HYDROLOGY

Principles and processes

To Mary and Kay

For their support and indulgence
as we pursued our passion for hydrology,
and the many hours we spent writing

"If there is magic in this world it is in water"

LOREN C. EISELEY (1957)
THE IMMENSE JOURNEY,
VINTAGE BOOKS, RANDOM HOUSE

HYDROLOGY

Principles and processes

M. Robinson, Honorary Fellow, Centre for Ecology and Hydrology, Wallingford, UK

R.C. Ward, Emeritus Professor of Geography, University of Hull, UK

IWA PUBLISHING

Published by IWA Publishing
Alliance House
12 Caxton Street
London SW1H 0QS, UK
Telephone: +44 (0)20 7654 5500
Fax: +44 (0)20 7654 5555
Email: publications@iwap.co.uk
Web: www.iwapublishing.com

First Published 2017

ISBN: 9781780407289 (Paperback)
ISBN: 9781780407296 (eBook)

Cover Image: Mountain landscape
Copyright: Biletskiy_Evgeniy
Stock photo ID:505823322
From istockphoto.com

CONTENTS

PREFACE

"Writing a book is an adventure. To begin with it is a toy and an amusement. Then it becomes a mistress, then it becomes a master, then it becomes a tyrant. The last phase is that just as you are about to be reconciled to your servitude, you kill the monster and fling him to the public."

WINSTON S CHURCHILL.

From a speech about authorship, National Book Exhibition at Grosvenor House, London, 2nd November 1949.

Thank you for picking up this book. Water is essential to all life and hydrology its study deals with all aspects of its occurrence on Earth. Earth, the 'Blue Planet', has three quarters of its surface covered by water and is the only body in the solar system on which water is known to exist in large quantities. Since ours is literally a water world, the growth of hydrology as both a practical study and a science is not surprising. Early knowledge of water developed almost exclusively through local attempts to manage and control it. However, the water which exists in such abundance on the Earth is unevenly distributed in both time and space, and its circulation, closely enmeshed with the circulations of the global atmosphere and oceans, is a vital component of the Earth's energy machine. Indeed much of the current impetus for the advancement of hydrology as a science comes from our increasing interest in and concern about climate variability and climate change and the key role of the global water circulation.

Water is essential to life but excessive variations bring disasters in the form of floods and droughts and its management and appropriation have already become the source of international tension and even potentially of regional conflict. Furthermore, as the world population continues to grow, the pressures exerted on and by water will also increase and over the next half-century large areas of the world will have insufficient water to meet their needs. Hydrology is therefore more important than ever before and certainly the need for a clear understanding of the operation of hydrological processes at all scales has grown significantly since the first hydrological textbooks were published more than 60 years ago.

Hydrological processes, and our improved understanding of their operation, dominate this discussion of the principles of hydrology. The structure of this book contains chapters in logical sequence of the hydrological cycle devoted to the major components of the hydrological cycle, each with up-to-date references and discussions. In addition, a concluding chapter has been added to draw together some of the ideas which are developed through the book and look to new developments and challenges ahead. There are web links for further information as well as Review Problems and Exercises at the end of each chapter.

We have included a wide range of recent references, as well as some seminal historical ones, as guidance for readers who wish to follow up specific topics in more detail. Inevitably, however, our selection is a personal one and in any case represents only a minute fraction of the growing stream of publications in the relevant journals. Fortunately, access to this great body of literature, is easier now than at any time, via libraries and in particular the Internet, and it is assumed that many readers will access additional information from these sources. Similarly, with the growth of the Open Data movement, access is more easily available to an increasing amount of environmental data including hydrology, and analytical tools for their study are readily available, such as spreadsheets (e.g. MS Excel and its free equivalent from Libreoffice.org) as well as more specialist numerical software.

Mark Robinson and Roy Ward
Summer 2016.

1

INTRODUCTION

"There is no life without water. It is a treasure indispensable to all human activity."

EUROPEAN WATER CHARTER, COUNCIL OF EUROPE, 6TH MAY 1968

1.0 INTRODUCTION

Water, the subject matter of hydrology, is both commonplace and unique. It is found everywhere in the Earth's ecosystem and it is essential to all known forms of life. The Earth is called the 'Blue Planet' because most of its surface (71%) is covered by water, although the total amount of water is under 0.5% of the Earth's total volume. About 97% (depending on the method of calculation) occurs as saline water in the seas and oceans. Of the 3% that is fresh water considerably more than half is locked up in ice sheets and glaciers, and another substantial volume occurs as virtually immobile deep groundwater that is not easily accessible. The really mobile fresh water, which contributes frequently and actively to rainfall, evaporation and stream-flow, represents only about 0.02% of the global total (see Figure 1.1).

A reliable source of water is *the* essential basis for human civilisation, and an integral part of the natural world. Without water every form of life on Earth would stop. Indeed, water makes up the bulk of most living things and it is a major medium for transporting energy, dissolved chemicals and sediments at scales from the molecular to the global. Moving water and ice are *the* agents of erosion and deposition. However, water creates risk to humans, in the form of floods and droughts, and polluted water harbours disease. Water resources for human exploitation must be allocated and used in a sustainable and equitable way to serve the competing and often conflicting requirements of agriculture, industry, households, power generation, navigation, flood protection and recreation, while

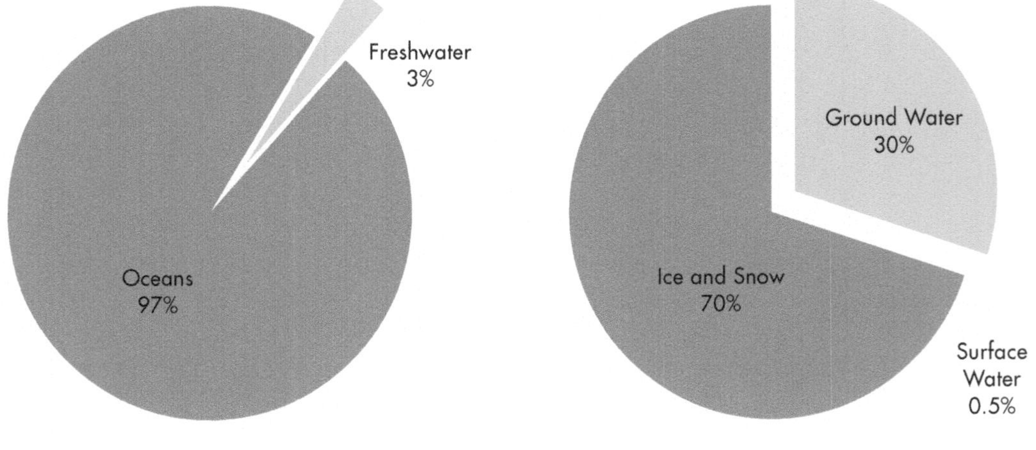

Fig 1.1: Global water constituents: Freshwater comprises only 3% of the total global water, and less than 0.5% of that freshwater is present as surface waters (Table 1.1).

maintaining a healthy environment. Taken for granted where plentiful, a prized possession where scarce, water is the only naturally occurring inorganic liquid, and is the only chemical compound that occurs in normal conditions as a solid, a liquid and a gas. The abundance of water is *the* feature that sets this planet apart from its fellows in the solar system.

Hydrology is defined formally as "the science which deals with the waters of the Earth, their occurrence, circulation and distribution on the planet, their physical and chemical properties and their interactions with the physical and biological environment, including their responses to human activity" (UNESCO, 1964). This book focuses on the *principles* and *processes* of hydrology which are concerned largely with *physical* or *environmental hydrology.* In this context, water is viewed in the same way as soil, vegetation, climate or rock, as an element of the landscape to be investigated and ultimately understood by means of rigorous, scientific examination and analysis. The science of hydrology overlaps with many other sciences. The study of precipitation and evaporation lie within the realm of the meteorologist, the storage of water in the soil and its percolation to groundwater are of direct relevance to agriculturalists, soil physicists and geologists. The flow of water in the river is the province of the hydraulic engineer. Only the hydrologist deals with these fragmented aspects of the **hydrological cycle** in its entirety (McCulloch, 1975). The aim of hydrology is to seek knowledge and understanding of the hydrological cycle (see Figure 1.3) in ways that lead to its safer exploitation, more reliable prediction and more effective control of water and water resources.

Hydrology is of fundamental importance in managing water, and seeks to understand the movement of water through the environment, and predict how water bodies will behave under different circumstances. At its broadest, hydrology encompasses all aspects of water as it moves through the water cycle, but is more usually taken to focus upon water on the land surface and in the soil profile, rather than in the air or the sea

Specifically, hydrology – and hydrologists help to provide a safer and better quality of life for people, and an enhanced environment for wildlife through:

- Securing water supplies for public use, including drinking water and sanitation,
- Ensuring the provision of water for food production, crops and livestock,
- Protecting against, and giving warning of, approaching floods and droughts to reduce their impact through economic and physical damage and loss,
- Protecting people and the environment from pollution and over-abstraction,
- Maintaining and improving aquatic habitats for wildlife, navigation, and recreation.

1.1 WATER – FACTS AND FIGURES

Water is present on Earth in three **phases**. In its liquid form, precipitation meets the basic water needs of humans, animals, and plants. Its runoff into streams sustains ecosystems and, along with percolation into aquifers, ensures long term storage and supply for human uses. The oceans are the world's primary source of water vapour that feeds precipitation. Atmospheric water vapour is a **greenhouse gas**, allowing much of the sun's shortwave radiation to pass through but absorbing the longwave radiation emitted by the Earth's surface, which results in the Earth's surface temperature being about 30°C warmer than it would be otherwise (Trenberth, 1992). In addition, water vapour may condense into clouds that reflect and absorb solar radiation, thus directly affecting the Earth's radiant energy balance. In water's frozen form, sea ice and snow cover tend to cool the planet by reflecting the incoming solar radiation. Glaciers, especially those at mid-latitudes, provide water storage and summer supply for both agriculture and urban areas around the world.

As the world population increases towards 10 billion by 2050, and climate change progresses, increasing stresses will be placed on water resources. The global distribution of fresh water over the globe is amazingly uneven in both space and time, and many regions of the globe currently suffer from water scarcity. Even is the case of Britain, a generally humid land, the spatial pattern of rainfall and runoff (greater in the North and West) is largely the reverse of the distribution of its population (higher in the South and East), creating problems of water supply. In fact by the World Bank criterion (<1,000 m^3 per head per year of available water) much of South East England can be classed as suffering from serious water stress.

Water is the only naturally occurring liquid most people see, and it is usually thought of as being a common, ordinary substance. But this colourless, tasteless, odourless fluid is far from being ordinary; water is one of the most extraordinary substances and defies many of the normal laws of physics and chemistry. It is its unusual attributes that make water of unique importance to life across the globe and to humanity.

1.1.1 THE SPECIAL CHARACTERISTICS OF WATER

Water combines two of the most common elements, but it does not behave like any other substance. Hydrogen and oxygen atoms form a strong **covalent bond** with electrons shared between them. Due to the distribution of electrons the oxygen side of the water molecule has a negative charge and the side with the hydrogen atoms has a positive charge. This means that the positive end of one molecule will attract the negative end of another water molecule to form a **hydrogen** or **polar bond** (Figure 1.2). This weak electrostatic attraction between molecules has only about 1/10 of the strength of a covalent bond within a molecule, yet determines most of water's unique properties:

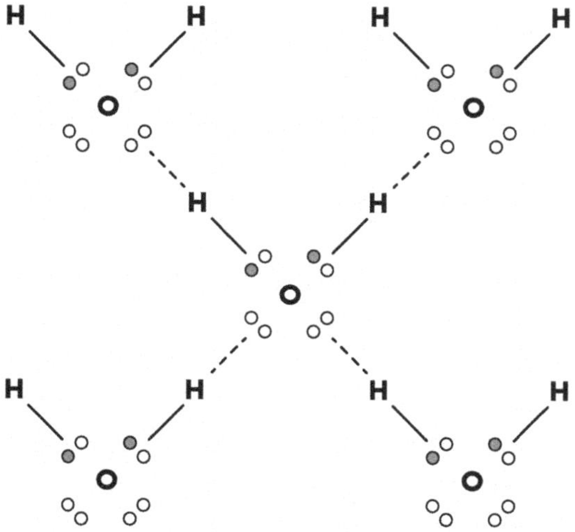

Fig 1.2: Water molecule showing the strong covalent bond (–) between the oxygen (**O**) and hydrogen (**H**) atoms in a water molecule, and the much weaker hydrogen bonds (---) with the oxygen atoms in adjacent water molecules.

Hydrogen bonds between water molecules result in water's cohesiveness - literally "sticking" to itself, giving it a high **surface tension**. Hydrogen bonds also form between water and a solid such as a soil particle. This is especially important to the movement of water in soil, and plants make use of **capillary rise** to draw water up through their stems to their leaves.

- Most substances shrink as they cool, but when water falls below 4°C it starts to expand and become lighter. That is why a water body such as the lake freezes from the top down rather than the bottom up, and why icebergs float. This insulates the lower liquid water, which helps aquatic life to survive through the winter.
- When water freezes its volume increases by about 9%; this is an important mechanism in rock weathering where liquid water percolates into rock cracks.
- Water has an exceptionally high **specific heat capacity** – the energy required to raise its temperature. This is about 5 times higher than dry soil, and so a water body will warm (or cool) much more slowly than the surrounding land. Its high **thermal inertia** minimizes temperature fluctuations and helps to keep the temperature of the planet stable. The ability of water to absorb, and later release, heat strongly affects the spatial pattern of climate as warm ocean currents transfer enormous amounts of heat from the tropics to higher latitudes.
- Changes of **phase** between liquid, vapour and solid (melting and freezing, evaporation and condensation) absorb or release more **latent heat** than most other common substances. This gives water an immense ability to store and transport heat around the globe, and atmospheric transfers of water vapour play a key role in the Earth's energy budget. Evaporation stores heat which is then released elsewhere with the condensation of water vapour.
- Due to its special molecular structure water is an almost **universal solvent**

that can dissolve almost anything. The hydrogen bonds between water molecules are continuously being broken down and remade several billion times a second (Hillel, 1991). Water dissolves a substance by forming hydrogen bonds with its molecules and once in solution water molecules surround individual **ions** of the substance (parts of molecules that become separate, electrically charged, entities) preventing them recombining. Water transports dissolved nutrients through the environment and is the medium in which life's key metabolic exchanges take place. But it may also carry pollutants, so water quality must be taken into consideration as well as quantity of water for human use.

The special properties of water described above are reflected in individual chapters of this book and include, for example, surface tension in soils (so they do not drain instantly after rain ceases), its large heat capacity resulting in its crucial role as a global energy transporter, and its dissolving power essential for transporting nutrients.

1.2 THE CHANGING NATURE OF HYDROLOGY

Hydrology is both an old and a new subject; old in that humans have been attempting to control and manipulate water out of practical necessity for many thousands of years; new in the sense that hydrology has only been studied as a separate academic discipline in its own right for less than a century.

The origins of hydrology can be traced back to the control and management of water at the start of civilisation. Indeed, the fact that water is essential to life and that its distribution and availability are intimately associated with the development of human society meant that it was inevitable that some development of water resources *preceded* a real understanding of their origin and formation. Some of the oldest civil engineering structures still in existence were built for the storage and supply of water. Hydrology as a modern scientific discipline had its origins in **hydraulics**, the study of flows within well-defined boundaries such as river channels.

With the increasing world population, industrialisation, climate variations and climate change, and the growing awareness of the fragility of the natural environment, the importance and security of water supply and wastewater treatment assume an ever higher profile in national and international strategies. This renders water prone to disagreements and disputes from headwater streams to the largest river basins (see Section 9.2.3). The risk of water wars rises with scarcity, and many have predicted that the wars of the 21st Century will be fought over water. Indeed, the word 'rival' derives from a Latin legal term *rivalis* referring to a person who shares with others the water of the same stream. Where basins and aquifers are shared by nations, equitable use of water resources is often the aim of the protocols and treaties that have been negotiated.

Archaeological discoveries and later documentary evidence emphasise the significant part played by the location and magnitude of water supplies in the lives of ancient people. The *Epic of Gilgamesh* written at least 1,000 years before the Bible's first books describes a Great Flood that is very similar to the Genesis story of Noah. There are similar accounts of a Great Flood found in central Turkey, in the Royal library of the Hittites, as well as in Hindu, Greek and Roman mythology (Barnett, 2015).

Some have attempted to attribute the origins of hydrology to a particular location or country, such as Greece, Egypt, Mesopotamia, China or South America. From the evidence available, however, it is much more likely that, from very early times, understanding, engineering skills and large-scale resource development progressed interdependently in multiple areas, especially those where water was a 'problem' - either because of its shortage or its over-abundance. Evidence of early structures to control water can be found in areas as wide apart as the Middle East and South America. There is evidence of large scale control of agricultural societies and a high degree of social organisation for food production and security in other warm and arid or semi-arid areas including the Yellow and Yangtze Rivers in China; and in South America irrigation canals were dug nearly 7,000 years ago in the Peruvian Andes (Ortloff, 2016).

About 10,000 years ago humans began mastering the skills and tools necessary for the beginnings of agriculture. Due to the critical importance of water many of the great early civilisations developed in the valleys of important river systems, including ancient Egyptians in the Nile Valley, the Sumerians in the Tigris-Euphrates plain of Mesopotamia (Iraq) and the Harappa in the Indus Valley of India/Pakistan. All three rivers rise in areas of high precipitation, and then flow through arid regions where rainfall is scanty and the inhabitants rely on streamflow for their water needs. Their experiences can provide valuable lessons for the present and the future.

In Egypt, rainfall is almost non-existent and the annual Nile floods derived from monsoon rainfall on the highlands of Ethiopia provide the only source of moisture to sustain crops. The Nile has a strong annual cycle of flows and this formed the basis of successful, large-scale, agricultural irrigation for more than 5,000 years. According to the Greek historian Herodotus, "Egypt is the gift of the River Nile". The level of the River Nile regularly began to rise in late July, reaching a peak in late September when the floodplain was inundated to its maximum extent. Then, as the waters receded they left a covering of nutrient-rich silt providing very fertile agricultural land. It was recognised that the higher the annual river peak level, the larger the area that could be irrigated on a particular reach of river and the bigger the expected harvest for an agriculture completely dependent on the river flood level watering the farmland. Accordingly the early Egyptian officials recorded the maximum water levels at many points along the river and compared them to the peaks in previous years to determine the amount of tax to levy from the farmers (see Photo 1.1). The earliest records can be traced back over 4,000 years to about 2,500 years BC when gauges were cut into the rocks in the Nile valley. But high floods could result in much destruction, a paradox the Egyptians understood well. If the river level was too high, it would damage the banks of the irrigation dykes, destroy villages on the plain and ruin the crops. It has been claimed that in flood conditions extremely good rowers were despatched, and rowing downstream with the current were able outpace the flood peak and provide a warning to the townspeople downstream (Biswas, 1967a). If correct, this was probably the world's first flood warning system. Although the Egyptians did not understand that the annual flooding was predominantly due to rainfall in the upper headwaters, they were able to use engineering skills to build simple canals dikes and reservoirs to manage the water and increase crop production. The remains of what is believed to be one of the oldest dams in the world, built between 2950 and 2750 BC, lies about 30 km south of Cairo (Biswas, 1970). The High Dam at Aswan was constructed in the 1970s to control the Nile floods, by impounding waters in Lake Nasser, which stretches 500 km upstream and has a storage capacity of about 170 km^3. This ended the annual flood risk and provided a more secure supply of water in Egypt when neighbouring countries suffered in severe droughts, but finished the natural deposition of fertile silt each year, so farmers must now use fertilizers for growing crops such as barley and wheat (see also Section 9.2.3).

The Sumerian civilisation arose about 5,000 years BC in the Tigris-Euphrates plain and created an impressive irrigation system of extensive canals, especially in the lowlands of southern Mesopotamia, but

Photo 1.1: The stepped Nilometer on Elephantine Island at Aswan in southern Egypt recorded flood levels at the first cataract (rapids) of the Nile. (Photo source: Jamie Hannaford)

they had no drainage system to carry off excess water and with high levels of evaporation of the irrigation water this led to the build-up of dissolved salts in the soil, eventually reducing agricultural yields severely. In the subsequent period the population in this area declined by nearly 60% from 2100 BC to 1700 BC (Thompson and Hay, 2004).

Although extending over a larger area than either of the other civilisations much less is known about the Harappan civilisation in the Indus valley region in northwest India and Pakistan, as their writing has not been deciphered. This had been a remarkably advanced culture nearly 5,000 years ago. Then around 4,000 years ago the people began to abandon their cities and the Bronze Age Harappan civilisation slowly disintegrated and disappeared, leaving no direct legacy in an area that now is now desert. Its agriculture had been sustained by monsoon-fed rivers flowing from the Himalayas and there is evidence from ancient lake sediments of a decline in precipitation that may have badly affected both inundation- and rain-based farming. About 4,100 years ago the summer monsoons began a rapid decline that lasted several centuries (Giosan et al, 2012).

All three civilisations arose on major rivers where rainfall was low and summer temperatures were high. The peoples had much practical knowledge of water storage, transfer and management for supporting their agricultural systems, and the link between irrigation agriculture and political control associated with large-scale management of water led to these societies being termed "hydraulic civilisations". However, their fortunes were different. The Harappan decline has been linked to changes in climate. The long-lasting success of the Egyptian civilisation, in contrast with the collapse of that in Mesopotamia, may instead be attributed to hydro-geomorphic factors (Hillel, 1991). Firstly, the annual flood of the Nile occurs in late September / early October, well after the spring harvest and the hottest time of the year, whereas in Mesopotamia the annual flood derives from snowmelt in the Armenian mountains leading to a spring peak, followed by high summer evaporation from the irrigated land tending to make the soil saline. Secondly, the flatter topography of the Tigris-Euphrates plain was conducive to extensive areas of irrigation with long canals and a widespread shallow water table, in contrast with the narrow valley of the Nile where the soil water table was controlled by the level of the river and tended to fall as the Nile declined after its peak, leaching salts from the soil back into the river (Hillel, 1991).

Major hydraulic works were slower to develop in more humid regions such as Europe which in any case could not be cleared of dense woodland until the development of iron cutting tools from about 1,000 years BC (Landes, 1998). The Romans undertook many water projects particularly supplying fresh water to urban areas, which they saw as essential to the health of a city. They built over 800 aqueducts, totalling 5,000 km. The largest single system was to supply Rome, the world's first megacity with a population of over 1 million. The first aqueduct to Rome, the *Aqua Appia*, was built around 300 BC, and by 4th Century AD there were 11 aqueducts totalling over 500 km in length, supplying the city with fresh water via a complex urban distribution system. In fact the water system provided more water than that to New York City in the mid-1980s (Hodge, 2002). Ancient Rome also had extensive drainage – the *Cloaca Maxima* or Great Drain, was one of the world's earliest sewage systems and discharged effluent into the River Tiber.

Limited measurements of water probably began with measurements of flood levels of the River Nile, and by rainfall depths in India in the 4th Century BC, while the earliest systematic official rainfall measurements were probably made in Korea and in China. The engineering achievements of early civilizations were based on close observation of nature, and by trial and error. Their ability to develop hydrological theory was, however, severely limited by their lack of technology to make precise measurements, particularly of time. The Romans, for example, recognised the importance of river velocity for determining flow rates, but lacked sufficiently accurate time measuring equipment to record speeds (Hodge, 2002).

There is clear historical evidence that relatively recent changes in climate have had a dramatic impact on human populations by its effect on harvests and disease. The 1300s marked the beginning of a five century climate shift known as the *Little Ice Age* (although it was not a true ice age) during which Europe and North America were subjected to much colder winters than during the 20th Century. The second decade of the 1300s was probably the rainiest in 1,000 years, and temperatures were exceptionally cold. Crops failed and it is estimated that the *Great Famine* of 1315–1322 affected an area of 1 million km^2 across northern Europe, when perhaps several million people died (Jordan, 2010). Environmental historians have also pointed to food poverty resulting from poor harvests as important factors in exacerbated existing tensions prior to the *French Revolution* of 1789, and the so-called '*Arab Spring*' beginning in the 2010–2011 winter when a once-in-a-century winter drought in China reduced global wheat supply and contributed to shortages and skyrocketing bread prices in Egypt, the world's largest wheat importer (Werrell and Femia, 2013). Similarly, the sea of mud endured in the trench warfare of World War I in northern France was made much worse by the exceptional rainfall; the 24 month period from mid-November 1914 to mid-November 1916 had 50% more rainfall than the long-term average (Brugge and Burt, 2015).

1.2.1 EARLY THEORIES ABOUT THE ORIGIN OF SPRINGS AND STREAMFLOW

It was not until the rise of Hellenic civilisation about 600 BC that people attempted to understand nature, just for the sake of understanding, breaking away from the tradition of attributing natural phenomena to divine action. Early Greek philosophers theorised about the source of rivers and rainfall, but failed to develop a complete understanding. Perhaps due to their observation in karst areas of Greece of large streams emerging from limestone cliffs, together with their knowledge that the Nile flowed through a desert area, they thought that rainfall was not adequate to account for springs and streamflow, particularly during long periods without rain. They postulated that rivers were fed either by sea water that by some means rose to great heights, and in the process changed from salt to fresh water (Thales, ~650 BC; Pliny 23–79 AD), or else by condensation of atmospheric waters within vast subterranean caverns which then fed underground rivers (Aristotle, 384–322 BC; Senecca, 3 BC–65 AD) (see Biswas, 1967b). In neither case was the mechanism responsible made clear.

It was not recognised that rainfall alone could account for streamflow, or that evaporation was the mechanism by which the atmosphere was replenished. And yet some of the ideas developed by the ancient writers were remarkably close to the truth as we now know it. For example, the Vedic texts in India, pre-800 BC, appear to show an understanding of the atmospheric portion of the hydrological cycle, with the sun breaking up water into small particles (i.e. evaporation) which are then removed by the winds before returning to the mother Earth as rain (NIH, 1990). Aristotle (384–322 BC) explained the mechanics of precipitation, and three centuries later the Roman author and civil engineer Vitruvius, recognised that springs originated from rain and snow; as did Al-Karaji, a Persian scholar of the late 10th Century AD, who expounded the basic principles of hydrology particularly groundwater, and was familiar with present-day concepts and principles inherent with the hydrological cycle (Pazwash and Mavrigian, 1981) long before they were identified in Europe. In Iceland the Edda poems of the 9th to 12th Centuries AD preserve Old Norse mythology dating back well over 1,000 years to the Iron Age, and indicate remarkable insights on the physical world including a description of the hydrological cycle which showed recognition of the important roles of evaporation from the sea, condensation and cloud formation, and rainfall over the land (Bergström, 1989).

Throughout the Middle Ages, and indeed until comparatively recent times, the search continued for an explanation of springs, streamflow and the occurrence and movement of groundwater. However, the hypotheses put forward were either based on guesswork or mythology or else were biased by religious convictions; few, if any, were based on the scientific measurement of the relevant hydrological factors. There was little further advance in understanding until the Renaissance (14th–17th Centuries). Leonardo da Vinci (1452–1519), studied water flow and discussed hydrology in a book written around 1500 (Pfister et al, 2009) and had somewhat confused ideas about the hydrological cycle, based on drawing an analogy with the circulation of blood around the human body. He recognised that water was constantly circulating and returning from rivers to sea and back again and insightfully wrote that "all the sea and the rivers have passed through the mouth of the Nile an infinite number of times", but he lacked the basic knowledge of the entire hydrological cycle, and assumed, like earlier Greek philosophers, that sea water was drawn up to the top of mountains in a similar way to the flow of blood up to the human head. He did however, have a much better understanding of the principles of flow in open channels than either his predecessors or contemporaries. Bernard Palissy (1510–1590) was the first to challenge the Hellenic concepts and state categorically (counter to Plato and Aristotle) that rainfall was the source of springs and rivers (Deming, 2005).

1.2.2 DEVELOPING UNDERSTANDING OF THE HYDROLOGICAL CYCLE

It was not until near the end of the 17th Century, however, that genuine *experimental evidence* was used to underpin plausible theories about the hydrological cycle. The greatest advances came largely through the work of three men: Pierre Perrault (1608–1680) and Edmé Mariotte (1620–1684), whose work on the Seine drainage basin in northern France demonstrated that, contrary to earlier assumptions, rainfall was more than adequate to account for river flow; and the English astronomer, Edmund Halley (1656–1742) who carried out evaporation measurements and concluded that evaporation from the oceans was sufficient to account for river flow. However, unlike Perrault and Mariotte he clung to the belief that subterranean condensation in the interior of mountains, rather than precipitation, maintained river flow. Nevertheless, because Perrault, Mariotte and Halley were contemporaries and undertook hydrological research of the modern scientific type, they are often regarded as the three founders of modern hydrology (UNESCO/WMO/IASH, 1974).

Subsequently, John Dalton (1766–1844) combined lysimeter measurements of evaporation with discharge estimates for the River Thames and a large number of raingauge measurements to estimate the water balance for England and Wales (Dalton, 1802a). He was able to prove that precipitation is equivalent in quantity to the sum of evaporation and river flow, thus correctly describing the hydrological cycle, and rejecting once and for all the notion that condensation within mountains made a substantial contribution to river flow (Oliver and Oliver, 2003).

The interdependence and continuous movement of all phases of water, i.e. liquid, solid and gaseous, form the basis of the **hydrological cycle** (Figure 1.3), which is driven by solar radiation causing heating and evaporation. Water vapour in the atmosphere condenses and gives rise to precipitation. The processes of streamflow, groundwater flow and evaporation ensure the never-ending transfer of water between land, ocean and atmosphere, followed by its return as precipitation to the Earth's surface. The uninterrupted, sequential movement of water, implied in this simplified description of the hydrological cycle, is rarely achieved. For example, falling or newly fallen precipitation may be returned to the atmosphere by evaporation without becoming involved in streamflow, soil water or groundwater movement. Much precipitation falls into the oceans, while in the terrestrial portion of the cycle not all of this precipitation will reach the ground surface because some will be intercepted by vegetation or by the surfaces of buildings and other structures, and will from there be evaporated back into the atmosphere.

Until the 20th Century advances in hydrology were largely the preserve of engineers, particularly hydraulic engineering applications related to the design and management of water infrastructures that underpin a reliable and hygienic lifestyle. As

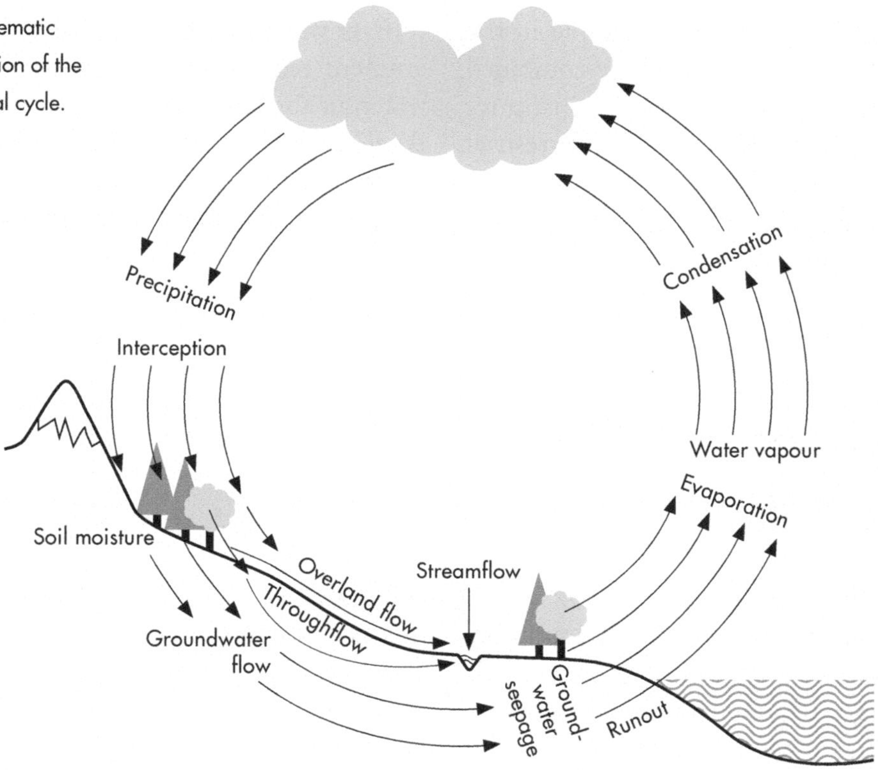

Fig 1.3: Schematic representation of the hydrological cycle.

these structures were largely completed in the developed world, hydraulic engineering started to lose its importance, and with increasing attention to environmental and social aspects it was natural that hydrology would separate from its engineering origins, and over time develop as an independent Earth science (Koutsoyiannis, 2014).

In this context the science of hydrology is still a relatively young discipline. The American Geophysical Union (AGU), for example, only established a separate hydrology branch in 1930, and in the UK the founding meeting of the 'Hydrological Group' of the Institution of Civil Engineers was held in 1963 (Hall, 1964). This marked the emergence of British hydrology as a scientific discipline, a position subsequently greatly strengthened by the creation of the Institute of Hydrology in 1968 (Robinson et al, 2013) and the formation of the British Hydrological Society (BHS) in 1983. Hydrological organisations formed in other countries include the New Zealand Hydrological Society founded in 1961 and the Canadian Water Resources Association in 1968. The National Institute of Hydrology (NIH) based in Roorkee in India was established in 1978. Throughout the development of the science of hydrology its progress has been closely interrelated with developments, including technological advances, in other sciences (McCulloch, 1988). To a large extent this is still the case, and the need for a multidisciplinary approach is certain to increase in the foreseeable future. For example in the UK, the Institute of Hydrology combined with other national research institutes dealing with freshwater and terrestrial ecology to form the Centre for Ecology and Hydrology (CEH) in 1995. The following year the National Rivers Authority (NRA) was amalgamated with other government regulatory agencies dealing with water, air and land to form the Environment Agency (EA), to provide a comprehensive environmental body able to tackle the increasingly complex and multidisciplinary problems of environmental protection.

1.3 QUANTIFYING THE GLOBAL WATER BALANCE AND HYDROLOGICAL CYCLE

Despite the continually changing global distribution of water vapour, evaporation, precipitation and runoff, and the associated changes in soil water and groundwater storage, the global water balance is, in effect, a 'closed system'. Virtually no water is lost from the system and the occasional releases of water from volcanic activity, or additions of water from comets, are minute. Furthermore, over quite short periods of time, the input of water vapour into the atmosphere by evaporation is wholly accounted for by the condensation-precipitation process. Thus although in one sense the global water balance is very simple, quantifying it has proved a difficult and elusive task due to sparse and often inadequate data.

Understandably, therefore, there is a broad range of estimates of the main global water storages and fluxes in the global water balance, depending upon the data used and the assumptions made (Speidel and Agnew, 1988). Of the relatively recent attempts, the most widely accepted estimates are probably those of Igor Shiklomanov and his colleagues at the State Hydrological Institute in St Petersburg (Shiklomanov, 1993, 1997; Shiklomanov and Rodda, 2003), building on earlier work including Lvovitch (1973) and Baumgartner and Reichel (1975).

Estimated values of global water stores are shown in Table 1.1. These estimates must be treated with caution because of the difficulties of monitoring and quantification at the macroscale. For example, the volumes of the ocean basins and of the major ice sheets depend upon sea bed and sub-ice topography. Similarly, reserves of deep groundwater are difficult to assess. Shallow groundwater storage is more accessible and mostly easier to estimate, although the proportion of useable, non-saline, water is still far from certain. Atmospheric water vapour content is normally monitored either by radiosonde balloons or from infrared spectrometers in weather satellites.

At any one time the amount of water held in the atmosphere is about 13 km^3, which is about 10 days' rainfall. The total water in lakes and wetlands amounts to about 190,000 km^3. The soil holds less than

20 km^3, while there is about 5,000 km^3 as readily accessible (i.e. shallow) freshwater groundwater. These volumes are dwarfed by the 25,000 km^3 water stored in icecaps and permanent snow cover, 20,000 km^3 as inaccessible or saline groundwater, and over 1,300,000 km^3 held in the oceans. These storage values alone can be misleading, as the available renewable water resource is not only a function of the total store size, but also of the *rate* of water flux; i.e. the **residence time** that water remains in a store is crucial (Oki and Kanae, 2006). Thus, for example, the global average annual precipitation is about 1,000 mm, about 35 times the amount held in the atmosphere at any one time, and total annual river water abstractions are several times the quantity of water stored in the worlds' rivers. Similarly, the importance of soil water is out of all proportion to its small total amount in the global water balance because its average residence time is relatively short, so that soil water is turned over several times a year, multiplying its effective contribution. Conversely, icecaps and glaciers are giant water reservoirs, and in glacial periods can lock up enormous additional amounts of water on land such that the sea level fell by over 100 metres, exposing areas of the continental shelves (e.g. the former 'land bridge' linking Britain to mainland Europe), whilst in warmer periods the water released raised the sea level. In addition, of course, the rate hydrological activity can vary in time; thus in hot deserts, rainfall is spasmodic and so too are other processes such as evaporation and streamflow, which can take place only for a short period during and after rainfall, to be followed by a long period of virtual inactivity, apart from a slow redistribution of groundwater at depth below the surface

Studies of the main components of the hydrological cycle have usually considered them in isolation. More recently, global sets of multiple components have been

	VOLUME (1,000 KM^3)	% TOTAL WATER	% FRESH WATER	TYPICAL RESIDENCE TIME
Oceans	1,340,000	96.5	–	10^3–10^4 yrs
Icecaps, glaciers	24,500	1.8	~69.	10–10^4 yrs
Groundwater – total	23,400	1.7	–	Months–10^4 yrs
Groundwater – fresh	10,500	0.75	~30.	Months–10^4 yrs
Lakes, swamps	190	0.015	0.4	Months–10 yrs
Soil water	17	0.001	0.05	100 days
Atmosphere	13	0.001	0.05	10 days
Rivers	2	0.0002	0.005	15 days

Table 1.1: The world water balance. Estimated magnitude of the stores and typical residence time. All numbers are rounded to avoid implying spurious accuracy (Based on Shiklomanov and Rodda, 2003; Oki and Kanae, 2006; Trenberth et al, 2007).

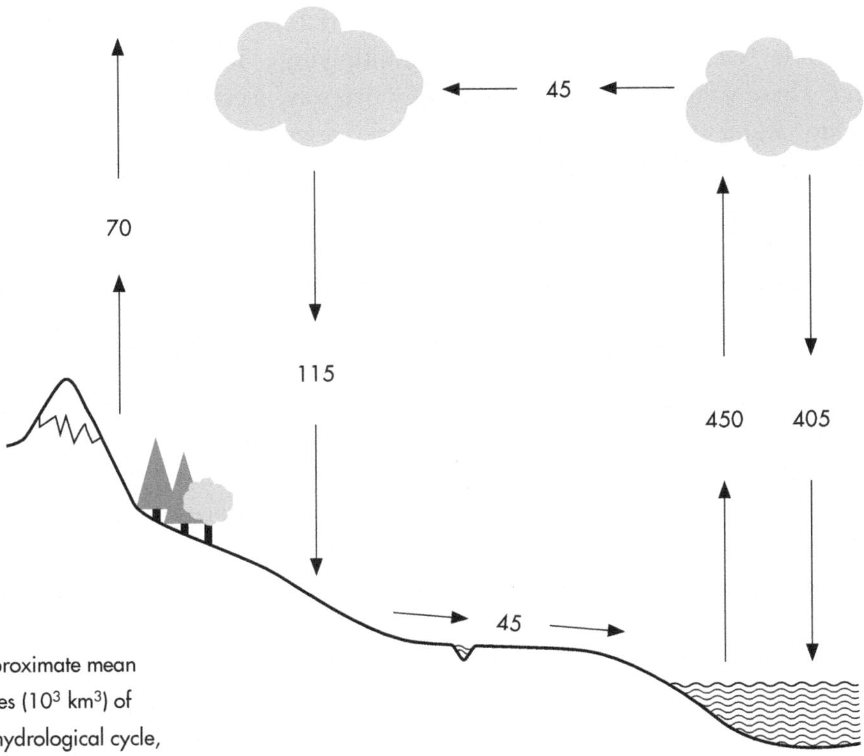

Fig 1.4: Approximate mean annual fluxes (10^3 km^3) of the global hydrological cycle, 2000–2010. Based on Rodell et al (2015).

assembled (e.g. Neale and Cosh, 2012). The increasing sophistication and availability of remote sensing products (see Box 1.1) means that they may have the potential to provide all the components of the water cycle, although at present there are too many uncertainties for remote sensing products alone to be able to provide a complete global water balance (Sheffield et al, 2009; Trenberth et al, 2007). A more realistic approach is to use a *combination* of measurements and modelling to reduce uncertainty about the magnitude of fluxes of water and energy, and provide a more secure picture of the hydrological cycle (e.g. Trenberth et al, 2007; Rodell et al, 2015).

Figure 1.4 shows mean annual fluxes obtained by combining ground-based and remote sensing observations with model estimates optimized by forcing the 'closure' of the water and energy budgets. The numbers are instructive, but values are approximate as there are still big uncertainties, particularly for precipitation onto the oceans and for evaporation. Groundwater flow into the ocean is also difficult to quantify, and although streamflow, is much larger and easier to monitor, its uneven distribution, both spatially and seasonally, complicates the assessment of its role in the global water balance. The amount of precipitation falling on land is about 115,000 km^3 per year. About 55% of this amount is evaporated by forests and natural landscapes and 5% by rainfed agriculture. The remaining 45,000 km^3 per year is converted to surface runoff (feeding rivers and lakes) and groundwater (feeding aquifers). These are called **renewable** freshwater resources (FAO, 2015). The majority of ocean evaporation is precipitated back to the oceans and the remainder transported to the continents. Conversely, the total precipitation onto land exceeds terrestrial evaporation, and this excess provides water to the soil and plants, streams, lakes and underground aquifers. River flow makes up most of the compensating flux of liquid water from the land to ocean.

These global values do not indicate the very uneven distribution. Clearly, the vast majority of the water that evaporates comes from the oceans, but there will be important regional differences in the contribution of land and the oceans to evaporation. Thus there is little evaporation from desert areas and very high evaporation from oceans in tropical regions. Tropical rainfall comprises more than two-thirds of global rainfall, and together with that of temperate regions generates almost all of the world's discharges to the sea. About 20% of global runoff comes from the Amazon Basin alone. In temperate climate regions approximately a third of precipitation goes to each of evaporation, runoff and groundwater recharge. In a semi-arid climate about half returns to the atmosphere by evaporation, 30% becomes runoff and 20% recharges groundwater. In an arid climate about 70% of the very low precipitation is lost to evaporation, 30% becomes runoff and only 1% recharges groundwater (UNESCO, 2009).

Ocean currents redistribute the excess runoff received by some ocean basins compared with others, and the imbalances between ocean evaporation and precipitation. For example, the Atlantic Ocean has a deficit in the annual precipitation-evaporation balance which is not made up by the inflow of land water. The remaining difference is compensated by inflows from the Arctic and Pacific oceans, which have a surplus in the precipitation-evaporation balance (Speidel and Agnew, 1988). In this way the hydrological cycle can be seen as a self-regulating system that moves water from one state or reservoir to another in complex cycles and the whole process enables life to survive on land because the net transfer of atmospheric water from the oceans to the land as additional precipitation helps to replenish soil water, lakes and streams. Hydrological processes form part of an interactive earth surface-atmosphere system, and it is therefore likely that future major advances in hydrology will be closely associated with scientific progress in our understanding of the atmosphere and the biosphere. Also, increasingly, the cycle is interrupted and modified by human activities. Indeed the natural cycle is more complex than shown here, with local and regional water cycles (see Figure 9.4), and much of the precipitation that falls on land actually comes from land-based sources as 'recycled precipitation' (see Section 9.2.2).

Links to Global energy budget

The global energy budget drives the atmospheric circulation that determines our weather and climate. Water has an important role in the global energy budget in several ways. Only about 25% of the energy comes from direct solar energy. The majority of the energy is transferred between the surface and the atmosphere by the evaporation-condensation cycle. As water changes from liquid to vapour and back to liquid, enormous quantities of heat are stored, carried by the atmosphere and ocean currents around the globe, and then released. Water vapour is the principal greenhouse gas in the Earth's atmosphere, an order of magnitude greater than CO_2, and has a key role in influencing the temperature of the planet (see Section 9.3). Clouds have a complicated effect on climate, being responsible for over half of the reflection of solar radiation back into space, but like water vapour they also trap longwave radiation. Soil water content plays an important role in partitioning solar radiation reaching the ground between sensible and latent heat fluxes.

In the past, hydrologists focussed their attention on the relatively small amount of fresh water occurring either as rivers, lakes, soil water and shallow groundwater, or in the vegetation cover and the atmosphere. Increasingly, however, it is recognised that the oceans play a dominant role in the global water and energy budgets and that large-scale perturbations of the hydrological system may result from changes of sea surface temperature, such as those associated with El Niño (see Section 2.5.3) or from modifications of the thermohaline ocean circulation which may result from the increasingly rapid break-up of major ice-sheets in both the northern and southern hemispheres.

1.4 THE NATURE OF HYDROLOGICAL PROCESSES

The global hydrological cycle provides a useful introductory concept, but it is of limited practical value to the hydrologist concerned with understanding and quantifying the occurrence, distribution and movement of water in a specific area. The natural unit of study for hydrologists is usually a drainage basin or catchment area which receives only water from precipitation falling within the topographic boundary, and from which the only outflows are evaporation and streamflow. The continuity equation, in the form of the hydrological or **water balance equation**, may be applied, so that, whether a small experimental plot of a few square metres or a large river basin or group of basins:

$$\text{Inflow} = \text{Outflow} \pm \Delta\text{Storage} \qquad (1.1)$$

If this equation can be solved, a quantitative assessment of the movement of water through the drainage basin becomes possible.

Research basin studies have played a major role in developing our understanding of hydrological processes. Three main bodies are responsible for coordinating international hydrological investigations: the International Association of Hydrological Sciences (IAHS), which is part of the International Union of Geodesy and Geophysics (IUGG), the World Meteorological Organisation (WMO) and the UN Educational, Scientific and Cultural

Organization (UNESCO), which are both agencies of the United Nations (see also Box 1.2). Together with other agencies, they are active in promoting a number of major international collaborative programmes. The first internationally coordinated programme of hydrological experiments was the IAHS International Hydrological Decade (1965–74). This was a remarkable example of international cooperation which facilitated major advances in hydrology and in the assessment of surface and groundwater resources. It was followed by UNESCO's long-term International Hydrological Programme (IHP) which aims to find practical solutions to specific problems of countries in different geographical locations and at different stages of technological and economic development.

Although the drainage basin is used as an example of a specific hydrological unit in the preceding section, hydrological processes are investigated over a very wide range of spatial and temporal **scales**. At one extreme, **microscale** investigations are exemplified by studies of the movement of soil solution through the interstices of the soil matrix or of the evaporation characteristics of individual plants growing in controlled environment chambers. At the other extreme, some of the emerging problems of environmental change associated with large-scale forest clearance or climate change, for example, will be resolved only by a better understanding of the hydrological system at the **macroscale**, i.e. global or regional, rather than at the **mesoscale** level of the drainage basin.

Such large-scale studies require international, as well as interdisciplinary, collaboration and depend increasingly on large data networks, on data derived from satellite-based remote sensing techniques, and on GIS and computerised data-bases (Marsh et al, 2015). As well as a greater concentration, in the future, on global-scale issues, hydrologists must also be prepared to deal more effectively than before with the enormous **spatial diversity** both at the Earth's surface and in the overlying atmosphere, a diversity that exists across the entire range from micro- to macroscale occurrences. Just as soils, geology and land use vary spatially so do the components of the water balance. Especially crucial is the relation between precipitation and potential evaporation, which varies enormously from one area to another and over time, both seasonally and in the long term. Accordingly, hydrologists must increasingly take a long-term and large-scale view in order to meet the considerable challenges posed by, for example, the impact of climate variability and climate change on hydrological processes and on the availability of water resources.

The great range of scale and of spatial diversity at which hydrological processes operate poses severe problems for the physical hydrologist, not least in the sense that it is virtually impossible to argue from the particular to the general using a simple deterministic notion of causality. It is also important to recognise that the small volume of mobile fresh water is itself distributed unevenly in both space and time. Wetland and prairie, forest and scrub, snowfield and desert, each exhibits different regimes of precipitation, evaporation and streamflow. Each offers different challenges of understanding for the hydrologist and of water management for the planner and engineer, and each poses different benefits and threats to human life and livelihood as between the developed and the developing world.

Accordingly, Beven (1987, 2014) saw the need for two developments in hydrology. Firstly, hydrological predictions should be associated with a realistic estimate of levels of uncertainty that can then be incorporated in the decision-making process, thereby permitting some allowance for hydrological processes that are not properly understood. Secondly, given the difficulty of scaling up theory to larger scales due to the inherent complexity of hydrological systems, we need an inherently stochastic, macroscale theory which can accommodate the spatial integration of heterogeneous nonlinear interacting processes, including preferential pathways, to provide a rigorous basis for predictions.

The discussions of hydrological systems and processes in this book show that hydrology has made great progress on these issues, although there is still some way to go. Many of the factors that were identified by Dooge (1988) as having hampered the development of hydrological theory in the past are rapidly receding in importance. Thus, approaches once considered as rivals are now seen to be complementary. Hydrologists are talking to other disciplines and increasingly working in multi-disciplinary teams with ecologists and socio-economists, as well as with civil engineers, agriculturalists and water chemists. Hydrologists now have the means, and the confidence, to work with 'real' rather than 'ideal' problem situations, and fully recognise the need to consider aspects of water quality as well as its quantity.

1.5 WHY HYDROLOGY MATTERS TODAY

This book is a scientific and not an engineering text, but we should be ever mindful that hydrology is more than just an intellectual exercise. The practical and theoretical advances in hydrology, reflected in this book, provide an essential basis for meeting the growing challenges posed, for example, by increasing environmental pressures resulting from population growth, climate variability and climate changes, including those triggered by the 'greenhouse effect'. Such pressures have the potential to greatly affect water resources, reducing even further the already paltry per capita availability of water in many countries of Africa, Asia and the Middle East (e.g. Biswas, 1996).

Although hydrology is concerned with the study of water, especially atmospheric and terrestrial fresh water, its emphases have changed from time to time and vary from one practitioner to another. The uneven distribution of water and human population is one of the reasons for the emerging global water crisis. Two-thirds of the world's population live in areas that receive only one quarter of the world's annual rainfall (OST, 2005). In contrast the most water-rich areas of the world, such as the

Amazon and Congo River basins are sparsely populated. Over the last century, fresh water abstraction for human use has increased at more than double the rate of population growth (Fox, 2013). Global water demand is largely influenced by population growth, urbanization, food and energy security policies, and macro-economic processes such as trade globalization, changing diets and increasing consumption. UNESCO (2015) predicted that by 2050 global water demand will be 55% higher, mainly due to growing demands from agriculture, manufacturing, thermal electricity generation and domestic use.

Virtually every component of the drainage basin hydrological system may be modified by human activity. These include: land cover changes such as afforestation, deforestation, and urbanisation, the widespread development of irrigation and land drainage; and large-scale abstraction of surface and groundwater. In recognition of humanity's profound impact on the planet, scientists have adopted the new term **Anthropocene** to denote the human-driven epoch of the planet (e.g. Monastersky, 2015).

SUGGESTED DISCUSSION TOPICS

For additional information see publications of UNESCO and FAO.

1.1 Less than 1% of the world's freshwater (0.008% of the total) is sufficient to fill all the world's rain clouds, lakes, swamps and rivers. About 70% of the available freshwater used by humans each year is used for agriculture; withdrawal is highly dependent on both climate and the place of agriculture in the local economy; it ranges from 22% of the total water withdrawal in Europe to 82% in Africa. (FAO, 2014).

1.2 Half of the urban population in Africa, Asia, and Latin America suffers from diseases associated with inadequate water supply and sanitation (Vorosmarty et al., 2005).

1.3 Bottled water is 500–1,000 times more expensive than tap water. A plastic bottle containing the most commonly available French water is transported approximately 650–700 km to reach shops in Britain.

1.4 The global circulation of water by the hydrological cycle, powers most of the other natural cycles and conditions the weather and climate (Vorosmarty, 2009)

1.5 About 70% of our bodies are made from water. The average human contains nearly 50 litres of water and must replace about 5% of it each day for vital bodily functions. The primary reason for the vast increase in life expectancy in developed countries in the 20th Century was not the development of new medicines, but rather the provision of clean drinking water and high levels of sanitation (Cutler and Miller, 2005).

1.6 The largest part of all water use by humans is for agriculture – 69% vs 23% for industry and 8% for households, and thus the largest opportunity for increasing water use efficiency comes from reducing agricultural use. Many irrigation systems lose up to 60 –80% of their water (WRI, 1996, 2000).

1.7 Many countries with low water availability compensate by importing a large amount of their food (See also Section 9.2.4). It takes 2800 litres of water to grow a kilogram of rice, and 50 glasses of water to grow enough oranges to make one glass of orange juice. (UNESCO, 2006).

1.8 Rainfall may be 'free', but it costs money to collect and store it, purify it for drinking water supplies and it is expensive to pump it to where it is needed. In many parts of the world water is wasted because it is not well priced (See also Section 9.2.5). The majority of the world's irrigation systems pay an annual flat rate irrespective of the amount of water used. More sensible pricing would increase the water use efficiency and help secure future supplies (UNESCO, 2009).

1.9 Several ancient civilisations collapsed due to mismanagement of the land and reductions in the rainfall, as is happening today in parts of the Sahel region across Africa.

BOX 1: SOME IMPORTANT REMOTE SENSING SATELLITES

AQUA Launched in May 2002, to study the Earth's water systems the satellite has six different instruments (including visible, infrared and microwave) and is named for the large amount of information being obtained about water in the Earth system.
http://www.aqua.nasa.gov/

SMAP (Soil Moisture Active Passive) mission to map global soil moisture.
Launched in January 2015 SMAP's radar transmits microwave pulses to the ground and measures the strength of the signals that bounce back from Earth (9 km resolution) , while its radiometer measures microwaves that are naturally emitted from Earth's surface (40 km resolution). Unfortunately, after three months operation the radar stopped transmitting leaving just the passive microwave working.
http://www.nasa.gov/smap

GPM (Global Precipitation Measurement) uses multiple satellites to measure rain, snow and other precipitation data every three hours. The GPM Core Observatory launched in February 2014 carries an advanced radar/ radiometer system to measure precipitation from space and serve as a reference standard to unify measurements from a constellation of research and operational satellites.
http://pmm.nasa.gov/GPM

OCO 2 (Orbiting Carbon Observatory) space-based global measurements of atmospheric CO_2 to characterize sources and sinks on regional scales. It can also quantify CO_2 variability over the seasonal cycles year. Launched in July 2014, the satellite observatory tracks the large-scale movement of carbon between Earth's atmosphere, its plants and soil, and the ocean, from season to season and from year to year.
http://www.nasa.gov/oco-2

GRACE (Gravity Recovery and Climate Experiment) measures tiny changes in the Earth's gravitational field by making accurate measurements of the distance between two satellites, using GPS and a microwave ranging system. Launched in March 2002, GRACE provides information on variations in water stored on and below the land surface in soils, aquifers, snowpack and river flows at a scale of about 150,000 km^2 and monthly time resolutions.
http://www.nasa.gov/mission_pages/ Grace/

BOX 2: INTERNATIONAL WATER INSTITUTIONS

Because water enters into so many facets of human life and endeavour the range of organizations and bodies for which water is a concern is virtually endless. This section can give only a very superficial view at the international level - the structures at the national and local levels are even more complex and differ from nation to nation.

The organisations concerned with water, especially freshwater, are many and varied, covering an enormous range of topics including sport and recreation, religion, policy making, economics, health, agriculture, industry, energy, transport, law, environment, science, natural hazards and climate change. These bodies vary widely in their scope and scale, and may change in remit over time. Some overlap and may even have competing interests. The WMO Hydrological Information Retrieval Service, INFOHYDRO, provides information on organizations dealing with hydrology and details of networks of hydrological observing stations.

Prof John Rodda, former Director Hydrology and Water Resources at the World Meteorological Organization describes it as follows: The simplest way of seeking order is to determine whether a body, organisation, association or group is governmental or non-governmental. The United Nations Organization is the principal international governmental body. The UN has a variety of interests in water; many are expressed in the Millennium Development Goals. But there are nearly 30 bodies and agencies of the UN which have their own programmes in water, separate from the UN itself. The World Health Organization (WHO), the World Bank (WB), the International Atomic Energy Agency (IAEA), the UN, Food and Agricultural Organization (FAO) and the World Meteorological Organization (WMO) are amongst these bodies. UNESCO, for example, has been undertaking an International Hydrological Programme for 50 years, and WMO plays a major role in coordinating the international collection of hydrological data and in hydrological modelling. These bodies meet in a committee called UN Water to co-ordinate their programmes and to promote the publication of the three-yearly World Water Assessment Report. Regional international organizations include the European Union and the Arab League. All these bodies are funded by governments, each member government contributing to the budget of the organization to an agreed scale, with the United States usually making the largest contribution. The emergence of climate change as a very serious global concern brought about the formation of the WMO/UNEP Intergovernmental Panel on Climate Change in 1988. A number of IPCC's regular assessment reports and their supplements have highlighted the impact of climate change on water resources, widening the number of bodies with concerns for water.

International non-governmental organizations tend to be much more varied than governmental ones. They are usually not-for-profit voluntary bodies and have fewer funds, some are concerned with a single issue and often they tend to be less visible. However a number frequently hit the headlines, such as when a disaster strikes, working alongside UN and national bodies to relieve its impact. These are generally aid, assistance and humanitarian organisations such as OXFAM, Medicines Sans Frontiers and CATHOD. The World Wildlife Fund, Greenpeace and the International Union for the Conservation of Nature mount programmes where protection of the aquatic environment is the focus. Consumer interests in water are represented in Consumers International, while scientific organisations with watery remits are headed by the International Council for Science which has a number of subsidiary bodies concerned with water, such as the International Geographical Union. Some form the International Union of Geodesy and Geophysics, notably the International Association of Hydrological Sciences, the International Association of Meteorology and Atmospheric Sciences and the International Association of Cryospheric Sciences. The World Water Council brings together a heterogeneous mixture of governmental and non-governmental bodies, both international and national and organises the three-yearly World Water Forum.

REFERENCE COMPILATIONS AND RESOURCES

The International Association of Hydrological Sciences (IAHS) produces the *Hydrological Sciences Journal*, as well as the *Proceedings of IAHS* (formerly *IAHS Publications*) of the International Symposia the Association organises each year – a valuable source of up-to-date scientific results and discussions. http://iahs.info/Publications-News.do

Many key papers in the development of hydrological science and understanding are dispersed amongst publications devoted to engineering, geography, geology, soil science, agriculture and physics, which may not all be easily accessible. Accordingly, some of the most influential papers have been brought together in the series of *Benchmark Papers in Hydrology*. Published by IAHS each volume includes a specially commission expert commentary and interpretation. Volumes include:

1. Streamflow Generation Processes
2. Evaporation
3. Groundwater
4. Rainfall-Runoff Modelling
5. Riparian Zone Hydrology and Biogeochemistry
6. Hydro-Geomorphology, Erosion and Sedimentation
7. Forest Hydrology
8. Isotope Hydrology
9. Palaeohydrology

Google scholar is a valuable online resource for tracing references, and for checking citations. But access to documents depends on the paywall structure as some articles are pay to view. http://scholar.google.com

Grey literature, such as unplublished theses, reports and conferences may be accessed at: www.greynet.org

USEFUL WEBSITES

Websites can be an invaluable source of up-to-date information but unlike peer-reviewed research journals and books there is no quality control of the accuracy of information available online. If you are in any doubt remember the 3 W's: Who is behind it? Why was it set up? When was the material written?

Below, and at the end of each of the following chapters, are some reputable well-established and evidence-based sites with good general sources of hydrological information. But by its nature this information is dynamic, both through updating of contents as well as changes in web page addresses and links. The web address links were checked correct in late-2016, but may change in the future.

Intergovernmental organizations and International research bodies

AQUASTAT (FAO Global information system on water and agriculture):
http://www.fao.org/nr/water/aquastat/main/index.stm

British Hydrological Society (BHS):
http://www.hydrology.org.uk/science_of_hydrology.php

European Environment Agency (EEA):
http://www.eea.europa.eu/themes/water

Flood Risk Management Research Consortium:
http://web.sbe.hw.ac.uk/frmrc/index.htm

Global Institute for Water Security (University of Saskatchewan):
www.usask.ca/water

Global Runoff Data Centre (WMO):
http://wwwbafg.de/grdc.htm

International Association of Hydrological Sciences (IAHS):
http://iahs.info/

International Water Management Institute (IWMI):
http://www.iwmi.cgiar.org/

UNESCO:
http://en.unesco.org/

Water Information:
http://www.unesco.org/new/en/natural-sciences/environment/water/

Freshwater:
http://www.unesco.org/new/en/natural-sciences/environment/water/

UN Water Facts:
http://www.unwater.org/downloads/Water_facts_and_trends.pdf

UN ENVIRONMENT PROGRAMME (UNEP):

Freshwater Portal:
http://freshwater.unep.net

Global Environment Monitoring System (GEMS):
http://www.unep.org/gemswater/

World Health Organization (WHO):
http://www.who.int/en/

World Meteorological Organization (WMO):
https://www.wmo.int

INFOHYDRO:
http://www.wmo.int/pages/prog/hwrp/INFOHYDRO/infohydro_index.php

National Research Bodies

Centre for Ecology and Hydrology (UK):
http://www.ceh.ac.uk/

Meteorological Office (UK):
http://www.metoffice.gov.uk/

CSIRO (Australia):
http://www.csiro.au/en/Research/LWF/Areas/Water-resources

National Hydrology Research Centre (Canada):
http://www.ec.gc.ca/

National Institute of Hydrology (India):
http://www.nih.ernet.in/

National River Flow Archive (UK):
http://nrfa.ceh.ac.uk/
NOAA's National Weather Service (USA):
http://www.noaa.gov/
UK Groundwater Forum:
http://www.groundwateruk.org/Image-Gallery.aspx
US Environmental Protection Agency:
http://www.epa.gov
US Geological Survey – Water Resources of the USA:
http://www.usgs.gov/water/

OPEN SOURCE DATA (UK) AND SOFTWARE TOOLS

These websites are particularly subject to alteration as government policy changes; such as the formation of UK Research and Innovation (UKRI) to take on the duties of the Research Councils, including the Natural Environment Research Council (NERC).

UK Open Data policy for public sector data:
http://www.nationalarchives.gov.uk/doc/open-government-licence/version/3/
Environment Agency Data Advisory Group (EADAG):
http://eadag.org/
Environment, Food & Rural Affairs:
https://defradigital.blog.gov.uk/category/open-data/
Public Sector Open Data:
https://data.gov.uk/
Natural Resources Wales:
http://lle.wales.gov.uk/
Environmental Data Scotland:
http://www.sepa.org.uk/environment/environmental-data/
NERC Open Research Archive (NORA) of its research centres' publications:
http://nora.nerc.ac.uk/
UK Public Sector research publications and data:
http://gtr.rcuk.ac.uk/
Environmental Data (UK):
https://data.gov.uk/data/search?theme-primary=Environment
WATCH Forcing Data (Global):
http://www.eu-watch.org/data_availability

Open Source word processing, spreadsheets and presentations software include:
LibreOffice www.libreoffice.org and Apache **OpenOffice** http://www.openoffice.org/

Application programming interfaces (API):
Real time flood monitoring
http://environment.data.gov.uk/flood-monitoring/doc/reference

2

PRECIPITATION

"Water is essential to life on Earth and precipitation is the hydrologic cycle's key process by which the fresh water supply gets regenerated."

MATTHIAS STEINER, US NATIONAL CENTER FOR ATMOSPHERIC RESEARCH

2.1 INTRODUCTION AND DEFINITIONS

Precipitation is a major factor controlling the hydrology of a region. It is the main input of water to the Earth's surface and knowledge of rainfall patterns in space and time is essential to an understanding of soil moisture, groundwater recharge and river flows. Precipitation data are more readily available, for more sites and for longer periods, than for other components of the water cycle. In some parts of the world precipitation data may constitute the only directly measured hydrological record (Perks et al., 1996). The study of precipitation is thus of fundamental importance to hydrology, and this chapter concentrates on those aspects of its occurrence and distribution that are of direct relevance to the hydrologist. More detailed investigation of the mechanisms of its formation is the domain of the meteorologist and climatologist, and it is assumed that the reader will refer to meteorological and climatological texts for a more extensive treatment of the subject (e.g. Barry and Chorley, 2010; Hewitt and Jackson, 2009).

The meteorologist is concerned to analyse and explain the mechanisms responsible for the distribution of precipitation, an interest ceasing when the precipitation reaches the ground. The hydrologist is interested in the distribution itself, in how much precipitation occurs and in when and where it falls. That is, the form in which precipitation occurs, its variations in both space and time, and the correct interpretation and use of the measured data.

Precipitation occurs in a number of forms, and a simple but fundamental distinction can be made between liquid and solid forms. Liquid precipitation principally comprises **rainfall** and **drizzle**, the latter having smaller drop sizes and lighter intensity. In contrast to these forms, which may play an immediate part in the movement of water in the hydrological cycle, solid precipitation, comprising mainly **snow** may remain upon the ground surface for a considerable time until the temperature rises sufficiently for it to melt. For this reason solid precipitation, particularly snow, is discussed separately in Section 2.7.

Other types of precipitation may be important locally. For example, in some semi–arid areas the main source of moisture may be dew, formed by cooling of the air and condensation of water vapour by cold ground surfaces at night. In coastal or mountain areas fine water droplets in low cloud or mist may be deposited directly onto vegetation and other surfaces. In practice, although it is not strictly correct, the terms 'precipitation' and 'rainfall' are often applied indiscriminately and interchangeably to any or all of these forms.

2.1.1 WATER VAPOUR

An **air mass** is a body of air with relatively uniform temperature and humidity derived from long contact with the underlying ground surface. However dry the air may appear to be, it always contains some moisture as water vapour molecules (Burt, 2012). A **vapour** is a gas that is below its 'critical temperature' and so may be easily condensed or liquefied by a comparatively small change in temperature or pressure.

The amount of water vapour in the air varies over time and can be expressed by the **vapour pressure**, which is the partial pressure of the water vapour, and is usually expressed in units of hectopascals (hPa) and are numerically identical to millibars (mbar). Vapour pressure is only a very small part of the total air pressure – typically about 2.5 hPa out of a total atmospheric pressure of about 985 hPa (Trenberth, 1992).

The amount of water vapour that the air can hold increases with temperature until the air becomes saturated. This maximum amount increases approximately logarithmically with increasing temperature. Once this maximum amount is exceeded, for example by cooling, then condensation may occur. This temperature is known as the **dew point**. The degree of saturation may be expressed as the **relative humidity** of the air, which is the ratio of the actual water vapour pressure of the air to that at saturation for the same temperature. Since the saturation vapour pressure depends upon temperature the relative humidity falls as the temperature rises and increases as the air cools. At vapour pressures below saturation the air is unsaturated and, if conditions are suitable, it can absorb additional moisture though evaporation (see Section 4.2). For each 1°C rise in air temperature, the amount of water it can hold increases by approximately 6%.

The total amount of water vapour in the atmosphere represents only a minute proportion of the world's water budget. At a given moment the atmospheric water accounts for less than 0.001% of the world's total stock of terrestrial, oceanic and atmospheric water (see Table 1.1), and yet this small amount serves as a continuing source of supply for precipitation. The global average atmospheric vapour amounts to about 25 mm of liquid water which, given the average annual precipitation over the whole globe is about 1,000 mm, represents only about 9 days' average supply. This *mean* value hides a great variation. Some water may be carried up into the stratosphere where it could remain for up to 10 years. At the other extreme, some water that is evaporated into the lower levels of a thunderstorm cloud may be precipitated out within an hour.

The vertical profiles of air temperature and pressure exert an important influence on precipitation. The variation of air temperature with height is known as the **environmental lapse rate (ELR)**. This averages a decrease of about 6°C per km, but it can vary greatly between places and through time, and this will have an influence upon the behaviour of air masses subject to a lifting mechanism. As a parcel of air ascends it expands due to decreasing atmospheric pressure, and the energy used in expansion causes it to cool. If there is no mixing, and hence no exchange of heat between the ascending air and its environment (an **adiabatic** process) this reduction in temperature is approximately 1°C per 100 m, which is known as the **dry adiabatic lapse rate (DALR)**. However, if the air cools sufficiently to become saturated, its **dew point** temperature, latent heat of vaporisation will be released as some water vapour condenses into droplets. This acts to offset part of the cooling, so that the rising air cools at a slower rate, the **saturated adiabatic lapse rate (SALR)**, which is typically about 0.5 °C per 100 m. The latent heat released may thus enhance the vertical motion. If a saturated air parcel is warmer (and hence lighter) than the surrounding air, and the environmental lapse rate is greater than the SALR (or the DALR if the parcel of air is not saturated), it will continue to rise and the air is termed **unstable**. Eventually the air parcel reaches a point at which it has cooled sufficiently for clouds to begin to form. This is called the **condensation level**. The ascent of the parcel will normally stop when its temperature is equal to that of the environment and it is no longer buoyant, and most convective cloud tops will be at this level. The intensity of precipitation generated by a cloud is often related to cloud height: deeper clouds tend to produce more intense rainfall. The changing properties of moving air parcels may be expressed on a **tephigram** and their stability and any condensation level can be determined (Figure 2.1). More detailed descriptions of tephigrams are given in meteorological texts including Hewitt and Jackson (2009).

2.1.2 CLOUDS

A **cloud** is a visible mass of minute water droplets or ice crystals suspended in the atmosphere, and appearing as a white or grey drifting body. Individual clouds may vary in lateral extent from tens of metres to hundreds of kilometres. At any given time approximately half of the Earth's surface is covered by clouds, and they have a very important effect on the radiation balance. Due to their high reflectance, or **albedo**, clouds cast back incoming solar radiation and re–reflect terrestrial radiation, helping to keep the Earth warm. Cloud droplet diameters are generally in the range 1–100 microns (μm) and a cubic meter of air could contain 100 million droplets. The amount of water in a unit

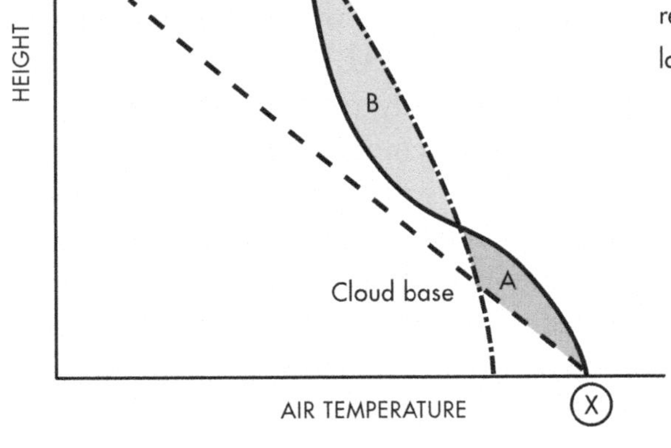

Fig 2.1: Tephigram showing the variation in air temperature with height. If a parcel of unsaturated air (X) is transported upward by turbulence, it cools at the DALR until at the dew point temperature, it becomes saturated and cloud formation commences. With continued ascent, the air will cool at the SALR. Shading: A = energy required to cause the air to rise; B= energy released from latent heat, providing buoyancy. See text for details.

volume of cloud can vary greatly depending on the type of cloud, but median values are about 0.1–0.2 g m-3 and theoretical maximum value is about 5 g m-3 (Pruppacher and Klett, 1997; Strangeways, 2007).

Clouds form when air becomes saturated, either by evaporation of water into the air, or more commonly by cooling of the air by upward motion. Water droplets condense onto aerosol particles that act as condensation nuclei. At temperatures below freezing, water vapour molecules may be converted directly to ice crystals by sublimation. **Condensation nuclei** typically range in diameter from 0.001–10 μm, and come from various sources, including smoke, dust, pollen, marine salts, pollutants and bacteria.

2.2 PRECIPITATION MECHANISMS

Although the atmosphere may have cooled sufficiently to produce clouds, precipitation will not occur unless condensation nuclei are present and there are suitable conditions for the growth of water droplets or ice crystals. These processes are discussed in detail elsewhere (e.g. Pruppacher and Klett, 1997; Wallace, 2006), and the following is a brief summary. Air may be cooled in a number of ways, for example by the meeting of air masses of different temperatures or by coming into contact with a cold object such as the ground. The most important cooling mechanism, however, is due to the uplift of air. As air is forced to rise, its pressure decreases and it expands and cools. This cooling reduces its ability to hold water vapour until at the dew point temperature the air becomes saturated and condensation occurs.

Since cloud appearance (shape, structure, patterns and transparency) express air movements, different types of cloud are associated with different weather conditions. Some may be associated with dry weather or only light rain, whereas others are indicative of heavy intense rainfall. Furthermore, it is common in weather systems for several types of clouds to occur simultaneously – at different altitudes, changing through time and at different parts of the storm. This visual information has been used in weather forecasting for over 2,000 years (e.g. NIH, 1990). The classification developed by Luke Howard 200 years ago (**Cirrus** – 'fibrous' with feather-like appearance, **Stratus** – 'sheets' with a layered structure and large horizontal extent and **Cumulus** – 'heaped' with a large vertical and limited horizontal extent) has been incorporated with cloud height: low (<2,000 m), medium (2,000–6,000 m) and high (>6,000 m) into modern classification schemes, such as the International Cloud Atlas (WMO, 1975, 1995a; Meteorological Office, 1982).

For clouds to result in precipitation there must be a mechanism to provide a source of inflow of moisture. Only when water droplets or ice crystals grow to a certain size are they able to fall through the rising air currents as precipitation. Since

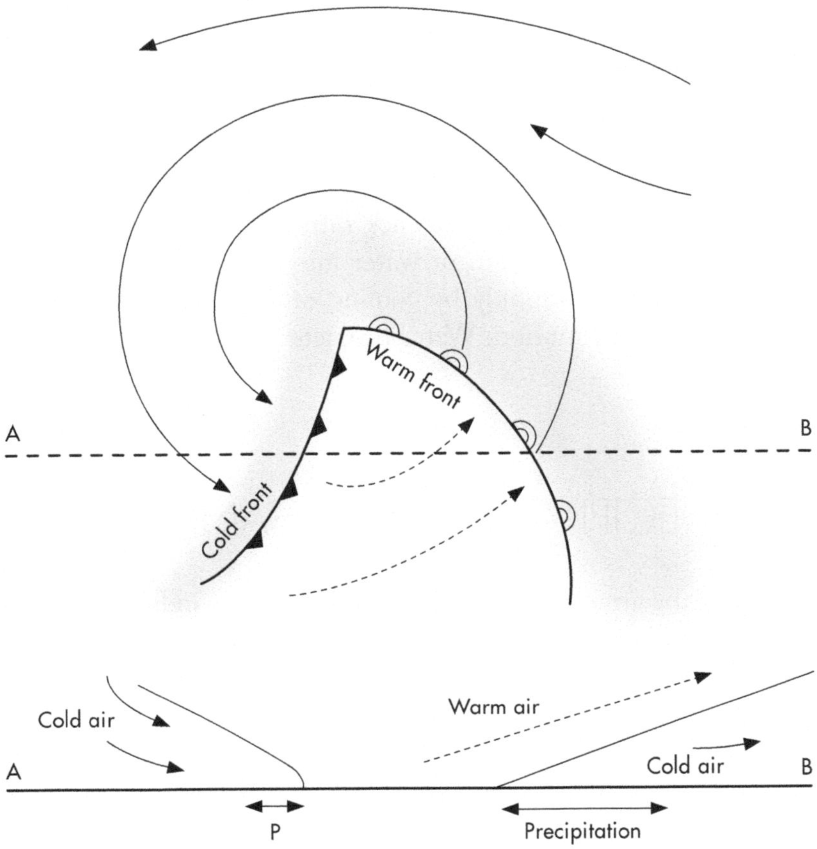

Fig 2.2: Cyclonic storm system for the Northern Hemisphere, showing the occurrence of fronts and resulting precipitation.

uplift is the major cause of cooling and precipitation, the following three–way division of precipitation according to the meteorological conditions causing the vertical air motion (Frontal/Cyclonic, Convectional, and Orographic) can be used in a very general way.

2.2.1 FRONTAL AND CYCLONIC PRECIPITATION

Precipitation outside of the tropics is often the result of large-scale weather systems (typically >500 km across) with precipitation occurring along the narrow boundaries, or **fronts** between the air masses, together with associated low pressure, **cyclonic**, systems where there is convergence and uplift of air. Cyclonic systems comprise air masses rotating anticlockwise in the Northern Hemisphere and clockwise in the Southern Hemisphere (i.e. the direction of rotation is the same as the Earth's rotation about the corresponding pole). In the case of frontal precipitation, warm moist air is forced to rise up and over a wedge of colder denser air. This may be either at a warm or cold front, which in broad terms may be distinguished in terms of the resulting precipitation (Figure 2.2). Cold fronts normally have steep frontal surface slopes that give rise to rapid lifting and heavy rain of short duration.

In contrast, warm frontal surfaces are usually much less steep, giving more gradual lifting and cooling, leading to less intense rainfall of longer duration. Over western Europe, warm fronts are more common in winter, when the westerly air moving over the Atlantic is warmer than the continental air to the east, whilst in summer the conditions are reversed and cold fronts are more common. Cyclonic systems are responsible for much of the cloud cover and most of the precipitation outside of tropical areas. They usually have relatively weak vertical air motions and typically produce moderate rain intensities of fairly long duration. In tropical areas, due to the greater heating, the resulting precipitation may be much more intense, and short–lived.

Satellite observations have revealed the important role of narrow zones (~300 km) of high water vapour fluxes at about 1–2.5 km altitude which act as a conveyor belt transporting large quantities of atmospheric moisture and heat thousands of km from the subtropics to the mid-latitudes (Lavers et al, 2011). At any given time there may be four or five of these **atmospheric rivers** across the mid-latitudes, and they account for about 90% of the poleward atmospheric water vapour transport. The exact mechanisms operating are still uncertain, and is the subject of active research (Dacre et al, 2015).

2.2.2 CONVECTIONAL PRECIPITATION

Convectional or convective rainfall results when heating of the ground surface by the sun causes warming of the air, and locally strong vertical air motions occur. If the air is **thermally unstable** (see Section 2.1.1) it continues to rise and the resulting cooling, condensation and cloud formation may lead to short-term, locally intense, precipitation. Such rainfall is dependent on heating, and moistening of the air from below, and is most common in tropical regions, although it occurs widely in other areas too, especially in the summer. In tropical cyclones, cloud cells may form spiralling bands around the central vortex, giving rise to prolonged heavy rain affecting large areas (Barry and Chorley, 2010).

Over warm continental interiors and tropical oceans slow–moving **mesoscale convective systems (MCS)** may produce appreciable amounts of rainfall, and in parts of central USA are responsible for a significant proportion of the summer growing season rainfall (Maddox, 1983; Houze, 2004). They comprise clusters of thunderstorm cells embedded in a much larger region of stratiform cloud shield several thousand km^2 in extent. Due to their extensive size and often long duration they can be very significant for flood hydrology (Smith and Ward, 1998). They are much rarer in maritime regions, particularly the mid–latitudes, such as the British Isles, where there may be only about one per year, although it has been suggested that such systems may have been responsible for some of the largest floods recorded in Britain (Austin et al., 1995). During the Boscastle storm of August 2004 in south-west Britain nearly 200 mm of rain fell in 6 hours (Burt, 2005). This was the result of a MSC with a sequence of convective storms channelled along the north Cornwall coast (Warren et al, 2014).

2.2.3 OROGRAPHIC PRECIPITATION

Orographic rainfall results from the mechanical lifting of moist air over barriers such as mountain ranges or islands in oceans, and is analogous to warm air being forced upward at a cold front. It may not be as efficient in producing precipitation as a convective or cyclonic system, but the lifting can induce convectional instability which may be more important than the orographic uplift itself. Typically more rain falls on windward than leeward slopes, since as the air descends it warms and the cloud and rain reduces. This effect can be seen along the western coast of northern Scandinavia and in the northern and western highland areas of the British Isles. On a somewhat smaller spatial scale, it is sometimes found that orographic effects may be translocated downwind, so that the largest falls are not recorded on the hill tops but some distance downwind. The intensity of orographic precipitation tends to increase with the depth of the uplifted layer of moist air.

2.3 GENERAL SPATIAL PATTERNS OF PRECIPITATION

In large storms the amount of precipitation may be several times greater than the average water content of a column of atmosphere, indicating that large–scale lateral inflows of moist air must play a key role in the distribution of precipitation, transferring large masses of moist air from areas of high evaporation to areas of high precipitation.

The large variations in the amount of precipitation, both in time and in space, are of considerable interest to the hydrologist. There is, for example, a great contrast between some of the driest deserts of the world which receive rainfall perhaps only once in 20 years and places such as Bahia Felix in Chile which on average has rain on 325 days per year (van der Leeden et al., 1990). The average annual precipitation over the land areas of the globe has been estimated to be about 720 mm, and may be contrasted with places such as Mount Waialeale in the Hawaiian Islands which receives about 12,000 mm annually and Cherrapunji in Assam, India, where over 26,400 mm were recorded in one year, and 3,720 mm fell in one 4-day period (Dhar and Nandargi, 1996).

The great mobility of the atmosphere means that the sources of water vapour may be hundreds or thousands of kilometres from the area where that water vapour is precipitated. Consequently, it would be extremely difficult to find a link between, say, changes in land use in one place and changing precipitation at another location. It has been estimated that only about 10% of the precipitation over Eurasia originates as evaporation from the land surface of the region, the remaining 90% is transported into the region from surrounding areas (Brubaker et al., 1993). Nevertheless it has been shown by using isotopes of rain and river water that about 30% of the rainfall in the Amazon Basin is 'recycled' from evaporation within the region, rather than advected water vapour from outside of the catchment (Lettau et al, 1979; Salati et al, 1979). The Andes form an effective western barrier to advected water.

While the behaviour and pattern of individual storms may be complex and variable, broad areal patterns of precipitation exist when averaged over long periods. This is the essential difference between **weather** as the day–to–day state of the atmosphere and **climate** which is the normal or average course of the weather.

2.3.1 GLOBAL PATTERN OF PRECIPITATION

The average water vapour content of the atmosphere, expressed as a precipitation equivalent, is about 25 mm. The overall distribution of atmospheric moisture over the globe is well related to the areal pattern of evaporation and transport by winds (Peixoto and Oort, 1992). Values decline systematically from the equator to the polar regions, and also vary seasonally, increasing in summer due to greater heating and evaporation. On the other hand, the pattern of world precipitation is not well related, being instead is closely dependent on the processes causing vertical motion in the atmosphere which produce condensation and precipitation. Rainfall is most abundant where air rises and cools, and least where it sinks. In broad terms, the greatest rainfalls occur in equatorial areas associated with converging trade wind systems associated with the **Intertropical Convergence Zone (ITCZ)** and monsoon climates, where annual precipitation may exceed 3,000 mm. High moisture contents and warm temperatures lead to abundant convectional rainfall. Almost two-thirds of global rainfall occurs in the tropics. There is a secondary rainfall maximum in the mid–latitudes (40–65°) due to the occurrence of polar fronts and associated cyclonic disturbances. The lowest precipitation, often less than 200 mm y^{-1}, occurs in:

a. High latitude polar areas, due to descending air masses and the low water content of the extremely cold air, and
b. Subtropical areas, which include many of the world's largest deserts, where high pressure cells give rise to descending, drying air.

This simple general pattern is modified by a number of other factors. Evaporation from the oceans (especially subtropical oceans) is the main source of global atmospheric moisture; evaporation from continents generally provides only a small proportion of precipitation over land (see Table 1.1). As a result, precipitation tends to decrease with distance from the sea, resulting in areas of extremely low rainfall near the centres of most of the major land masses. In coastal areas, precipitation is generally greater over land than over the nearby sea, due to the greater mechanical and thermal overturning of the air. Mountain ranges tend to accentuate precipitation amounts, particularly where the prevailing air movement is onshore. Many of the world's rainiest places are on the windward side of mountains close to the sea – for example the Pacific side of the Rocky Mountains in N America, and the village of Cherrapunji in India on the steep Khasi Hills overlooking the Bay of Bengal.

2.3.2 REGIONAL PRECIPITATION

When regions such as the North America, Europe or the British Isles are considered in detail, the orographic influence is far more apparent, dominating the annual and, to a lesser extent, the seasonal distributions. The pattern of precipitation across Europe (Figure 2.3) is strongly influenced by the extensive ocean to the west, the distribution of mountains and the predominant direction of rain–bearing winds (from the west). Moist air from the Atlantic results in the highest precipitation (over 1,000 mm y^{-1}) on west coasts and mountain ranges, including western parts of the British Isles, Norway, the Iberian Peninsula, the Pyrenees, Italy, the Dalmatian coast of the Balkan mountains, and the Alps. The lowest precipitation (under 500 mm y^{-1}) falls in southern and eastern areas in the lee of mountain barriers, such as Sweden and Finland downwind of the Scandinavian mountains, central and south–east Spain, north–east Italy, and eastern Greece. The lowlands of western and central Europe

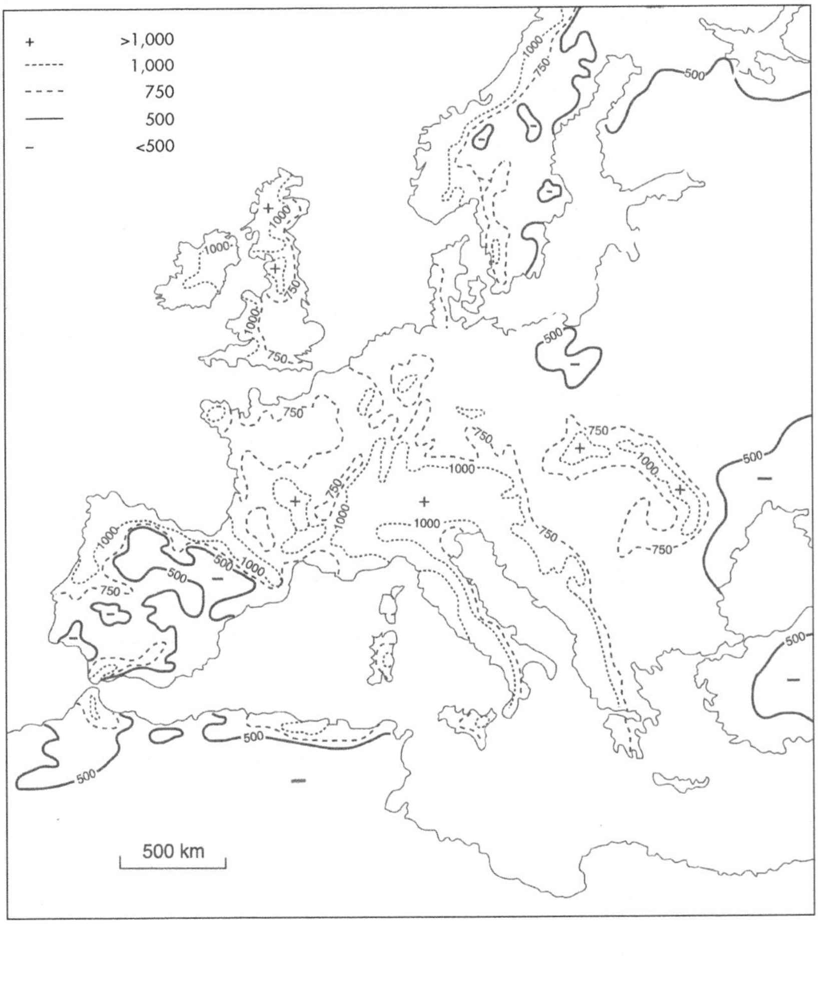

Fig 2.3: Generalised pattern of annual average precipitation over Europe (mm).

generally have a fairly even distribution of about 500–750 mm y^{-1}.

There is a winter precipitation maximum in western coastal areas (British Isles, Norway, north–west France) and in Mediterranean areas (Iberian Peninsula, Italy and Greece), with a summer maximum over much of central Europe due to summer heating and intensified convective activity. The climate of Europe is described in texts such as Wallen (1970) and Martyn (1992).

The UK lies in the latitude of predominantly westerly winds. The wettest areas are in the west as they are nearest the tracks of rain-bearing winds and they are also the most mountainous parts of the country (Figure 2.4). They typically receive 3–5 times as much precipitation as the drier areas. The short-term weather over the UK is strongly influenced by the position of the mid-latitude **jet stream**. This consists of ribbons of very strong winds 10–15 km above the surface of the Earth, flowing

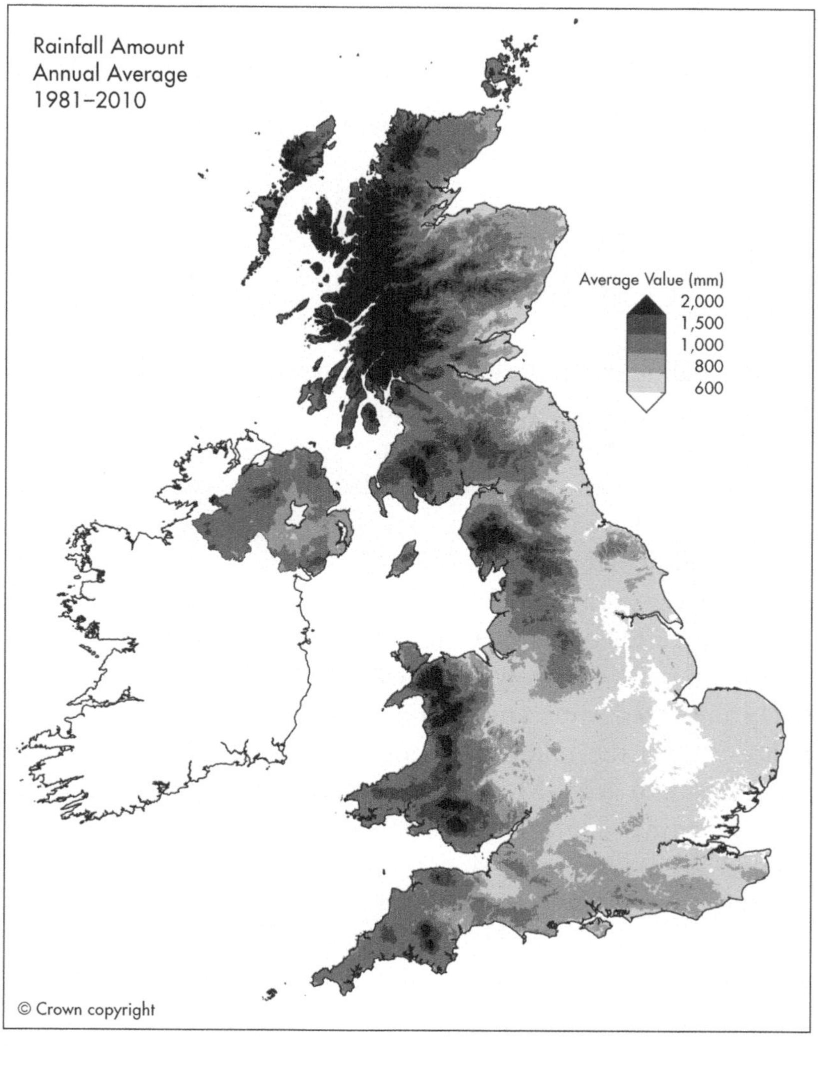

Fig 2.4: Average annual precipitation (1981-2010) (Met Office National Climate Information Centre)

along the boundary between the polar and mid-latitude air masses which move weather systems around the globe. The position of the jet stream over the UK determines the source of air and type of weather – if it is significantly to the south, then the UK will experience colder than average weather, whilst if it lies to the north, then the country will experience warmer than average conditions.

Unlike the rain in many other countries, particularly in the tropics, average hourly rainfall rates in the British Isles are quite low, typically ranging from 1–2 mm h^{-1}. Short-term rates can of course be much higher, and an hourly total of 10 mm is not uncommon, and 20–25 mm h^{-1} may be expected to occur once in 5–10 years. Greater falls are usually associated with heavy thunderstorms. Getting a month's rainfall in 24 hours at a single point in the UK is not very unusual – occurring perhaps once every 4–8 years on average. Most are caused by convective summer storms and tend to be localised. It would be unusual, although not unprecedented, for this heavy rain to occur over a wide area. A notable example occurred central-southern England and the south Midlands on 20 July 2007 when 147 mm of rain was recorded by one gauge, and the area receiving over 100 mm on that day extended over more than 3,500 km^2 (Prior and Beswick, 2008).

2.4 PRECIPITATION MEASUREMENT

Before dealing in detail with rainfall variations in time and space, and methods of analysing aspects of its magnitude and frequency, it is appropriate to briefly review the different means of measuring and recording precipitation, and to discuss some of their problems and limitations.

Of the different forms of precipitation (rain, hail, snow, etc.) only rainfall is extensively measured with any degree of certainty, and so the following sections deal primarily with rain. Methods specifically for measuring snow are dealt with in Section 2.7.2. The measurement of rainfall comprises two aspects: first, the point measurement accuracy at a gauge and, second, the use of the catches at a number of gauges to estimate areal rainfall.

2.4.1 POINT MEASUREMENT

A raingauge is basically an open container to catch falling raindrops or snowflakes over a known area bounded by the raingauge rim. The amount of rain collected may be measured by manually emptying a **storage raingauge**, usually at daily or longer intervals, and noting the amount of accumulated water, or else by using a **recording raingauge** which automatically registers the **intensity**, or rate of accumulation of rainfall. The most common type of recording gauges are **tipping bucket raingauges (TBR)** which digitally record the number of bucket tips in a set time period in increments, typically 0.1 to 0.5 mm, of rain. Short–period rainfall data are necessary to understand rainfall interception losses, limits to soil infiltration rates and catchment runoff hydrographs. For urban runoff studies, rainfall depths over only a few minutes' duration are necessary. Automated gauges became increasingly common from the 1980s, and they now account for about 30% of gauges operated by the UK Met. Office. Strangeways (2007) and Habib et al (2013) describe some of the different types of gauges and their advantages and limitations. Guidelines on procedures for collecting and processing raingauge data are provided by BSI (2012) and WMO (2008).

The first written account of rainfall measurement is from India over 2,000 years ago (NIH, 1990), and the first raingauges in Europe date from about the seventeenth Century (Biswas, 1970). In the 1670s, in northern France, Pierre Perrault used a raingauge to prove, for the first time, that the annual rainfall to a small catchment was adequate to account for the observed streamflow. Nevertheless, many problems remain in the collection of accurate rainfall data. The major problem is under-catch due to wind turbulence around the gauge (Sevruk, 1982). This may be due both to the exposure of the site and to the type of raingauge itself.

Controlled experiments in wind tunnels show that a raingauge acts as an obstacle to the wind flow, leading to turbulence and an increase in wind speed above the gauge orifice. The result is that precipitation particles that would have entered the gauge tend to be deflected and carried further downwind (Sevruk et al., 1989). This effect is even more pronounced in the case of snowflakes. Errors due to turbulence increase with wind speed and with reducing drop size, and so will be greater in temperate areas, such as Britain, than in some tropical areas, due to smaller raindrop sizes, higher wind speeds and the occurrence of snowfall. Catch efficiency is independent of raingauge diameter above 10 cm width (Strangeways, 2007). In all cases the volume of water collected has to be converted to a depth by applying the horizontal area enclosed by the rim of the gauge or by using a specially graduated measuring cylinder for that gauge diameter.

Raingauge measurements are sensitive to changes in the immediate environment surrounding the gauge. General advice on **raingauge siting** is provided in texts such as BSI (2012; Part 1), and WMO (2008). A site should not be over–exposed and subject to strong winds, nor should it be unduly sheltered by nearby obstacles. As a general rule the gauge should be at a distance of at least twice (and preferably four times) the height of any obstacle. But these guides also recognise that some degree of shelter is needed.

The most direct way of reducing wind losses is to place the raingauge in a pit so that its rim is level with the ground (BSI, 2010). If properly sited and surrounded by an anti–splash grid, this type of gauge provides the most accurate measure of the amount of rainfall that would have reached the ground if the gauge had not been present. However, this design has not been widely adopted because a pit is prone to fill with leaves or drifting snow, and on poorly drained ground it may fill with water. A more widely adopted approach is to add a shield around the rim to reduce turbulence. Different types include the rigid Nipher

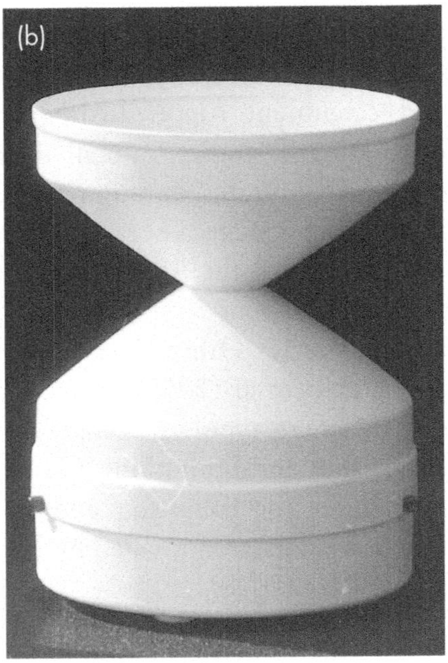

Photo 2.1: Raingauge designs to reduce wind losses include (a) Pit gauge (with standard gauge behind), (b) aerodynamic gauge (Photo sources: Ian Strangeways).

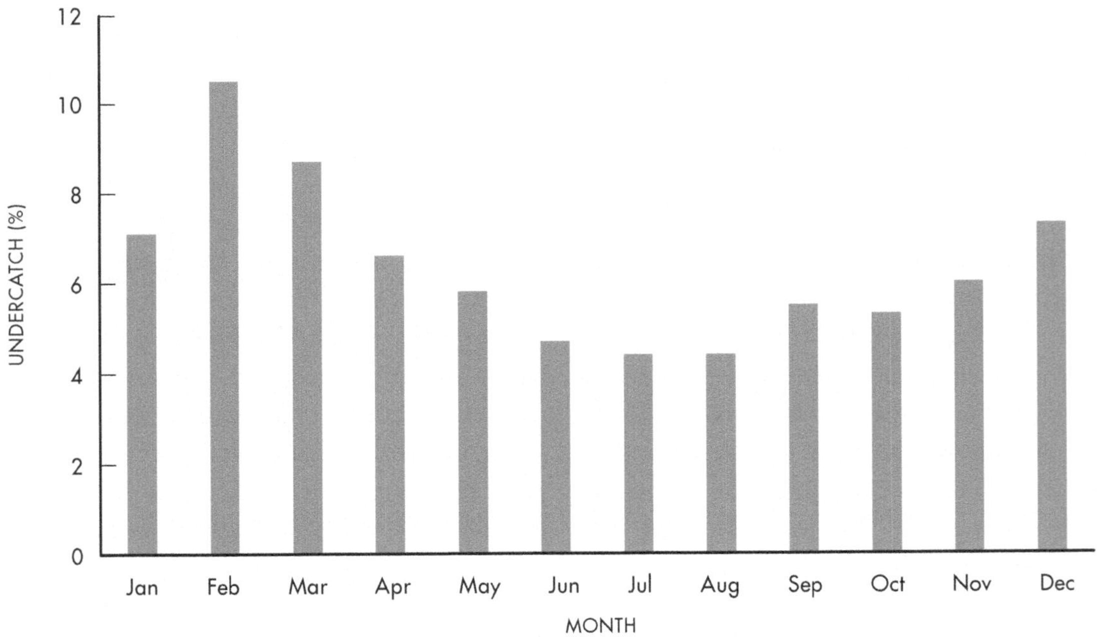

Fig 2.5: Average standard raingauge under-catch at Wallingford, S England as a percentage of the ground level gauge catch, 1969–2007, reflecting the greater winter wind speeds and larger summer raindrop sizes (Rodda and Dixon, 2012).

shield and the flexible Alter shield comprising metal strips which can move in the wind, and so restrict the accretion of snow. Such shields have, however, only been partially successful, and snow catches are still subject to considerable error. A promising alternative where a ground level gauges is not practicable, is an aerodynamically shaped raingauge, designed to interfere less with the flow of the wind (e.g. Strangeways, 2004). This can be a useful alternative, if not quite so efficient (Strangeways, pers comm, 2016).

Consideration of wind losses must be made when interpreting rainfall records. Early raingauges were often placed on a roof or a high wall to be safe from human or animal interference, thus inadvertently increasing the potential under-catch. This practice is still common in many developing countries. The first continuous record of daily rainfall in Britain, for example, was made in the 17th Century, using a gauge on the roof of a house (Craddock, 1976). Similarly, from 1815 to 1852 the main raingauge at Oxford University's Radcliffe Observatory was on a roof 7 m above the ground (Wallace, 1997). The larger under-catches in exposed upland areas led to the belief that rain increased as it fell through the atmosphere to lower altitudes, resulting in one case to the abandonment of plans in 1838 to build an upland reservoir to supply water to Oldham in northern England (Binnie, 1981).

Legates and Willmott (1990) used records of nearly 25,000 raingauges to compute global precipitation, and attempted to allow for gauge under-catch using correction methods devised by Sevruk (1982). They estimated that under-catches amounted to 10% globally, varying from 40% near the poles, due to snowfall, to under 5% in the tropics. Even larger errors may occur in individual storms. The under-catch also varies between seasons, being largest in winter when drop sizes are smaller and wind velocities are higher (Rodda and Dixon, 2012) (Figure 2.5).

Over 50 different types of national precipitation gauge are currently in use, with rim heights varying from 0.2 to 2 m (Sevruk and Klemm, 1989; Strangeways 2007). Some of the most widely used gauges are summarised in Table 2.1. In Britain and Australia, for example, the standard gauge has its orifice at 305 mm (1 foot) above the ground. In contrast, some countries prone to heavy snowfall have adopted rim heights of 1 or 2 m. This may lead to complications with apparent jumps in precipitation values at national boundaries (Groisman and Easterling, 1994). To provide a basis for comparison the World Meteorological Organization (WMO) proposed an Interim Reference Precipitation Gauge; but it, too, suffers errors from the effect of wind. In fact there are good physical reasons for the continued use of a variety of gauges; taller, wider gauges are more suitable in areas with much snow, whilst shorter gauges or ground level gauges are more appropriate in windy areas where rainfall predominates. Many countries are reluctant to change from their traditional type of gauge as this could introduce inhomogeneity into rainfall records, leading to problems in their use for studies of climatic variations and in evaluating long–term average values. The addition of Alter wind shields to raingauges at sites in the western USA in the 1940s, for example, created discontinuities in the

COUNTRY OF ORIGIN	GAUGE NAME	ORIFICE AREA (CM^2)	RIM HEIGHT (M)	ESTIMATED NUMBER WORLDWIDE	AREA USED (KM^2)
Germany	Hellman	200	1.1	30100	10250
China	Chinese	314	0.7	19700	10880
UK	Mk2/Snowdon	127	0.3	17800	10400
Russia	Tretyakov	200	20.	13500	25340
USA	Weather Bureau	324	1.1	11300	12560
India	Indian	200	0.3	11000	3290
Australia	Australian	324	0.3	7600	7940
France	Association	400	1.0	3000	5000
France	SPIEA	400	1.0	1800	4710

Table 2.1: Different types of widely used storage raingauges. (Based on data in Sevruk, 1982 and Sevruk and Klemm,1989).

records (Groisman and Legates, 1994). Details of precipitation gauge changes in a number of countries are given by Sevruk and Klemm (1989) and Groisman and Easterling (1994). These considerations demonstrate the crucial importance of **metadata**, records of site changes, which include instrumentation and site exposure to be able to separate true climate trends from local site effects. The processing and quality control of rainfall data are described in various meteorological texts including Burt (2012) and Strangeways (2003).

Particular measurement problems

Particular measurement challenges are faced when measuring rainfall in forested areas, in very steep terrain and during very intense rainstorms.

About 30% of the Earth's land surfaces are covered by forests, and it is not always possible to measure precipitation in clearings. It may be necessary to install a raingauge on a tower at tree top (canopy) level. It is possible for catches very similar to ground–level gauge values to be achieved in this way because the airflow disturbance of the gauge is similar to the roughness of the forest canopy, but this is very dependent upon the height of the gauge relative to the canopy (Jaeger, 1985; Robinson et al, 2004). If it is too high it will experience wind-induced under-catch, whilst if it is too low it may suffer from over-catch due to drip from adjacent branches, or under-catch due to sheltering. It is particularly difficult to define a suitable height for the gauge rim where the trees are of irregular height or the topography is uneven. The level of the gauge must also be regularly raised in line with growth of the trees.

In very steep terrain, not only is rainfall spatially very variable but a standard gauge with horizontal rim will tend to under-catch precipitation in winds blowing upslope, and over-catch in winds blowing downslope.

The best solution may be a ground level gauge with its rim inclined parallel to the ground slope. The effective catch area must then be converted to a horizontal standard by dividing by the cosine of the slope angle.

The measurement of extreme rainfall events is of special importance to the hydrologist studying rare floods of great magnitude. Under such severe conditions, the performance of the instruments, rather than wind effects, may be the major problem (Sevruk and Geiger, 1981). Rainfall amounts may exceed the capacity of storage gauges, and high intensity rainfall can cause recording gauge mechanisms to jam, or to lose accuracy due to the finite time taken for tipping buckets to tip (~0.5 seconds). Calder and Kidd (1978) provide a dynamic calibration equation for tipping bucket gauges. An alternative is to use a weighing gauge, which will have less moving parts and may be able to cope with both solid and liquid precipitation (Grust and Stewart, 2012). However, some may not give stable readings in intense storms when the weight of accumulated rain in increasing rapidly. An international intercomparison of 26 types of recording gauges under field conditions (Lanza and Vuerich, 2009) found a similar performance between *calibrated* tipping bucket gauges and those weighing gauges that had good *dynamic stability*.

2.4.2 AREAL RAINFALL

Even if raingauges provide accurate point measurements, they are only representative of a limited spatial extent. Hydrologists often need to estimate the volume of rainfall over a catchment area and require an adequate number of measurements in order to assess the spatial variation. This may be achieved with a network of raingauges alone, or by using additional information from remote sensing by weather radar or satellites.

Design of raingauge networks

A network of raingauges represents a finite number of point samples of the two-dimensional pattern of rainfall depths. The UK has one of the highest densities of raingauges in the world with an average of one gauge per 80 km^2 (Allott, 2010); yet the total collecting area of all the gauges in the UK is less than the size of one standard football pitch! Comparable areal values (km^2 per gauge) for a range of countries include Germany (90), France (120), Netherlands (130), China (470), India (790), Australia (1,010) and US (1,040) rising to over 8,100 for Saudi Arabia and 47,400 for Mongolia (WMO, 1995b). The figures are not stable and there has been a general reduction in hydrometric networks in recent years (Mishra and Coulibaly, 2009), due partly to save costs and partly in response to an increase in methods of remote sensing. Thus in the UK, for example, the number of raingauges has fallen by nearly 50% since a peak in the 1970s (Eden, 2009), see Figure 2.6.

The accuracy of areal precipitation estimates will increase as the gauging network density increases. But a dense network is difficult and expensive to maintain, and so a number of general guidelines for gauge density have been produced. The World Meteorological Organization (Perks et al., 1996) evaluated the adequacy of hydrological networks on a global basis for the Basic Hydrological Network Assessment Project and gave the following broad guidelines for the *minimum* gauge density of precipitation networks in various geographical regions: one raingauge per 25 km^2 for small mountainous islands with irregular precipitation; 250 km^2 per gauge for mountainous areas; 575 km^2 elsewhere in temperate, Mediterranean and tropical climates, and 10,000 km^2 for arid and polar climates.

Of course, many other factors are likely to be important, including type of topography and climate characteristics. In estimating the areal pattern of rainfall from a given gauge network, errors will occur due to the random nature of storms and their paths relative to gauges, and the spatial variability of precipitation. More gauges will be required in steeply sloping terrain and in areas prone to localized thunderstorms rather than frontal rainfall. The density of gauges required also depends upon the time scale of interest; shorter period rainfall intensities (e.g. hourly) are generally much more variable than daily or annual totals.

The accuracy of areal rainfall estimation depends on both the total number of gauges and their spatial distribution. Raingauges may be sited *a priori* within a classification of 'domains' representing classes with different ranges of geographical and topographic characteristics – such as altitude, distance to the sea, ground slope and aspect – which are thought likely to influence rainfall. Alternatively, a large number of gauges may be used initially to identify the predominant areal pattern of precipitation, and the number subsequently reduced. The areal distribution of gauges might also reflect the intended use for the network; thus if the main purpose of precipitation measurement is for runoff studies, then one approach to network design would be to locate gauges in the wettest areas that contribute most to runoff.

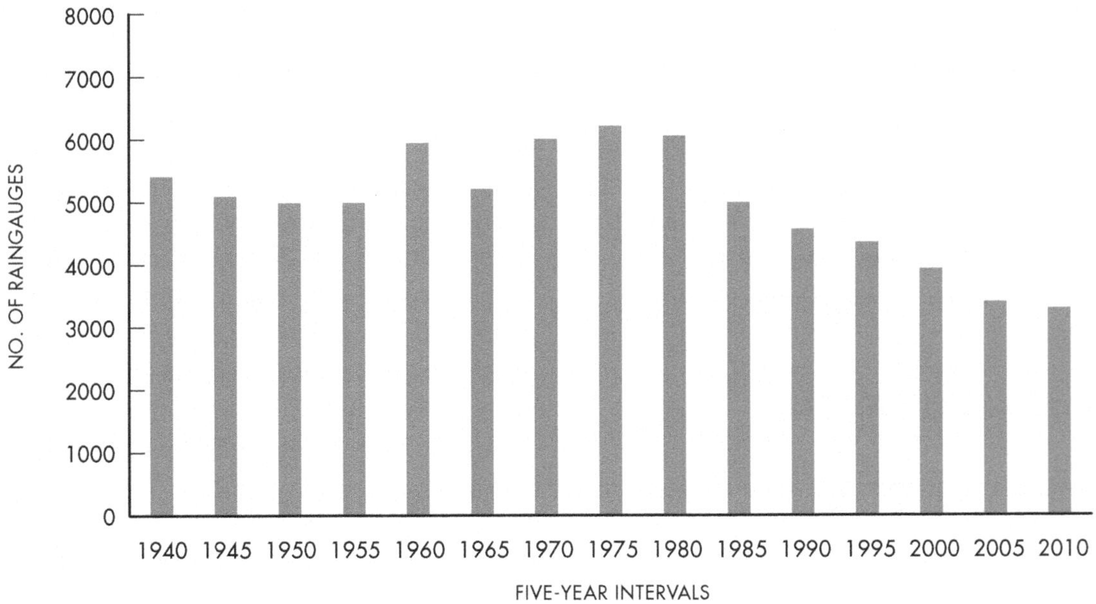

Fig 2.6: Changing raingauge network in Britain over time (redrawn from Eden, 2009).

Kriging is a statistical method that uses the variogram of the rainfall field (i.e. the variance between pairs of points at different distances apart) to optimize the gauge weightings to minimize the estimation error. It has the advantage that it can be used to generate a map of the standard error of the estimates that indicates where additional gauges would be of most benefit. IH (1999) used the technique to estimate areal rainfall, to indicate the degree of redundancy in a gauge network and to identify locations where additional gauges might be most useful (Figure 2.7).

If an area has no existing gauges to indicate the spatial distribution of precipitation, it may be necessary to transpose information on rainfall variations in time and space from a similar area to help to design a preliminary network. Various techniques, including multiple regression and kriging, have been used to quantify the statistical

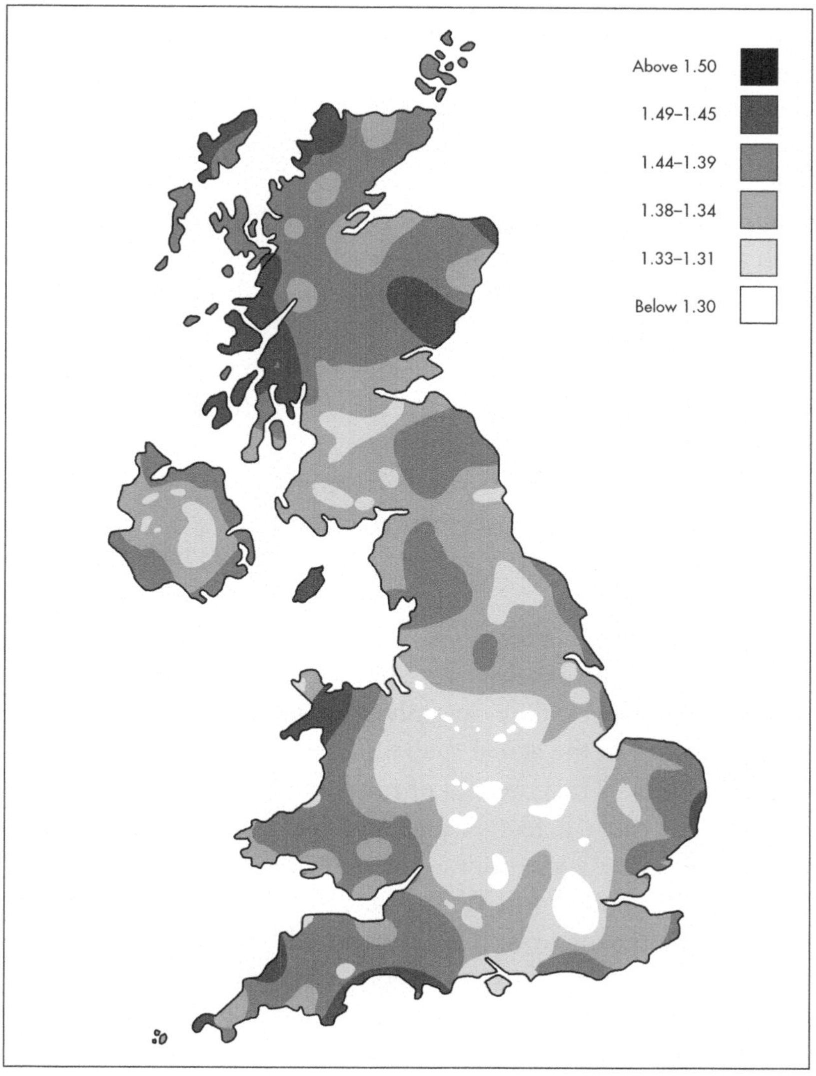

Fig 2.7: Standard deviation of estimates of 1-hour median annual maximum rain depth, RMED (mm) across the UK, produced by kriging. (Reproduced and simplified from Vol 2 Fig 7.5 of Flood Estimation Handbook, IH, 1999).

structure of rainfall patterns, and to identify a more effective network design (Periago et al., 1998). In using any method based on the correlation between gauges it is of paramount importance to ensure the homogeneity of record of each gauge. Unrecorded changes in siting or exposure may weaken correlations between gauges, resulting in networks that are denser than necessary. A number of gridded spatial datasets are routinely available (see Section 2.6.1).

It must be recognised that there are additional social and economic constraints and considerations in determining network density. A raingauge or any other network exists to serve certain objectives. The minimum network is one that will avoid serious deficiencies in developing and managing water resources on a scale commensurate with the overall level of economic development and environmental needs of the country (WMO, 2008).

Over much of the globe raingauge networks are too sparse to capture the variability of precipitation in time and space. An alternative approach is offered by remote sensing by ground-based weather radar and satellite sensors.

Weather radar and satellites

However great the density of existing networks of raingauges, they can only give an approximation to the actual spatial pattern of precipitation. The potential of **radar** (an acronym for **ra**dio **d**etection **a**nd **r**anging) to show weather patterns was first noted during the Second World War and considerable development work has been carried out in a number of countries to apply it for monitoring storm rainfall. Advances in data processing, as well as communications and display technology were needed before the full potential of meteorological radar could be exploited. The major advantage over rain gauges alone is that the radar samples and averages the rainfall in a volume of many millions of cubic metres of the atmosphere (Austin et al., 1995).

There are several ways of measuring rainfall using radar, including Doppler radar and attenuation, but the most widely adopted approach is based on radar echo, or reflectivity. The weather radar does not measure rainfall but uses a calibration relation between radar reflectivity and rainfall rate. This relation is not constant, and varies with factors including the concentration of drops, their size distribution and the pattern of vertical wind velocity. Marshall and Palmer (1948) analysed extensive experimental results on rainfall rates and drop size distributions, which is the basis of an empirical relation between rainfall rate, R (mm/h), and radar reflectivity, Z (mm^6/m^3) of the form:

$$Z = AR^B \tag{2.1}$$

The values of the parameters *A* and *B* are typically of the order of 200 and 2.6 respectively, but these can vary widely; the main reasons include:

- Variations in rain drop–size distribution, e.g. cumulonimbus have larger size drops than layer clouds;
- The presence of hail, snow or melting snow which have a much higher reflectivity than raindrops, producing a 'bright band' with artificially high estimates;
- Growth or evaporation of raindrops below the radar beam height;

- Ground echoes ('clutter') due to hills and tall buildings blocking the beam; and
- Attenuation of the signals due to heavy rain along the beam.

For many years this unpredictable variation in the relation between radar reflectivity and rainfall rate prevented the use of radar for quantitative rainfall measurement. The major advance enabling radar to be used for quantitative precipitation estimates came from the use of measured rainfall rates for 'real-time' adjustment of the Z:R relation. This lumps together all sources of radar error and deals with them in a single process. The radar precipitation field is continually 'calibrated' or 'adjusted' by point raingauge observations, while retaining the areal pattern observed from the radar.

Calibrating raingauges can be used in real-time to apply a varying correction factor to radar estimates over periods as short as a few minutes. The calibration factor can be determined in real-time using telemetry gauge information, and the nature of its temporal changes can be used to automatically identify the rainfall type.

The UK Met. Office operates a network of 15 radars across the UK with about 1,000 real-time recording gauges. The merged gauge and radar data are better than either of the individual methods alone. To achieve it in 'real-time' requires automated quality control of the gauge data and merging with the radar data within 15 minutes, and be available within 30 minutes of the end of the accumulation period. Quality-controlled rainfall is made available on a 1 km grid (Harrison et al., 2012)

Each radar emits short pulses of electromagnetic waves in a narrow beam. Between each pulse, the radar station serves as a receiver as it listens for return signals from particles in the air. The time for an echo to reach the radar indicates the distance from the radar. Each radar completes a series of scans every 5 minutes at from four to eight elevations above the horizontal (0.5 to 4 degrees, depending on the height of any surrounding hills). The lowest beam is preferred for rainfall estimation, since there is least opportunity for raindrop growth or evaporation before reaching the ground, but a higher beam may have to be substituted on compass bearings where hills and tall structures obstruct the lower beam. The scans are combined to produce 5-minute rainfall estimates. Each scan gives good quantitative data for 1 and 2 km grid resolutions within 75 km of the installation (the radar network provides this resolution over 85% of England and Wales), and on a 5 x 5 km grid for distances 75–210 km from the radar. Beyond 210 km the radar estimates are not considered to be sufficiently accurate, since the beam becomes too high and diffuse, and the return signal becomes too weak. Weather radar calibration in the UK is discussed by Harrison et al (2012) and across Europe by Huuskonen et al (2014).

The techniques of radar measurement have developed enormously with hardware advances in radar technology and data processing capacity and with software developments in real–time calibration. Radar is now widely used in operational systems. At the European level there are over 200 operational weather radar systems with a median separation of approximately 130 km (Huuskonen, et al, 2014). In the USA the NEXRAD (**Next** generation weather **rad**ar) has been operating for over 25 years and following the completion of an upgrade in 2013 has 160 Doppler radars. As well as precipitation estimates they are able to provide detailed information on wind patterns and the internal structure of storm clouds (Serafin and Wilson, 2000; Huuskonen, et al, 2014).

Urban areas are characterised by high spatially variability and fast runoff responses, with very short lag times between rainfall and potential flooding. They are particularly sensitive to high intensity downpours that may be of very limited extent. Very short wavelength (X-band) dual polar radar can provide the very fine scale rainfall measurements needed to predict hydrological response and potential flooding in urban catchments (Ten Veldhuis et al, 2014).

Raingauges or radar coverages are inadequate over much of the Earth's surface, comprising the oceans, most of the desert and semi–desert regions, most major mountainous regions and extensive humid regions in the tropics. Satellite techniques can provide more uniform and spatially continuous information than ground instruments, and are the only systematic means of estimating precipitation for the entire globe (Kidd and Levizzani, 2011). Satellite techniques were first developed for convective rainfall in the tropics and subtropics, and that is where they have been most widely applied. Tropical rainfall comprises more than two-thirds of global precipitation.

Satellites provide observations of clouds, not rainfall, and so are not as accurate as raingauges and radar for estimating rainfall depths, which should be used wherever possible. Using satellites, precipitating clouds must be inferred from cloud types and the way in which they alter through time. Rainfall rates are estimated indirectly from the albedo and radiation emissions of cloud tops, providing information on cloud extent and temperature. Most operational programmes using satellites for rainfall monitoring are based on visible and infrared wavelength radiation while research has centred on the use of passive and active microwave techniques (Kidd and Levizzani, 2011). Visible radiation is most strongly related to the albedo of highly reflective surfaces such as clouds. High brightness implies a greater cloud thickness and probability of rainfall. Thermal infrared imagery is largely dependent on temperature. Since temperature varies with altitude, this may be interpreted as indicating the cloud top height. Low temperatures imply high cloud tops and large thickness of clouds, with a greater probability of rainfall. In practice, neither bright clouds nor cold clouds necessarily produce rainfall, and the best approach is to use both types of information together (Kidd and Huffman, 2011). Rain is more likely in clouds that are both cold and bright.

Precipitation can also be inferred from microwave techniques because it is primarily attenuated by precipitation-sized particles. The Earth naturally emits low levels of microwave radiation, and the absorption of these passive microwaves is related to the total amount of liquid water in the atmosphere. Active microwave techniques are the most direct method of precipitation estimation. However, as with all radar systems, they rely on the interpretation of the backscatter of radiation from precipitation. This is broadly proportional to the number of precipitation-sized particles and therefore the precipitation intensity. This relationship is not constant, and furthermore the background reflectivity of land areas also varies, with such factors as topography, ground wetness and vegetation, while that of the oceans varies with waves on the surface.

The TAMSAT project has produced rainfall estimates over tropical Africa for over 30 years (beginning 1983) by combining thermal infra-red imagery (approx. 4 km grid) with local raingauge data from about 1,000 sites provided in cooperation with African national meteorological agencies (Tarnavsky et al, 2014).

Blended techniques use a combination of microwave data with visible and infrared wavebands (Roca et al, 2010). The Tropical Rainfall Measuring Mission (TRMM) which began in 1997 was a joint US-Japanese satellite mission to study the distribution of rainfall and latent heat transfer over the tropical and subtropical oceans and continents (Liu et al, 2012). It carried a rain-measuring package combining a spaceborne radar yielding information on the intensity and distribution of the rain, rain type and storm depth, a Visible and Infrared Scanner measuring clouds, and a passive microwave sensor to quantify the water vapour, cloud water, and rainfall intensity. The TRMM was replaced in 2014–15 by the Global Precipitation Measurement (GPM) project. This extends the observations to higher latitudes and has dual-frequency radar providing more accurate and detailed rainfall measurements.

Ground-based measurements (raingauges and radar) and satellite techniques are complementary. The former are more appropriate for areas smaller than about 10,000 km^2 (the typical extent of an individual surface radar system), while satellites are better for larger areas and where ground measurements are sparse or lacking. Raingauge measurements are still essential for calibration and checking purposes of radar and satellite estimates.

2.5 TEMPORAL VARIATIONS IN PRECIPITATION RECORDS

Variability is an intrinsic feature of the Earth's climate. Point precipitation records exhibit great changes from hour to hour, week to week and even from year to year. This variation is far larger than that of any other component of the hydrological cycle. Evaporation, for example, is strongly controlled to the radiation output from the sun and the wetness of the ground, while streamflow represents a much moderated pattern of the precipitation inputs. In principle, the pattern of precipitation is deterministic, being related to the geographic or **synoptic** weather conditions and the properties of the air masses. Considerable advances have been made in numerical prediction models of weather systems for forecasting purposes but, in practice, for many hydrological purposes, the analysis of rainfall data is often based on the statistical properties of observed rainfall time series.

Variations in precipitation records may incorporate three time series components: stochastic, periodic and secular. **Stochastic** variations result from the probabilistic or random nature of precipitation occurrence, and may be so great that they effectively dominate the time series. **Periodic** or cyclic variations may be related to the diurnal and annual astronomical cycles. Finally, **secular** or long–term variations of climatic change may incorporate both cyclic and trend characteristics.

2.5.1 STOCHASTIC VARIATIONS

The great variability in rainfall totals can be explained by the frequency distribution of rainfall, since only a small proportion of storms or rain days in a year may provide a disproportionate amount of the total rainfall (Figure 2.8). The presence or absence of only a small number of storms may therefore have a considerable effect upon the total precipitation. The variability of annual rainfalls is much greater for areas with low average annual precipitation, where rain may fall only occasionally, than for, say, equatorial regions where rain may occur nearly every day. Thus, estimates of water resources in arid and semi–arid areas are particularly sensitive to short precipitation records.

In addition to rainfall amounts, the time intervals between storms are of great interest to the hydrologist, especially in the drier parts of the world. The importance of the time interval depends upon the storage capacity and the depletion characteristics of the particular system of interest, such as a column of soil or a water supply reservoir. Due to differences in the vapour inflows and the mode of uplift of the air between passing weather systems there is a tendency for rainy days to cluster in groups. This tendency for serial correlation, or **persistence**, has often been described using Markov chain analysis (Essenwanger, 1986).

The timing and magnitude of individual storms are largely stochastic in nature, and a number of studies (e.g. Essenwanger, 1986) have represented variables such as the time interval between storms, the storm durations and the precipitation depths by statistical frequency distributions (see Figure 2.9). The pattern of rainfall during a storm, however, will be largely deterministic since it depends upon the weather system. Convectional rainfall is usually of higher intensity and shorter duration than rain from frontal systems. In general, convective and frontal type storms tend to have their peak rainfall rates near the beginning, while cyclonic events reach their maximum intensity nearer the middle of the storm. The temporal profile of rainfall is almost infinitely variable, depending not only on the rainfall type, and the state of development or decay of the rainfall system as it passes over the rainfall measurement point, but also on the speed of movement of the system. If enough storms are examined for a particular site then their shapes can be summarized statistically.

2.5.2 PERIODIC VARIATIONS

These are regular cyclic variations with rainfall minima and maxima recurring

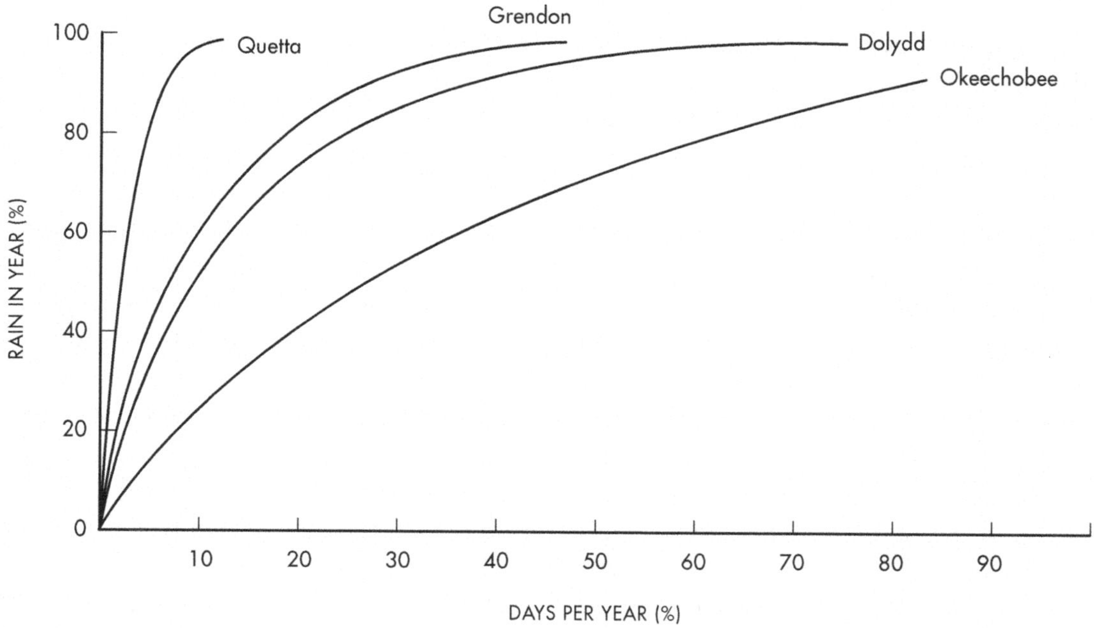

Fig 2.8: Percentage of annual rainfall occurring in a given percentage of the year for Grendon (lowland Britain, 630 mm y^{-1}), Dolydd (upland Britain, 1780 mm y^{-1}), Quetta (N Pakistan, 200 mm y^{-1}) and the unusual situation at Okeechobee (S Florida, 1330 mm y^{-1}) where the more even rain distribution results from its mix of both synoptic disturbances and sea breezes. (Based on data from CEH, and Burpee and Lahiff, 1984)

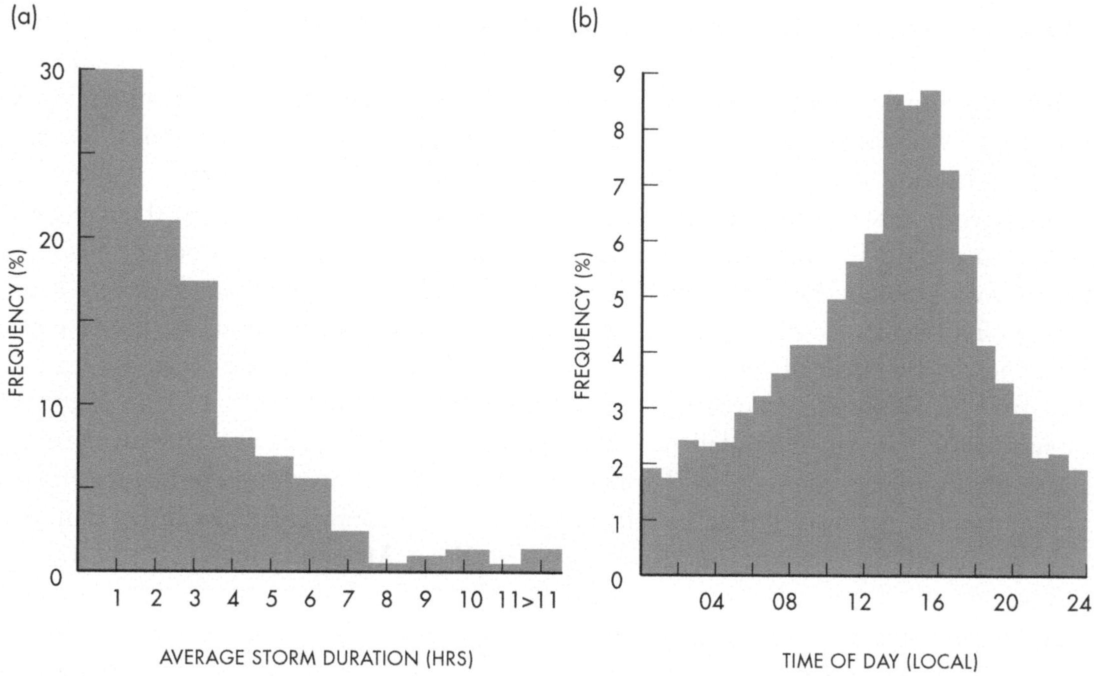

Fig 2.9: Frequency distribution of a) storm duration and b) time of occurrence of rainfall during the day for a site in central Amazonia (Lloyd, 1990).

after approximately equal time intervals. The best known are the daily and annual cycles leading to more rain over land at the warmest times of the day and during the warmest seasons – when the water vapour content of the air is high and thermal convection is strongest – and least rain around dawn and in late winter.

Diurnal or within–day variations occur in areas where a large proportion of the rainfall derives from convective storms generated by local surface heating. This pattern is most often found in warm tropical continental climates although a preference for certain times of the day has been noted in many other areas. Peak convective activity in the middle of the day tends to be self-supporting (due to the release of latent energy) often leading to heavy thunderstorms and can last into the evening. In Quetta in northern Pakistan, for example, 80% of the annual rain falls between 2pm and 8pm (Rudloff, 1981). The strength of such convectional rainfall patterns will vary through the year with changes in the degree of radiant heating and convection, and variations in evaporation and hence the vapour content of the air. The pattern of an afternoon maximum may be modified by the interaction of land and sea breezes near the coast and by the effect of topography. For most parts of the world, however, there is no systematic pattern of rainfall over the course of the day. A far more widespread cycle in rainfall patterns is that associated with the changing seasons of the year.

The annual cycle is the most obvious weather cycle and results from the regular seasonal shifts in the zones of atmospheric circulation in response to the changes in the heating patterns accompanying the migration of the zenith sun between latitudes 23°N and 23°S. Near to the equator these movements usually result in two maxima, while in tropical areas there is often a distinct summer rainfall maximum. Such changes in rainfall are most pronounced in areas on the fringes of arid zones, such as the Mediterranean region, where cyclonic depressions bring rain in the winter but the summers are typically dry. In Europe, to the north and east of the Alps, reflecting the increasing continentality, the greater proportion of rainfall occurs in the summer half–year. Over much of Asia there is a marked dominance of summer rainfall associated with the summer monsoon. The annual rainfall regimes at locations in different climatic zones of the world are described in standard climatology textbooks (e.g. Martyn, 1992).

The most famous seasonal rains are the **Asian monsoon**, and almost 1 billion people are affected by this annual rainfall that brings essential water for drinking and, farming, but can also bring about damaging floods, and the intermittent failures of the Asian monsoon has been the cause of terrible famines. The monsoon occurs due to temperature differences between land and sea. Summer warming of the ground makes it warmer than the ocean waters. As air heats over the land it becomes lighter and rises, drawing cooler wetter air in from the sea. When this air meets the Himalayas it rises and cools leading to heavy rainfall. The **West African monsoon** is the result of the seasonal shifts of the Intertropical Convergence Zone and the great seasonal temperature and humidity differences between the Sahara and the equatorial Atlantic Ocean. It migrates northward from the equatorial Atlantic in February, reaches western Africa on or near June, and then moves back to the south by October.

Even in temperate climates, without a strong seasonal rainfall pattern, there may be fairly regular seasonal weather sequences termed **singularities**. Certain types of weather tend to occur at the same time each year with a greater frequency than expected for a random distribution. Thus, in Britain and large parts of continental Europe, there are certain dates when the probability of rainfall is significantly higher or lower than the average. The reason is that the general circulation of the atmosphere is strongest in winter, when global heat gradients are greatest, and this often results in cyclonic activity and high rainfall. Circulation strength decreases to a minimum in spring (March to May) which for many parts of north–west Europe is the period of lowest rainfall. Singularities in circulation types and rainfall regimes have also been noted in the Mediterranean region (Kutiel et al., 1998). Although singularities do not provide definite forecasts, they can provide a reasonably sound guide, with a physical basis to likely periods of unsettled weather and a higher risk of heavy rainfall.

2.5.3 SECULAR VARIATIONS

It is clear that climate has changed in the past (Section 1.2) and even excluding human influences (Section 9.3) there is no reason to suppose that it will not change in the future. Analyses of historical, botanical and palaeoclimate records indicate that parts of Europe and North America experienced a warmer climate than at present during the 'Medieval Warm Period' between the 10th and 13th Centuries, at a time when, for example, grain cultivation in Norway extended north of the Arctic Circle. Towards the end of the 13th Century the climate began to become colder, and in Europe the frequency of severe winters and cool wet summers increased. The period from the early 16th Century to the mid-19th Century has been called the 'Little Ice Age'.

Nevertheless, until relatively recently hydrologists assumed the principle of **stationarity** – namely that any changes occur so slowly that the historical record is a suitable base for characterising current and future conditions (e.g. Figure 2.10). This is of great significance to hydrology, since most design procedures (including flood estimation and water resource assessments) are based on this assumption.

Whilst attempts to find generally applicable cycles of precipitation have been largely unsuccessful, investigations have demonstrated numerous examples of non–cyclic secular variations of precipitation, and some studies have considered the consequent effects on runoff. Karl and Knight (1998) analysed US daily rainfall data from nearly 200 gauges for the period 1910–96, and discussed this in relation to an observed increase in river flooding. They concluded that precipitation had increased by about 10% over this period. This was partly due to an increase in the number of days in each year with rainfall, and more importantly to an increase in the frequency of very heavy falls (>50 mm per day).

Investigations have emphasised that many non–cyclic variations in precipitation are caused directly by a combination of geographical and climatological factors. Observed changes in rainfall amounts have been attributed to shifts in the global wind circulation resulting in changes in the paths of rain–bearing winds. An increase in the vigour of the mid–latitude westerly airflow circulation from the mid–1970s resulted in a consistent pattern of increased rainfall over the western uplands of the British Isles (Mayes, 1996). Osborn et al (2000) examined the frequency of heavy falls at 110 gauges in the UK and found there had been an increase in winter events, while in summer there had been a decrease. This is in line with the general trend in the UK since the middle of the 19th Century of decreasing precipitation in summer and an increase in winter (Strangeways, 2007). Fowler and Kilsby (2003) analysed records from 1961-2000 for more than 200 raingauges across the UK and found there had been an increase in the frequency of multi-day prolonged heavy rainfall in the (wetter) northern and western regions. Subsequently Kendon et al (2014) looked at the implications of possible climate change scenarios on sub-daily rainfall and found a likely intensification in short-duration summer convective rainfall intensities that could lead to an increase in flash flooding.

Of particular note has been the recognition of the importance of the **El Niño** phenomenon, the first climate feature to be shown to depend upon coupled interactions of the dynamics of *both* ocean and atmosphere. This greatly modifies the energy balance of the Pacific – the world's largest ocean – and a huge energy store. In some years the normally westward flow of

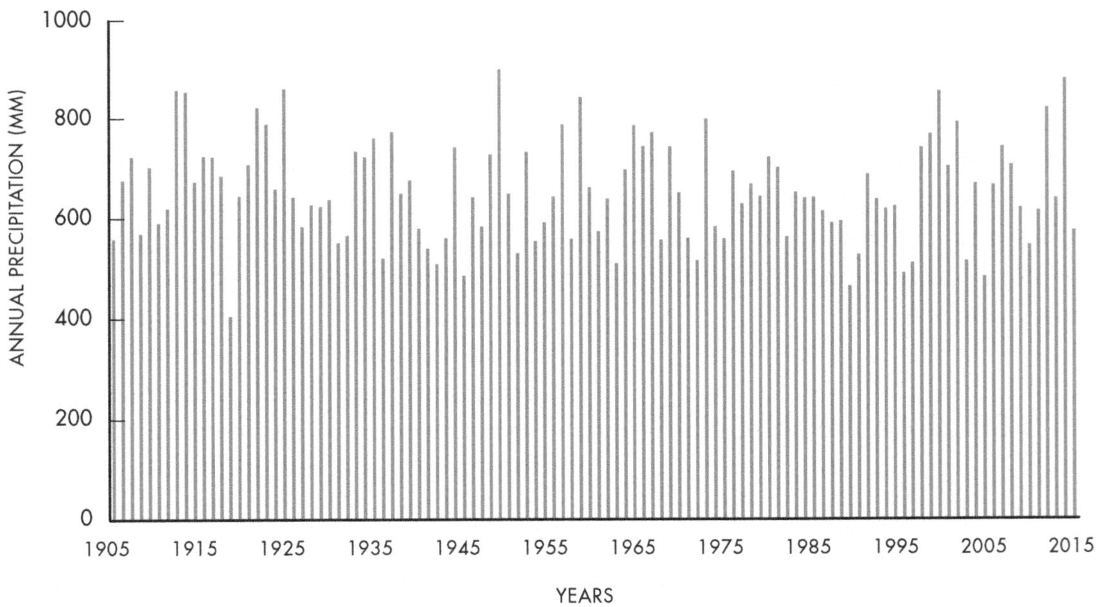

Fig 2.10: Annual precipitation at Reading, Berkshire, 1908–2015 shows no sign of a long-term trend, although recorded air temperatures were rising over this period. Data provided by University of Reading, Department of Meteorology (Brugge and Burt, 2015).

the atmosphere and the ocean in the equatorial Pacific is reversed, resulting in significantly increased rainfall in the eastern Pacific, especially the normally dry coastal zone of Ecuador and Peru. Conversely, in the western Pacific, there may be drought in Indonesia and northern and eastern Australia. The El Niño typically develops in May-August, peaking in December to April (when it has most effect on precipitation) and then declining in the following May to June. This result from the Southern Oscillation, a major shift in air pressure between the Asian and east Pacific regions. The intensity of this change varies, but it is particularly strong on average every 2 to 7 years. There are also long-distance atmospheric **teleconnections** – with impacts around the world including drier conditions in north east Brazil and a weakening or even failure of the monsoon in India and eastern Africa. Thus, the very strong El Niño of 1982–83, which caused flooding along Peru's normally dry northern coastline, has been linked to the severe droughts and wildfires in Indonesia and Australia and intense storms in the southern USA causing over US $8 billion in global damage. The even more extreme El Niño of 1997–98 caused the loss of approximately 23,000 lives and an estimated US $35–40 billion in damages worldwide. The 2015–16 El Niño was one of the strongest on record and was to blame for floods in parts of South America including Paraguay, Peru and Argentina, and contributed to drought in southern Africa and parts of the Horn of Africa, the Caribbean and Central America, and Southeast Asian countries including Indonesia and Vietnam. Recent climate modelling studies highlight the increasing likelihood of such events due to greenhouse warming (Cai et al, 2014).

Dai et al. (1997) examined global datasets and found evidence that even in some middle latitude regions (including western and central Europe) precipitation is moderately affected by the ENSO (El Niño Southern Oscillation) events. Ogallo (1988) noted that there were some relationships between the Southern Oscillation and seasonal rainfall over parts of East Africa. There is evidence that the impact of the ENSO is felt in crop production, including grain yields in south Asia, Australia, East Africa and the North American Prairies (Hansen et al., 1997). Although due to the many economic, social and technical factors influencing crop yields it is very difficult to quantify the impact (Iizumi and Ramankutty, 2015).

While the El Niño impacts are mostly felt in the tropics there are several other pressure oscillations, but they have less global importance as they are not in the equatorial zone where most precipitation occurs. In the Northern hemisphere there are important changes in the atmospheric circulation, which is characterized by the North Atlantic Oscillation Index (NAOI) defined by the atmospheric pressure difference between Iceland and the Azores (Hurrell, 1995). When the pressure gradient is stronger this results in more frequent westerly circulation types, more Atlantic frontal systems and higher rainfall.

2.6 ANALYSIS OF PRECIPITATION DATA

In order to aid comparisons between different places, there is international agreement on the use of a standard period of record for the calculation of climatological 'normals', including precipitation. At present, this is the 30–year span 1961–90, which has been retained for comparison with more recent weather to highlight any observed warming trend. In practice it is not always possible or desirable to use this period, for example when records are short or when comparing current weather with the seasonal normal – what people have experienced in recent years. Arguably in a transient climate more weight should be given to more recent data (Fowler and Kilsby, 2003). Whenever comparisons are made between areas, efforts should be made to ensure that a common time period is used, and to specify the period of record chosen.

Several aspects concerned with the use and interpretation of precipitation data are of direct concern to the hydrologist. A basic requirement is to estimate the average rainfall over an area from a number of point measurements, or perhaps to determine the spatial pattern and movement of an individual storm, often from comparatively widely separated gauges. Hydrologists are also interested in the frequency of occurrence of rainfalls of different magnitudes, and so study the statistical properties of rainfall data. Finally, there is the special case of trying to estimate the largest rainfall that is physically possible over a given area, i.e. what is the Probable Maximum Precipitation?

2.6.1 CATCHMENT MEAN RAINFALL

An estimate of mean areal rainfall input is a basic requirement in many hydrological applications, including water balance and rainfall–runoff studies. There are many techniques for calculating areal rainfall from point measurements, including polygonal weighting, inverse distance weighting, isohyetal, trend surface analysis, analysis of variance and kriging. Singh (1989) provides a detailed discussion of 15 different methods. The selection of the most appropriate one for a particular problem will depend upon a number of factors, including: the density of the gauge network, the known spatial variability of the rainfall field, the time available and the expertise of the hydrologist,. In general, the accuracy of all the methods for estimating areal rainfall will increase with: a) the density of gauges, (b) the length of period considered and (c) the size of area.

The **isohyetal** method is potentially the most reliable of the standard methods of areal precipitation calculation, but it is subjective. It involves manually drawing **isohyets**, or lines of equal rainfall, on a map between gauges, making judgements based on local knowledge for factors such as topography, prevailing winds and distance from the sea. Areal precipitation is then computed by calculating the areas between the isohyets. This method uses all the data and knowledge about rainfall patterns in a particular area, but can involve a considerable amount of time to construct the maps and so it is only feasible for special studies of small catchments.

The simplest objective and automated technique suitable for large numbers of basins is to calculate the arithmetic mean of all the raingauge totals within the area of interest. This may be satisfactory for areas of flat topography with little systematic variation in rainfall and a uniform distribution of gauges. Such conditions are not generally found in practice and there is often a tendency for the distribution of gauges to mirror human populations. Thus, gauges are often most widely spaced in mountainous areas where rainfall depths and spatial variability are typically greatest.

The nearest neighbour **Thiessen polygon** method (Thiessen, 1911), has been widely adopted as a better method for calculating areal rainfall than the mean. It allows for a non–uniform distribution of gauges by assigning 'weights' to the measured depths at each gauge according to the proportion of the catchment area that is nearest to that gauge. The method may be carried out graphically, or can be programmed for computer application by superimposing a regular grid over the area and allocating each grid point value to the nearest gauge. The resulting rainfall surface is, however, a series of polygonal plateaus with sharp steps between them. A modification is to allocate a region or 'domain' to each gauge based on physical factors such as local meteorological conditions and topography thought likely to influence rainfall (see Section 2.4.2). British Standards (BSI, 2012, Part 4) recommends Voronoi interpolation which is a development of the Thiessen method and has the advantage for hydrological applications that it produces a much more realistic rainfall surface.

Other methods of estimating areal rainfall overlay the area with a regular grid. Rainfall is then estimated for each grid square and these are averaged to give the areal mean. A common method for computing depths at these grid points is to weight the values at the nearby gauges by the inverse square of their distance to the grid point (Essenwanger, 1986). The UK Met. Office has generated 5 km grids of a range of weather variables including rainfall from 1961 onwards (Perry et al, 2009). For many applications there is a need for finer spatial resolution and the Gridded Estimates of Areal Rainfall dataset (GEAR) provides 1 km gridded estimates of daily and monthly rainfall totals for Great Britain and N. Ireland from 1890 to the present (Keller et al, 2015). It is updated each year with a 2–3 year time lag whilst data from the contributing organisations are collated, quality controlled and analysed. Very importantly, given the changes in the size of the observation network over time, GEAR provides information on the distance to the closest gauge used to calculate the rainfall at each grid cell. Access to the data is free, and via the CEH Information Gateway https://gateway.ceh.ac.uk.

The preceding discussion is applicable to situations for which radar data are either not available or inappropriate. These include locations that are distant from radar installations or are very hilly (leading to obstruction of the beam), and to cases where information is required over very small areas (current national weather radar systems provide rainfall data on grid sizes of a kilometre or more).

2.6.2 STORM PRECIPITATION PATTERNS

Radar greatly improves the spatial (and temporal) interpolation between gauge observations and can give a detailed quantitative record of the movement of storm systems over large areas. In the absence of reliable radar estimates, storm cell movement has been studied for small areas (about 25 km^2) using dense networks of recording gauges, but it was very difficult to synchronize and validate recording rainfall data collected independently at very many sites. An operational network of weather radars provides detailed information on storm precipitation intensities and distributions in Britain, and further integration of networks of weather radars is planned for Western Europe. Figure 2.11 shows the synoptic weather map and weather radar during a large frontal storm that brought a band of heavy rain from central southern England to eastern Scotland.

Radar data are important for 'real-time' flood warnings and for monitoring storm cell patterns that are too small for raingauge networks. Current systems can produce a 1 km rainfall grid at 5 minute resolution, and developments in weather radar technology may soon be able to produce a much finer grid at 100 m grid and 1 minute resolution that will be valuable for urban flood design (Ten Veldhuis et al, 2014).

2.6.3 RAINFALL STATISTICS

The frequency of heavy rainfall is of interest to the hydrologist for a number of reasons including the design of hydraulic structures such as bridges, culverts and flood alleviation schemes. In Britain, designs are commonly based on rainfall depths with a **return period** (the *average* length of time between occurrences) of between 2 years and 100 years. The exceedance probability within a time period, generally one year, is the reciprocal of the return period. Thus, a storm with a 1% chance of occurring in any year has an annual exceedance probability of 0.01 and a return period of 100 years. The duration of the design storm rainfall that is selected will depend upon the design objectives. The critical storm duration may be several days for very slowly responding river catchments, but only a few hours for medium and small catchments. In the case of storm sewer design in fast–responding impermeable urban catchments, rainfall inputs over only a few minutes may be appropriate.

The choice and fitting of alternative frequency distributions to rainfall data are discussed in various texts (e.g. Sevruk and Geiger, 1981; Essenwanger, 1986, IH, 1999, vol 2), to which the reader may refer for further details.

To date due to the relatively short period of radar observations and the uncertainties and errors in deriving quantitative estimates of precipitation for many applications such as flood risk assessment conventional measurements from rain gauges are still preferred (Sebastianelli et al, 2013).

Point rainfall frequencies

Daily read raingauge records are the most commonly analysed rainfall data, largely because of their greater availability both in terms of the number of measurement points and the length of records, compared

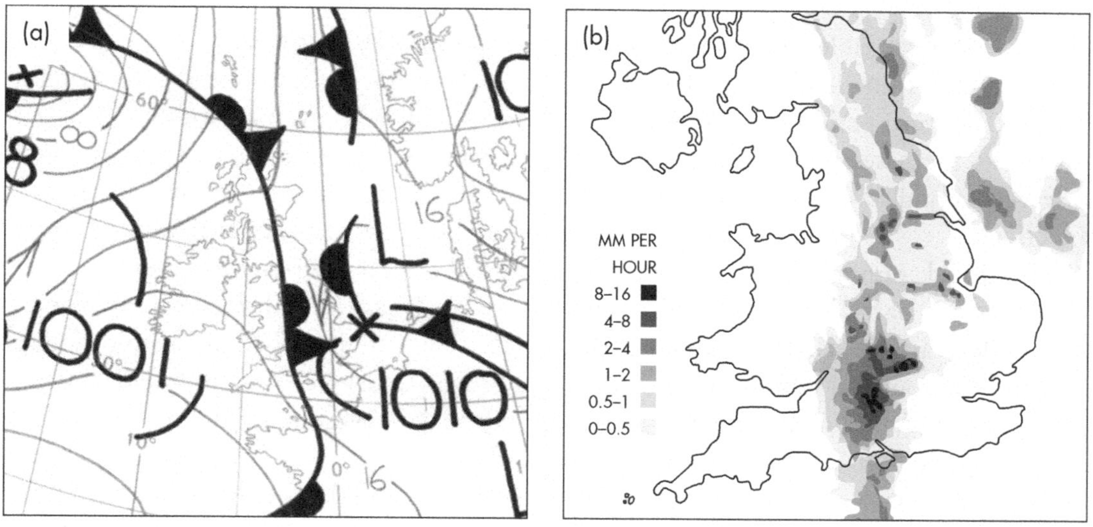

Fig 2.11: Storm of 3 June 2008 showing : (a) Traditional synoptic weather map, and (b) Calibrated hourly rainfall. © Crown Copyright 2016, adapted from image from Met Office'. http://www.metoffice.gov.uk/learning/library/publications/historical-facts.

with shorter time interval information (Sevruk and Geiger, 1981). Accordingly they were used in the UK Flood Studies Report (FSR) (NERC, 1975), which provides guidance on how daily falls of given return periods may be estimated for different durations at each site (NERC, 1975, IH, 1999 Vol 2). A problem with the use of daily data is that the time interval must correspond to a standard 'rain day'. For example, in the United Kingdom, storage raingauges are traditionally read and emptied at 0900 GMT each day, and consequently the precipitation from storms which span this interval is split between two rain days. Accordingly, the maximum falls in rain day periods are smaller than the maximum falls over a 24–hour duration, on average by 14% (Sevruk and Geiger, 1981) to 16% (Dwyer and Reed, 1995). The use of daily totalled rainfall can be even more misleading for the estimation of maximum short–duration intensities, since rain may only fall for a small part of the day.

Short–duration rainfall statistics have been studied by a number of investigators using data from recording raingauges. In a major national study of rainfall statistics in the 1970s the UK Meteorological Office analysed data from approximately 200 recording gauges, in addition to records from over 6000 daily read gauges (NERC, 1975, Vol. II). The UK rainfall extremes were reanalysed in the Institute of Hydrology's Flood Estimation Handbook (FEH) (IH, 1999; Stewart el al, 2015). This benefitted from a larger number of raingauges, but more importantly the FEH incorporated an additional 25 years of recording gauge data. The final dataset consisted of over 6,500 daily raingauges (a slight increase in the number used in the FSR), and over 900 hourly gauges (substantially more than the FSR).

The procedures enable the estimation of a design rainfall from 1 hour to 8 days duration, and a return period up to 1,000 years, as well as the assessment of the rarity of observed rainfall events at any location in the UK.

The FEH rainfall frequency analysis comprises two parts. Firstly, the estimation of an index variable, RMED, which is the median of the annual rainfall maxima of a given duration (and for an annual series has a return period of 2 years). Secondly, a rainfall growth curve is derived for the site of interest based on pooling data from all nearby gauges. This enables the index variable to be scaled up to the desired design return period. Figure 2.12 shows the computed 1–hour rainfall depth of 100–year return period.

Studies of storm rainfall depth, duration and frequency indicate that because very high rain intensities occur only rarely they contribute less overall to annual totals of rainfall than smaller, but more frequent falls. **Depth–duration frequency (DDF)** curves present rain depth and duration for different return periods. They are steeper for areas with convectional rain than for those with predominantly frontal storms characterized by longer, less intense rainfall. A new rainfall DDF model for the UK (termed FEH13) was completed in 2015, providing rainfall frequency estimates for durations between 1 hour and 8 days for return periods of up to 10,000 years (see Stewart el al, 2012, 2015). The FEH web service provides online access to the catchment descriptors and rainfall model outputs including the FEH13 rainfall model – see http://fehweb.ceh.ac.uk.

Areal rainfall frequencies

The hydrologist often needs to estimate the rainfall of a given return period over a catchment area rather than at a point. Since the T–year return period rain depth at a point is bound to be larger than the average rainfall of that return period over an extensive area, an **Areal Reduction Factor (ARF)** is often applied to a point value. Svensson and Jones (2010) provide a comprehensive review and discussion of the technique. Values of ARF are typically between 1 and 0.7, being smaller for larger areas and for shorter duration storms. For a given storm duration this ratio appears to be constant for different return periods and for different regions of the UK. Work incorporating weather radar data (Stewart, 1989), has shown that, in line with studies in other parts of the world, ARFs for a given area and duration decrease slightly with increasing return period.

Determining the Probable Maximum Precipitation (PMP)

From a consideration of the processes of precipitation generation, it is generally agreed by most meteorologists that there must be a physical upper limit to the amount of precipitation that can fall on a given area in a given time. The difficulty lies in estimating that amount, as such extreme events are rarely observed, and almost never measured properly. Current knowledge of storm mechanisms and their precipitation producing efficiencies is inadequate to enable the precise evaluation of extreme precipitation, and assumptions have to be made. Nevertheless, for the design of structures, such as a large dam, where failure due to an overtopping flood would have catastrophic consequences in terms of environmental or physical damage or the loss of life, it is necessary to estimate the precipitation with a very low risk of exceedance.

The upper limit to precipitation is termed the **Probable Maximum Precipitation (PMP)** and is defined as "the greatest depth of precipitation for a given duration that is meteorologically possible for a given storm area at a particular location at a particular time of the year, with no allowance made for long–term climatic trends" (WMO, 2009). The word 'probable' is intended to emphasize that, due to inadequate understanding of the physics of atmospheric processes, and imperfect meteorological data, it is impossible to define with certainty an absolute maximum precipitation. It is *not* intended to indicate a particular level of statistical probability or return period, although such associations are sometimes

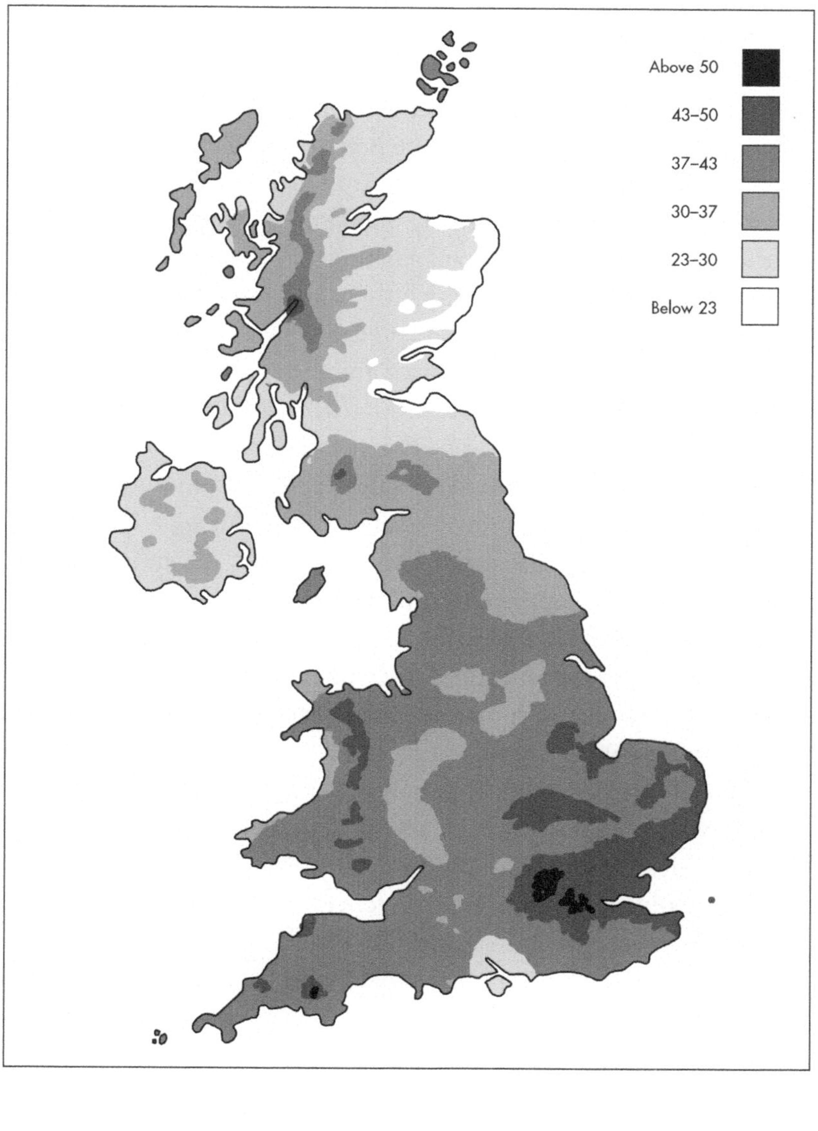

Fig 2.12: One–hour rainfall depth (mm) of 100 year return period across the UK. (Reproduced and simplified from Fig 11.6 in Vol 2. Flood Estimation Handbook, IH, 1999).

made. In recognition of the uncertainties in estimating the PMP, yet the usefulness of such a concept, the PMP has often been called a 'convenient fiction'.

There are two main approaches for estimating PMP (WMO, 2009). Firstly, the maximization and transposition of real storm events, and secondly the statistical analysis of extreme rainfalls. Storm maximization and transposition is the most widely used method, and involves the estimation of the maximum limit on the humidity concentration in the air that flows into the space above a basin, the maximum rate at which wind may carry humid air into the basin and the maximum fraction of the inflowing water vapour that can be precipitated. PMP estimates in areas of limited orographic control are normally prepared by the maximization and transposition of observed storms. In areas in which there are strong orographic controls on the precipitation, storm models have been used for the maximization procedure (Wiesner, 1970).

The maximization/transposition techniques require a large amount of data, particularly rainfall data, and involve subjective decisions regarding the maximum values assigned to the meteorological factors. Austin et al. (1995) used weather radar data with a convective storm model to estimate PMP in a more objective manner, for reservoired catchments in northwest Britain.

In the absence of suitable data it may be necessary to transpose storms across large distances despite the considerable uncertainties involved. In this case, reference to published values of maximum observed point rainfalls can be helpful. Worldwide maximum falls for various durations (WMO, 2009) are shown in Figure 2.13, together for comparison with maximum recorded falls in the UK.

The world maxima are dominated by tropical region rainstorms due to their plentiful supplies of moisture and heat energy. By comparison maximum falls are much smaller in the temperate areas such as the UK, having typically less than a third of the depth for a given duration. It is to be expected that areas of less rugged topography and a cooler temperate climate area would experience less intense falls than tropical zones subject to hurricanes or the monsoons of southern Asia. Thus, for example, La Reunion is a mountainous volcanic island in the Indian Ocean (rising to 3,000 m elevation) and subject to intense convective storms, and Cherrapunji is on a plateau about 600 m above the surrounding plain. In both cases the extreme falls were augmented by orographic effects. The highest recorded daily rainfall in the UK of 341 mm in December 2015 at Honister Pass (358 m elevation) was dwarfed by the record fall of over 1,800 mm recorded at La Reunion (gauge at over 2,500 m elevation) during a 1966 tropical cyclone.

From an analysis of the maximum rainfall in each year at several thousand gauges, Hershfield (1961) used a general formula for the analysis of extreme value data to relate the PMP depth (mm) for a selected duration to the mean (X) and standard deviation (σ) of the largest falls in each year:

$$\mathrm{PMP} = X + K\sigma \qquad (2.2)$$

Parameter K was originally set to 15 based on daily rainfalls, but it was found to vary widely between sites, and the method was subsequently modified to allow K to vary with the mean annual maximum, X, and the storm duration (Chow et al., 1988; WMO, 2009). Although the method is somewhat crude, it has the apparent advantages that it is easy to use, it is based on observed data and – since the processes in short, intense, thunderstorms are similar for different parts of the world – it could be widely applicable for such rainfall conditions. The disadvantages are that, like all statistical methods, its success depends upon the length and nature of the available record; and parameter K may depend on factors other than rainfall duration and the mean of the annual maxima. In particular, the frequency of thunderstorms varies widely. Consequently, this 'quick' approach should only be used in conjunction with other methods (Wiesner, 1970). In the final analysis, however, there is no completely objective way of assessing the level of a PMP estimate, and judgement based on an understanding of the meteorological processes is most important.

Droughts

As well as being concerned with the distribution and amount of above–average rainfall, hydrologists are also concerned with periods of **drought**. This is a regular and normal part of climate, and is not to be confused with **aridity**, which is a perennial state of very low average precipitation. Droughts are a normal, recurrent feature of climate. There are *many* different ways of defining a drought, depending on the particular purpose and disciplinary perspectives. A meteorologist may define a drought in terms of a departure from the 'normal' or average precipitation. But

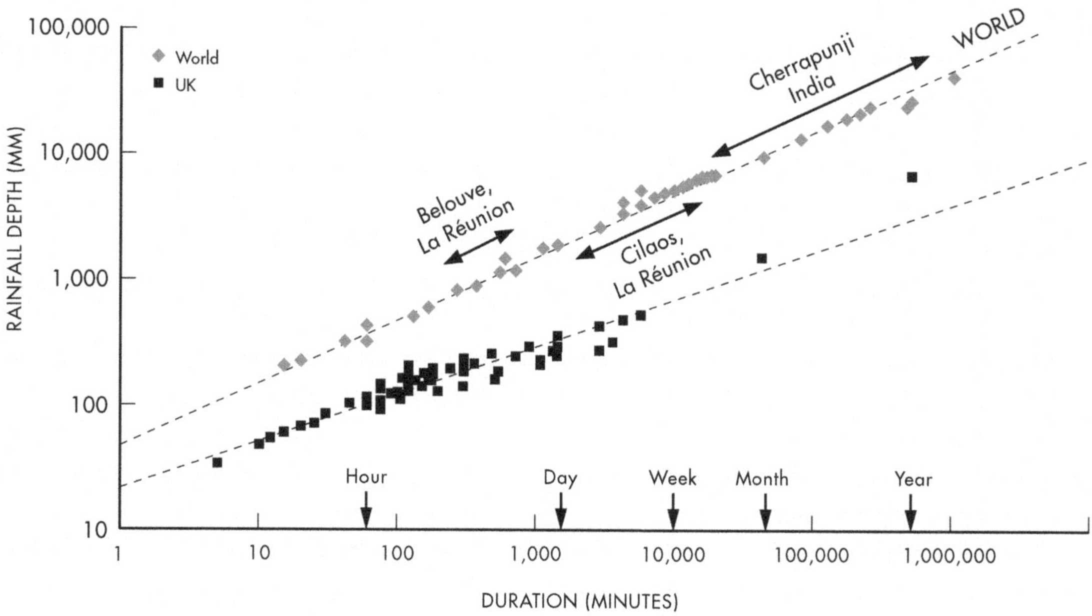

Fig 2.13: Maximum recorded point rainfalls of different durations across the world and the UK (Data sources: WMO, 2009; Burt 2005; UK Met Office, unpublished).

rainfall alone does not give the complete picture; other factors that intensify or mitigate drought impacts include temperature, sunshine, soil type and vegetation development, as well as the initial state of water reserves in soils, aquifers, streams and lakes. Farmers will be particularly concerned with the impact upon soil water contents during the growing season and hence upon plant development and crop yield. Water resource managers may distinguish between long duration groundwater droughts, which include one or more dry winters, and surface water droughts over a much shorter period. Thus, a dry winter (limiting aquifer recharge) followed by a wet spring (replenishing the surface soil layers) may result in a 'groundwater drought' without any soil water stress. Reduced flows in rivers may impact adversely on water quality by reducing the dilution of pollutants, increasing the risk of eutrophication and sedimentation of channels. Ecological impacts of a drought may be felt in terms of effects on plant communities and the ecosystems they support, both terrestrial and aquatic life.

Droughts may be characterised in terms of three essential characteristics: a) intensity or severity (usually measured as a departure of a climatic index from normal), b) duration, and c) spatial extent (Wilhite, 1993). Over 150 definitions of drought are described in the literature, but there is no consensus on which indicator best represents drought impact occurrence.

One of the most widely used is the Palmer Drought Severity Index (PDSI), which uses a simple water balance computation based on weekly or monthly rainfall and air temperature data. However, it is slow to detect fast-emerging droughts, and its calculations are based on arbitrary rules (Alley, 1984). The Standardised Precipitation Index (SPI) is based on the probability of precipitation over different time scales (McKee et al, 1993). The difficulty with any drought index is the choice of thresholds that define the *onset* and the *end* of a drought (Bachmair et al, 2015). Reed (1995) discussed some of the problems and limitations of drought assessment. Rodda and Marsh (2011) reviewed the 1975–76 drought in the UK and concluded that while its severity by some measures (including groundwater) was matched by several more recent droughts, it was nevertheless exceptional in terms of its impact on river flows, its areal extent and range of impacts.

The US is far ahead of Europe in providing continental-scale drought indices and forecast systems. The U.S. Drought Monitor provides a blend of numeric measures of drought (including the PDSI, SPI, satellite-based assessments of vegetation health, indicators of soil water status and snowpack amount) together with experts' judgment into a composite map every week (www.drought.gov). There is a move to more proactive risk management rather than crisis management with regard to droughts, but drought forecasting is difficult as the seasonal forcing functions are not well understood, and will vary over time and space. For example the El Niño may be good guide for drought forecasts in Australia, but not for other areas.

2.7 HYDROLOGICAL ASPECTS OF SNOW

Snow and ice account for just over 75% of the Earth's fresh water, although most of this is held as ice in Antarctica and Greenland, with a residence time of the order of 10,000 years. Of more relevance to the hydrologist is the fact that about 5% of the global precipitation falls as snow. This is, of course, not evenly distributed and, for example, snow accounts for about 15% of the annual precipitation in the contiguous United States. Hydrological interest in snow is concentrated in middle and higher latitudes and in mountainous areas, although even in Britain with its mild temperate climate, flooding due in part to melting snow is a regular occurrence in upland areas (Smith and Ward, 1998).

Snow has a great hydrological importance, as it is a very important 'on/off' switch affecting radiation budgets. It has a cooling effect on climate by increasing the albedo, and modifying the surface radiation balance and the near–surface air temperature, and it causes a great amount of energy to be expended on melting. Seasonal snow cover changes are known to affect global atmospheric circulation, and may have an important role influencing climatic change.

In arid and semi-arid areas bordered by high mountains, including semi-arid western United States, northern India and Iran, snowmelt is an important seasonal source of water (Section 7.9). The presence of snow on the ground is important due to disruption of travel and commerce, and seasonal flood risk may be increased by snowmelt. In addition to providing a store of water, snow cover can serve as protective insulation for soil and crops through the winter. For these reasons the hydrologist is normally interested in a number of aspects of snowfall – *where* it falls and *how much* has fallen. The timing of and speed of snowmelt runoff are discussed in Chapter 7.

2.7.1 DISTRIBUTION OF SNOW

Seasonal snow cover extends out from the permanent ice sheets of Antarctica and Greenland, over large areas in Asia, Europe and North America. Snowfall is the predominant form of precipitation when the temperature in the lower atmosphere is below 0°C, and ground temperatures below freezing are necessary for the deposited snow to remain unmelted.

Due to its high albedo, snow cover can be readily distinguished from snow–free ground using visible radiation reflectance. Remote sensing, from aircraft or satellites, enables the rapid mapping of the extent of snow cover over large areas. The area of annual snowfall varies from year to year between 100 and 125 million km^2 (Shiklomanov et al, 2002). Robinson et al. (1993) analysed snow cover records for the Northern Hemisphere for 1972–92, and found periods of greater snow extent in the late 1970s and mid–1980s, and much reduced snow cover in the 1990s, coinciding with increasing global air temperatures. Derksen and Brown (2012) examined subsequent data, and confirmed snow extent has been reducing by about 15% per decade since 1979.

It is, however, often difficult to distinguish snow from cloud cover using visible reflectance alone, without repeated photography over time to filter out the variable cloud pattern. This can be overcome by the use of passive microwave radiation emitted naturally by the Earth's surface. This can penetrate cloud cover and allow the mapping of snow extent unobstructed by weather effects. However, passive microwave emissions can be masked by liquid water in the snowpack, and due to the relatively weak microwave signal from terrestrial surfaces they have a low spatial resolution of about 25 km. Frei et al (2012) provide a review of a number of satellite based remote sensing techniques of snow cover extent.

2.7.2 AMOUNT OF SNOWFALL

It is often not sufficient to be aware of the presence of snow, and the hydrologist requires some measure of the quantity of snow. Usually of much greater importance than the depth of lying snow is the **water equivalent** of the snow, i.e. the equivalent water depth of the melted snow that is potentially available for runoff and for soil moisture replenishment. The depth of snow can vary greatly with topography and with drifting, resulting in spatial sampling problems that are even more severe than those already discussed for rainfall. For this reason depth and density measurements have traditionally often been made along predetermined snow courses selected to be representative of conditions over a wide area (US Army, 1956). Such snow courses are expensive to operate, particularly in remote terrain, and measurements generally cannot be carried out very frequently.

The difficulties in measuring the amount of snowfall in gauges are even greater than those of rainfall. Snowflakes are even more prone than raindrops to turbulence around gauges, resulting in severe under-catches. Although wind effects can be greatly reduced by using wind shields around the gauges, the errors are often still too great to be acceptable (Sevruk, 1982). The WMO initiated a comparison of the catches of some of the most widely used precipitation gauges with those of a Double Fence Intercomparison Reference gauge (DFIR) comprising a Tretyakov gauge within two concentric fence shields (Goodison et al., 1989). Undercatches by the standard gauges relative to the DFIR increased from only a few percent for rain in light winds, up to 50% or more for snowfall in strong winds. Eventually it may be possible to derive correction procedures to reduce the catch errors of standard gauges, but this would require continuous real-time information including wind speed and the discrimination between 'solid' and 'liquid' precipitation.

It may be better to directly measure the depth of snow lying at particular 'representative' locations and convert these to the snow water equivalent using the density of the snow. The density of freshly fallen snow is typically about 0.1, but varies from 0.05 to 0.20 (50 to 200 kg m^{-3}), depending on the temperature during the storm. The density increases over time due to settling and compaction under gravity as well as to any partial melting and refreezing of the snow pack, and may reach values of up to 600 kg m^{-3}.

One solution which provides much greater time resolution information is to measure the water equivalent directly by

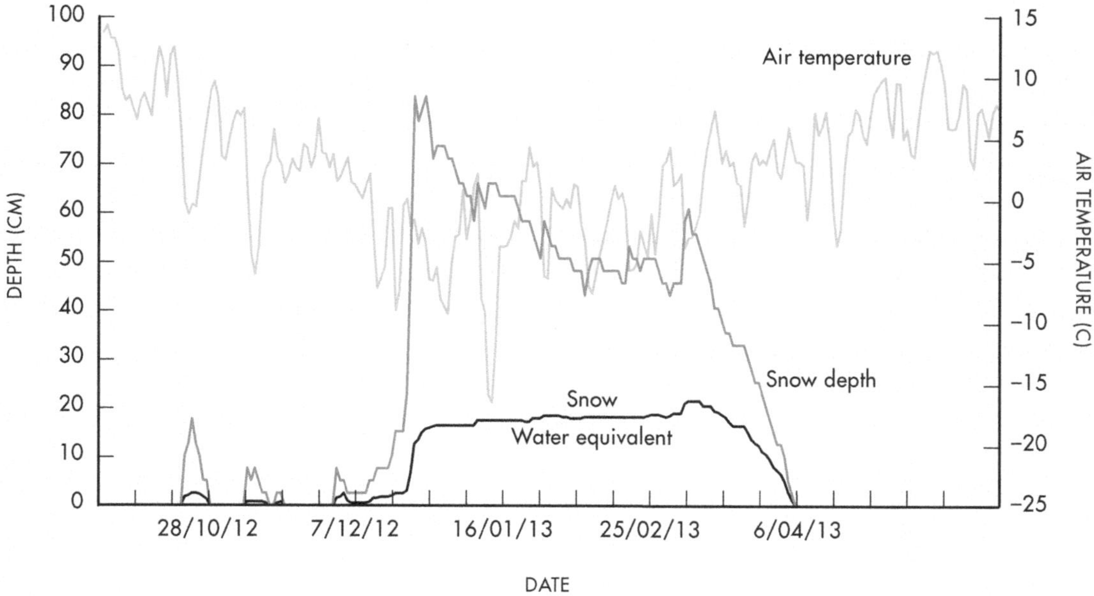

Fig 2.14: Snow time series in Sierra Nevada Mountains, N California, elevation 7000 feet. (Redrawn from data supplied by Snow Survey, Natural Resources Conservation Service, US Department of Agriculture).

automatically weighing the snow that falls onto a **snow pillow**, several metres in diameter, comprising a metal plate or a flexible bag filled with an antifreeze liquid. The overlying weight of snow is recorded as changes in pressure using a manometer or a pressure transducer. Choosing a representative site for the gauge is probably the most important factor to be considered. Snow pillows are widely used in the western USA, where the SNOTEL (SNOw TELemtry) network has over 600 automatic gauges to provide daily or sub-daily data from remote mountain basins (http://www.wcc.nrcs.usda.gov/snow/). Each site is equipped with a radar sensor for snow depth, a precipitation gauge that measures the total precipitation (both solid and liquid) using a pressure transducer, and a snow pillow. Figure 2.14 shows a series of snow records at a site over one winter.

Point measurements can provide only limited information on the uneven areal distribution of snow cover. Areal snow water equivalents must be estimated from point values, for example by correlations with topography, altitude and vegetation parameters. Remote sensing in combination with conventional snow surveying methods offers the opportunity to obtain quantitative information on the areal distribution of snow cover. Tait (1998) found reasonable agreements between passive microwave measurements, obtained by satellite, with ground data obtained from snowlines in the former Soviet Union and SNOTEL data from the USA, for different land covers, terrain and snow state based on surface temperature. The natural radioactivity of the Earth also provides a means of measuring the snow water equivalent over large areas. Natural gamma radiation from the upper layer of the soil will be attenuated by the overlying snow. Airborne measurements carried out when snow is lying on the ground can be compared with results when the ground is snow free to provide information on the snow water equivalent. However, the data collected with this method have a low temporal resolution and cannot be used for large scale studies. (Frei et al, 2012). The GRACE satellites measuring changes in terrestrial water storage can be used to estimate snowpack water equivalent where snow mass is the primary component, such as in winter Arctic river basins (Niu et al, 2007). This has a spatial resolution of about 100-300 km and a monthly sampling rate.

REVIEW PROBLEMS AND EXERCISES

2.1 Explain the importance of very short residence time of atmospheric water for hydrology.

2.2 Discuss the principal uplift mechanisms resulting in precipitation.

2.3 What is a cloud, and why do some types produce large quantities of rain whereas others produce little or none?

2.4 Explain why over a year some areas of the world receive no rainfall whilst others may receive over 12,000 mm.

2.5 Discuss some of the factors to be considered when choosing a method to estimate the rainfall over an area from a number of raingauges.

2.6 Discuss the importance of real-time calibration for weather radar.

2.7 Define and distinguish the following pairs of terms: return period and exceedance probability, aridity and drought.

2.8 What are the main problems in measuring snow water equivalent at a point, and over a catchment area?

WEBSITES

National Center for Environmental Prediction (NCEP) climate datasets:
http://www.esrl.noaa.gov/psd/cgi-bin/data/composites/printpage.pl

El Niño:
http://www.education.noaa.gov/Weather_and_Atmosphere/El_Nino.html

Global Precipitation Climatology Centre (GPCC):
http://www.esrl.noaa.gov/psd/data/gridded/data.gpcc.html

UK Flood Estimation Handbook web service:
http://fehweb.ceh.ac.uk

US Drought Portal:
www.drought.gov and http://droughtmonitor.unl.edu/

US SNOTEL network:
http://www.wcc.nrcs.usda.gov/snow/

Meteorological Services:

Bureau of Meteorology (Australia):
http://www.bom.gov.au

India Meteorology Department:
http://www.imd.gov.in/pages/main.php

Met Éireann (Ireland):
http://www.met.ie/climate-ireland/rainfall.asp

Météo-France:
http://www.meteofrance.com/accueil

Meteorological Office (UK):
www.metoffice.gov.uk/

Meteorological Service (New Zealand):
http://www.metservice.com/national/home

National Weather Service (USA):
http://www.weather.gov/

Nigerian Meteorological Agency:
http://nimet.gov.ng/

South African Weather Service:
http://www.weathersa.co.za/

3

INTERCEPTION

Interception: To stop, deflect or seize on the way from one place to another.

COLLINS DICTIONARY OF THE ENGLISH LANGUAGE

3.1 INTRODUCTION AND DEFINITIONS

When precipitation falls onto a vegetated surface, only a part may actually reach the ground beneath. Depending upon the nature and density of the vegetation cover a proportion of the rain may be **intercepted** by the leaves and stems of the vegetation canopy and evaporated back into the atmosphere. This water then takes no part in the land–bound portion of the hydrological cycle, and is termed the **interception loss**. The remaining water which reaches the ground beneath the vegetation is termed the **net rainfall** (Figure 3.1). The bulk of this comprises **throughfall** consisting of rain drops that fall through spaces in the vegetation canopy and water which drips from wet leaves, twigs and stems to the ground surface; a generally much smaller amount of water trickles along twigs and branches to run down the main stem or trunk to the ground as **stemflow**. A forest may have a shrubby understorey vegetation which itself will also have interception and stemflow components. A layer of leaf litter on the surface of the ground may also intercept some water.

The **interception storage capacity** for a given vegetation type is generally taken to be the amount of water held on the canopy at the end of rainfall once drainage has stopped, expressed as a depth over the plan area of the vegetation. This is discussed further in Section 3.3.

The interception process is important for a number of reasons. Firstly, the net rainfall beneath a vegetation canopy is

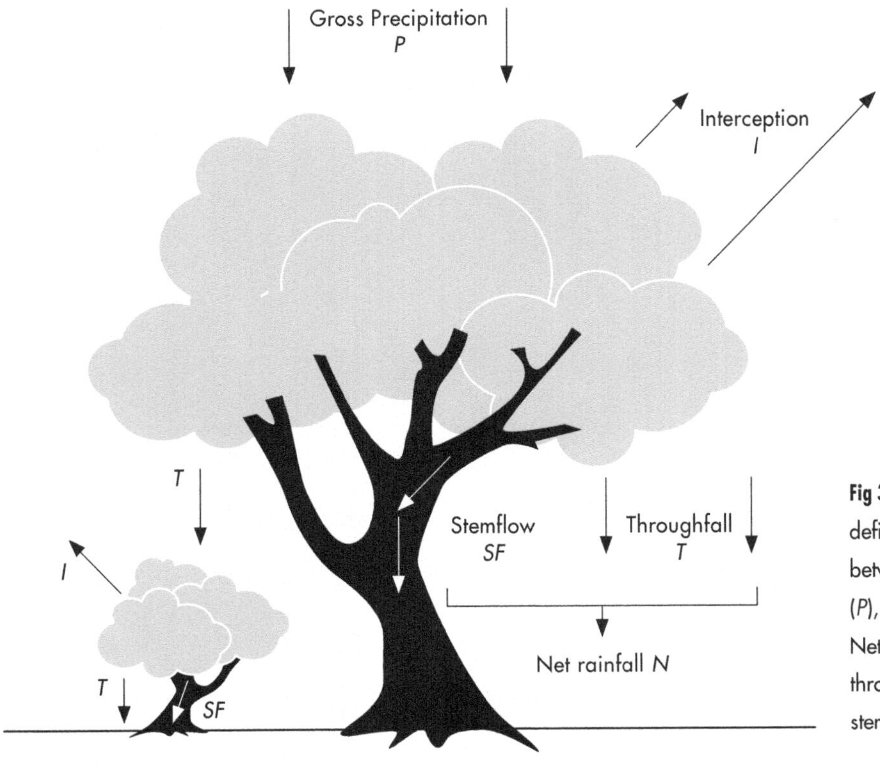

Fig 3.1: Interception loss (I) defined as the difference between gross precipitation (P), and net rainfall (N). Net rain is made up of throughfall (T) and stemflow (SF).

generally less than the **gross rainfall** falling onto the top of the vegetation canopy. In some cases the interception loss may be quite large and can have a significant impact on the water balance. Secondly, as a result of passing through a vegetation canopy the spatial variability of net rainfall is much greater than for gross rainfall. Throughfall, and dripping meltwater, are concentrated at the edges of the tree crowns, whilst concentrated drip close to the trunk, and the stemflow itself often result in high values of infiltration and soil moisture recharge, and even the initiation of minor rills and channels in the ground surface. Thirdly, the passage of rainfall through the vegetation foliage may lead to alterations in both the drop size, which can have implications for soil erosion, and it may also lead to significant changes in water chemistry (see Chapter 8).

Interception is discussed here, in a separate chapter to evaporation (Chapter 4), because although they have the same physics, interception loss can only occur when the vegetation canopy is wet. It is therefore much more dependent upon short–term variations in rainfall, and specifically storm duration and the dry intervals between them.

3.2 INTERCEPTION AND THE WATER BALANCE

Our understanding of the mechanisms that control evaporation and transpiration from vegetation and the role of interception in the water balance of catchment areas is now comparatively well understood, but for many years was the subject of debate and controversy.

Although there was considerable evidence that net precipitation under trees could be much less than in open ground (e.g. Horton, 1919) it was not clear if the intercepted water loss represented a *real* increase in evaporation. Some people argued that interception losses would be balanced by a corresponding reduction in the water uptake by the vegetation, which is suppressed or ceases when the foliage is wet. In that case the overall effect on the water balance would be neutral, or so small that it could be ignored. Even now, although the importance of interception is much more widely recognised, it still receives only a passing reference in some engineering and forest hydrology textbooks, yet as will be shown in this chapter, interception can be hydrologically very significant.

The neutral hypothesis emphasizes that interception losses are essentially evaporative; in any period of time only a finite amount of energy is available. This can be used *either* to evaporate water from within the leaf (transpiration), or to evaporate water from the surface of the leaf (interception). Thus, interception is at least in part balanced by a reduction in transpiration that would have occurred if it had not rained. Early experiments on grasses using weighing lysimeters (McMillan and Burgy, 1960) indicated that there is no difference between the evaporation of wetted foliage and the transpiration of unwetted grass adequately supplied with water. They concluded that grass interception loss is balanced by an equivalent reduction in transpiration. Thus, interception losses are an *alternative* and not an *addition* to transpiration, and so would have little, if any, effect upon the water balance of a catchment area. It was argued by analogy that a similar situation would exist for forests, but it was too difficult to verify this experimentally. This hypothesis of neutral effect was given substantial additional credence through its support by Howard Penman, the recognised authority on evaporation, who affirmed that "whilst energy was being used up to get rid of the intercepted water, the same energy could not be used to get rid of the transpired water" (Penman, 1967).

A growing body of evidence, however, indicated that this was not the complete answer. Catchment studies in Europe and the USA indicated that forestry reduced total streamflow relative to grassland (Engler, 1919; Bates and Henry, 1928; Keller, 1988; Swank and Crossley, 1988). Opponents argued that this was due to experimental errors, or to forests having deeper roots than shorter vegetation and so being able to extract greater quantities of water during times of soil water shortage. A crucial step forward in understanding was provided by Law (1958) who studied the water balance of a small (450 m^2) natural lysimeter set within a spruce forest. Measurements included both gross and net rainfall as well as surface drainage. The annual water use (gross rain minus drainage) was 50% greater than that recorded at either a grass covered percolation gauge or a nearby grass covered catchment. Furthermore the forest water use appeared to be significantly greater than the net radiant energy available for evaporation.

This result portrayed interception, not as an alternative to transpiration, but as a net loss of water. This raised the fundamental questions of how evaporation from the wetted surfaces of vegetation could take place at a significantly higher rate than the transpiration from unwetted vegetation, and whether significant evaporation of intercepted water could occur in circumstances when transpiration rates would otherwise be negligibly small, such as from wetted dead and dormant vegetation and during the winter and at night. Affirmative answers to these questions then required an explanation of the sources of energy for the additional evaporative losses.

Evidence accumulated rapidly to support the conclusion that intercepted water evaporates much faster than transpired water, and therefore much of the interception loss represents an additional loss in the catchment water balance. Rutter (1963) found that evaporative losses from cut wetted branches exceeded those from dry, living branches which lost water only by transpiration. More realistic field experiments by Rutter (1967) indicated that during the winter period the loss of intercepted water considerably exceeded the transpiration rate *in the same environmental conditions*. Results from small catchment studies showed that substantial increases in water yield resulted from the removal of forest vegetation (Hewlett and Hibbert, 1961; Hibbert, 1967) and that decreased yields resulted from the conversion of decidous hardwood forest to evergreen pine forest (Swank and Miner, 1968).

Subsequently a combination of theoretical analysis and field data collection have confirmed that precipitation intercepted by forest evaporates at a greater rate than transpiration from the same type of vegetation in the same environment (Murphy and Knoerr, 1975). The difference may be of the order of 2–3 times (Singh and Szeicz, 1979) or as much as 5 times the transpiration rate (Stewart and Thom, 1973). Both Singh and Szeicz (1979) and Stewart (1977) concluded that 68% of interception during daylight was additional to transpiration (i.e. 32% could have been compensated for by transpiration). Pearce et al. (1980) found that, if account is taken of the additional water losses resulting from the high night-time rates of evaporation of intercepted water, the net interception loss may be as high as 84% of gross interception. Net interception loss will increase as the proportion of night-time rainfall duration and amount increases. This means that, in many high rainfall areas, especially in maritime climates, where at least one-half of rainfall may occur at night, the importance of interception as an evaporative loss and the magnitude of the net loss may be much greater than in areas where rainfall is dominated by daytime convective activity.

In specific conditions other factors may result in additional net interception losses. For example, in some areas transpiration may be limited more by the availability of water than of energy. Then by increasing the amount of available water, interception would increase the total loss of water from a catchment area. The evaporation of water intercepted by dormant or dead vegetation and by a litter layer would certainly represent a net interception loss, the only factor involved in this case being the interception storage capacity and its depletion by evaporation.

The primary explanation of the higher evaporation rate from wet vegetation surfaces, and especially from wet forest canopies, relates to the relative importance of the two main resistances imposed at the vegetation canopy on the flux of water vapour into the overlying atmosphere. This will be discussed in more detail in Chapter 4, in relation to the Penman–Monteith equation for calculating evaporation. At this stage it will be sufficient to note that the **surface resistance** is a physiological resistance, imposed by the vegetation canopy itself on the movement of water by transpiration, and the **aerodynamic resistance** is a measure of the resistance encountered by water vapour moving from the vegetation surface as wet surface evaporation into the surrounding atmosphere. In dry conditions forest canopies probably have a slightly higher surface resistance than grass and other lower-order vegetation, but when the vegetation surfaces are wet this resistance to vapour flux is effectively 'short-circuited' and reduces to zero for all vegetation types (Calder, 1979). The aerodynamic resistance depends essentially on the roughness of the vegetation surface, which tends to be significantly greater for trees than for grasses and other short vegetation. The aerodynamically rougher canopies of forests generate more effective mixing of the air, which is the dominant transport mechanism for water vapour.

The additional energy required to maintain the higher rates of evaporative loss permitted by the dominating role of the aerodynamic resistance for wetted vegetation appears to be attributable to **advection** energy. This refers to the *horizontal* movement of energy in the atmosphere (i.e. parallel to the Earth's surface) in contrast to **convection** which is *vertical* movement. And whereas convection is usually a local phenomenon (see Chapter 2) advection is normally regional or global in extent. Rutter (1967) showed that in wet canopy conditions evaporation losses may not be controlled predominantly by the radiation balance but rather that the wet canopy acts as a sink for advected energy from the air. Importantly, he found that when intercepted water was being evaporated the foliage was measurably cooler than the surrounding air and that the resulting temperature gradient was sufficient to yield a heat flux to supply the energy deficiency. This hypothesis was subsequently confirmed in a number of investigations, mainly of forested areas (e.g. Stewart and Thom, 1973; Thom and Oliver, 1977), and it is recognized now that the advected energy may be derived both from the heat content of the air passing over the vegetation canopy (Singh and Szeicz, 1979) and from heat stored in the canopy space and the vegetation itself (Michiles and Gielow, 2008). Stewart (1977) measured gradients of temperature and water vapour evaporation during wet canopy conditions at a 20-minute time scale, and found that for 70% of these periods the evaporation exceeded the net radiation, the additional energy being derived from the air passing overhead. Michiles and Gielow (2008) measured the thermal energy stored in the biomass of a tall forest in Amazonia and found that in the night, and during and shortly after rainfall it could be an important energy source for the evaporation of water intercepted on the forest canopy. A good discussion of this subject is provided by McNaughton and Jarvis (1983).

Additional proof of the role of advection and stored energy in promoting wet canopy evaporation was provided by Pearce et al. (1980) who confirmed the evidence of high evaporation rates during the night when there was no other energy source. In this connection, it should be noted that the earlier studies of Singh and Szeicz (1979) and Stewart (1977) were carried out in a comparatively small forested area, surrounded by farmland, where large-scale advection of 'surplus' energy was to be expected. It was possible, though, that in the case of very extensive forests, as for example on the Canadian shield or the Amazon basin, where trees extend for many hundreds of kilometres, less surplus energy may be available when large areas are wetted. Localized, thunderstorm-type wetting would, however, still permit sensible heat to be released from the dry areas to boost evaporation in the wetted areas.

Simulation of the energy exchange between the atmosphere and a vegetation surface by Murphy and Knoerr (1975), however, indicated that energy balance modifications may also play a significant role. The integrated effect of interception from a forest stand was an increase in the latent heat exchange, at the expense of the long-wave radiation and sensible heat exchange, which varied according to relative humidity and windspeed conditions. As a result they concluded that enhanced evaporation of intercepted water *can* occur for forests of large areal extent where horizontal advection may be negligible.

Pereira et al (2009a) showed that the surface temperature of a wet tree crown depends on the available energy and wind-speed, and that a fully wet single tree crown in a sparse forest will behave like a wet bulb, allowing evaporation of intercepted rainfall to be estimated by a simple diffusion equation for water vapour which is not restricted by the assumptions of one-dimensional transfer models usually used at the stand scale.

van Dijk et al (2015) provide an excellent overview and discussion of some of the remaining gaps in knowledge. These include the measurement of evaporation from wet canopies during rainfall, when eddy covariance measurements tend to be unreliable, unmeasured heat input for evaporation from cooling of the soil and vegetation biomass, as well as horizontal advection from nearby dry areas and the impact of occasional large scale turbulent eddies under otherwise stable conditions.

Detailed studies have also been conducted on other types of vegetation cover, including, grasses and agricultural crops. Finney (1984) investigated the possible paths taken by raindrops falling on Brussels sprouts, sugar beet and potatoes, i.e. they may fall between the leaves, their properties remaining unaltered; be intercepted and redirected as stemflow; be intercepted and coalesce, to fall subsequently as drip; or be intercepted and shattered by impact with the vegetation and then be redirected as small drops between the leaves. He found that as the plants matured and their interception area increased, the resulting decrease in throughfall was accompanied by an increased stemflow and leaf drip and a reduction in soil detachment except at leaf drip points.

Rainfall simulator experiments with tussock grasses showed the way in which the plant structure, with its convergent leaf arrays, directed intercepted rainfall towards the base of tussocks (De Ploey, 1982).

3.3 MEASURING INTERCEPTION

The most common method of measuring interception loss (I) in the field is to compute the difference between the precipitation above the vegetation layer (P) and the net precipitation below the vegetation canopy, comprising the throughfall (T) and stemflow (SF) (Figure 3.1). Thus:

$$I = P - T - SF \tag{3.1}$$

Due to the difficulties of installing equipment underneath a vegetation canopy, this method has been used more for forest vegetation than for lower-order covers. Throughfall may be measured using funnel or trough gauges placed beneath the forest canopy and stemflow may be collected by small gutters sealed around the circumference of the trunk leading into a collecting container. Even then a number of problems may arise.

It has been found that throughfall depends upon canopy coverage and **leaf area index (LAI)**, i.e. surface area of leaves (one side only)/projected crown area, whether the trees are evergreen or deciduous, and the leaf surface smoothness. Leaf shapes and orientation can concentrate throughfall at drip points. Stemflow may be influenced by branch orientation and by the roughness of the bark.

Generally, gross precipitation is measured in open areas, but sometimes this is not possible and there may be problems

Photo 3.1: Interception sheet gauge under a regularly-spaced conifer plantation. The water runs downslope to a collector with a measuring gauge (Photo source: Mark Robinson).

due to the effects of the aerodynamic roughness of the vegetation cover on the catch of the canopy–level gauges (see Section 2.4.1). Additionally there are sampling difficulties imposed by the great spatial variability of throughfall and stemflow in tropical forests (Jackson, 1971). Work in Amazon rainforest (Lloyd and Marques, 1988) has revealed the magnitude of the sampling problems involved due to considerable localised concentrations of throughfall in drip points (Figure 3.2). Previous estimates of interception losses for tropical forests may be in error as a result of inadequate spatial sampling (Bruijnzeel, 1990) since many early investigations did not fully appreciate the great spatial variation in throughfall necessitating a large number of randomly sited gauges, frequently relocated, to make an accurate assessment.

A more satisfactory method of collecting net rainfall may be to use large sheet gauges (Photo 3.1) which collect both throughfall and stemflow, and provide a sampling area significantly larger than provided by funnel gauges or troughs (Calder and Rosier, 1976; Hall and Hopkins, 1997). There is also evidence that locations with high throughfall are close to those with low throughfall, so that monitoring adjacent places is an additional advantage (Lloyd et al, 1988). It also may be easier to operate one large sheet gauge than a large number of throughfall and stemflow gauges. However, large sheet gauges are not appropriate where information is required about the spatial variability of net rainfall, as may be the case in studies of soil moisture recharge, erosion or water quality. Whilst they work well for dense young plantation

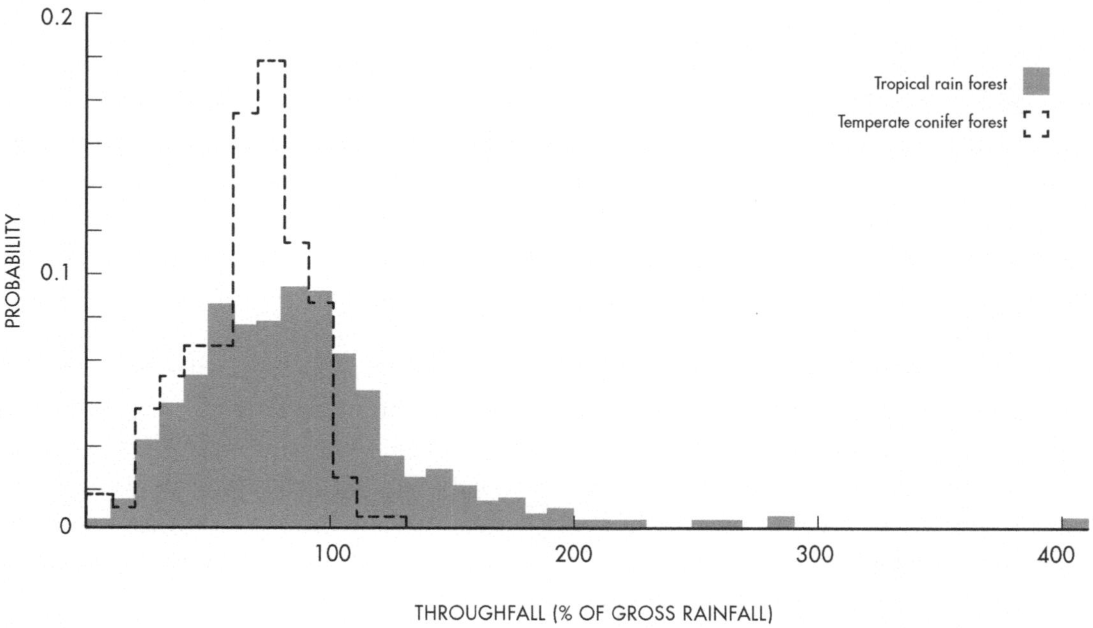

Fig 3.2: Comparison of throughfall catches in a network of collecting gauges expressed as a proportion of the gross rainfall for (a) Tropical rain forest, (b) Temperate conifer forest (Lloyd and Marques, 1988).

forests, sheet gauges are unsuitable where there is dense understorey vegetation, or where the trees are widely spaced and sheets would have to be excessively large (Lloyd and Oliver, 2015). Furthermore, if measurements are required over long periods the presence of the sheet may affect the tree canopy by cutting off the supply of net rainfall and nutrients to the ground.

Stemflow on trees may be measured by sealing flexible guttering around the tree stems, and leading the water into a collecting device. Although many early studies ignored stemflow, it is now known that it can be significant for certain species, and it may be particularly important for ecology and water chemistry as it focusses water in the soil around the trunk base and it often has high solute concentrations. There is some evidence that with increasing tree age stemflow reduces as a proportion of the gross rainfall (e.g. Johnson, 1990). This may result partly from their bark becoming rougher or covered with mosses and lichens and partly as older branches tend to become less steeply angled upwards.

Interception studies are more difficult for grasses and other lower vegetation for which other techniques may be possible. Small weighing lysimeters, for example, have been used to measure the wet–surface evaporation loss from heather (Hall, 1985; 1987), but care must be taken to exclude periods of transpiration loss.

Catchment water balance approaches have been used to measure indirectly the magnitude of interception loss. Some studies utilized small instrumented catchments to provide large–scale estimates of interception loss. Swank and Miner (1968), for example, reported that the effect of converting mature hardwoods to eastern white pine on two experimental catchments in the southern Appalachians was to reduce streamflow after 10 years by almost 100 mm. Since most of the water yield reduction occurred during the dormant season, it was attributed mainly to greater interception loss from the evergreen pine than from the deciduous hardwoods. Increases in water yield, also attributable largely to interception effects, were reported by Pillsbury et al. (1962) and by Hibbert (1971), after conversion of chaparral scrub to grass.

Finally, there has been a range of approaches to quantify interception storage capacity and its individual components. The simplest approach is to plot throughfall against gross rainfall and fit an upper envelope to the throughfall points. The line gives a negative intercept on the throughfall axis which represents the canopy storage capacity. However, the data usually have a large scatter due to the pattern of wetting and drying cycles within each rain event as well as the large experimental variance. An envelope tends to bias results upwards to larger points with errors or just natural variability due to various causes. In addition there is subjectivity in excluding the smallest storms for which there was incomplete wetting of the canopy. As with the interception loss, the measurements of actual canopy storage and the maximum storage capacity are all expressed as equivalent depths (usually mm) per unit ground area and not as physical thicknesses of water films on the foliage.

Direct experimental approaches include wetting vegetation and measuring subsequent changes in weight as the water evaporates. This may be done for short vegetation using a weighing lysimeter; this technique was used by Calder et al. (1984) to study the interception characteristics of heather. Rutter (1963) and Crockford and Richardson (1990) weighed tree leaves and branches and scaled up to the complete tree, whilst Teklehaimanot and Jarvis (1991) cut and suspended a tree from a load cell and spraying it with water and monitored the changes in its weight directly. The storage capacity can also be derived indirectly from the net and gross rainfall depths for a number of storms (Klaassen et al, 1998; Pereira et al, 2009b)

Herwitz (1985) determined the interception storage capacity of tropical rainforest leaf surfaces using a rainfall simulator and the interception storage capacity of the trunks and woody surfaces by immersing bark fragments in aqueous solutions. He combined these data with measurements of the leaf area index, calculated with the help of large–scale aerial photographs, and a woody area index (WAI), i.e. woody surface area/projected crown area, to determine the total interception storage capacity. He found storage capacity values of 2–8 mm, which are much higher than the 1–2 mm generally found by in situ methods for temperate forests (Table 3.1).

Note, it is important to distinguish between the storage capacity defined by Herwitz, as the *maximum* water that could be stored on a canopy while still draining, and the more widely used definition by Horton (1919), and used in the Rutter and Gash models (see Section 3.6). This is the

VEGETATION TYPE	CANOPY STORAGE CAPACITY 'S' (MM)		FREEFALL 'P' (PROPORTION)	
CONIFERS				
Corsican pine (*Pinus nigra*)	1.05		0.25	
Norway spruce (*Picea abies*)	1.5		0.25	
Sitka spruce (*Picea sitchensis*)	1.7		0.05	
Douglas fir (*Pseudotuga menziesii*)	1.2		0.09	
Heather (*Calluna vulgaris*)	1.1		0.13	
Grasses	1.3		–	
Tropical forest	1.1–4 .9		0.0–0.08	
DECIDUOUS FOREST	**IN LEAF**	**BARE**	**IN LEAF**	**BARE**
Hornbeam (*Carpenus betulus*)	1.0	0.65	0.35	0.55
Oak (*Quercus robur*)	0.85	0.3	0.45	0.8

Table 3.1: Typical values of canopy rainfall interception capacities for different vegetation types, expressed as an equivalent water depth over the ground area of the vegetation (Rutter et al., 1975; Shuttleworth, 1989; Hall, 1985).

minimum amount that just saturates the canopy, i.e. the amount left at the end of a storm in zero evaporation conditions when all drip has finished. The two measures are quite different.

Remote sensing may also be used to measure directly the amount of water held on a whole forest canopy. Calder and Wright (1986) used gamma–ray attenuation. A transmitter and receiver were suspended from two towers, 40 metres apart, and were raised and lowered to allow the beam to scan across different levels in the canopy. However, for safety reasons, this could not be used for long term unattended monitoring.

Interception losses can be obtained from eddy covariance measurements (see Chapter 4). This technique has been used in many studies including van der Tol et al (2003) over pine forest, Mizutani et al, (1997) over broad-leaved forest, Herbst et al (2008) for hedgerows, Cabral et al (2010) for Eucalyptus and Czikowsky and Fitzjarrald (2009) for Amazonian rainforest. In addition, large datasets are becoming increasingly available from measurements made in the Fluxnet and ICOS networks (Baldocchi et al, 2001). The strengths and weaknesses of using eddy covariance under wet conditions are discussed in van Dijk et al (2015). In particular, flux measurement during rainfall require special scrutiny, and there are also unresolved issues concerning horizontal advection from nearby dry areas, and the possible mechanical removal of splash droplets by infrequent large-scale turbulence under stable atmospheric conditions.

3.4 FACTORS AFFECTING INTERCEPTION LOSS FROM VEGETATION

If rain falls onto a dry vegetation canopy the interception loss is usually greatest at the beginning of the storm and reduces with time. This largely reflects the changing state of the **interception storage** of water on the vegetation canopy, i.e. the ability of the vegetation surfaces to collect and retain falling precipitation. At first, when all the leaves and twigs or stems are dry, the available storage – the **interception storage capacity** – is at a maximum, and a very large percentage of precipitation is prevented from reaching the ground. As the leaves become wetter the weight of water on them eventually overcomes the surface tension by which it is held and further additions from rainfall in excess of evaporation fall as water droplets from the lower edges of the leaves. There is a widespread assumption based on apparent 'common sense' that the interception capacity of trees is greater than that of shorter vegetation and grasses. In fact as Table 3.1 shows that canopy storage capacities are very similar, and the capacities for some grasses are actually higher than some forest values! However, as shown in Section 3.5.2, grass interception losses certainly will not be greater than those from the forests.

It must be remembered that condensation and raindrop formation high in the atmosphere does not necessarily mean that the air near the ground is also saturated. Overall the evaporation during rainfall may account for up to half the total interception loss (Gash, 2016). It must be remembered the temperature and water holding capacity of the atmosphere at the altitude at which raindrops form, may be very different to that at the vegetation canopy. A considerable amount of water may be lost by evaporation from the leaf surfaces during rainfall, so that even when the initial interception storage capacity has been filled, there is some further fairly constant retention of falling precipitation to make good this evaporation loss. Indeed, during long continued rains, the interception loss may be closely related to the rate of evaporation, so that meteorological factors affecting the latter are also relevant to this discussion. While rain is actually falling windspeed is a factor of real significance; evaporation tends to increase with higher windspeed, so that during prolonged periods of rainfall the interception loss is greater in windy than in calm conditions. This observation may, however, be less applicable in short duration rainfall during which high windspeeds reduce the interception storage capacity by prematurely dislodging water collected on vegetation surfaces, and so partially outweighing the greater evaporative losses. Other variables include differential vapour pressure and net radiation (see Pereira, et al, 2009a).

The duration of rainfall is another factor that influences interception by determining the balance between the reduced storage of water on the vegetation surfaces, on the one hand, and increased evaporative losses, on

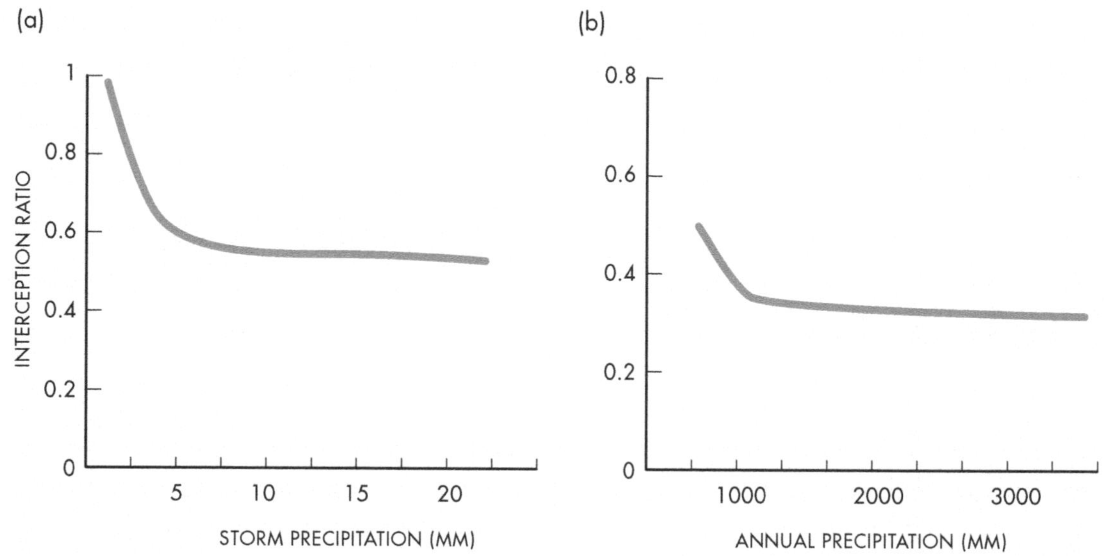

Fig 3.3: Changing interception ratio (loss as a proportion of rain) with increasing: (a) storm rain depth, and (b) annual precipitation (based on data in Calder, 1990).

the other. Data collected during classic work by Horton (1919) and in numerous subsequent investigations showed that the absolute interception loss increases with the duration of rainfall, although its relative importance (i.e. interception as a proportion of rainfall) decreases. Since the amount and duration of rainfall are often closely related, many investigators have related interception losses and rainfall amount. When a storm commences, the losses are high as the initial interception storage capacity is filled; then they increase more slowly with evaporation of intercepted water during further rain as the storage capacity is replenished. The rate of this evaporation is generally lower than the rainfall, so that the relative size of interception losses will tend to decrease as the amount of rainfall increases. This is illustrated in Figure 3.3a in which the interception ratio (i.e. interception loss/precipitation) is plotted against storm precipitation amounts in an area of tropical forest in Puerto Rico. The relationship also holds good for annual conditions, as is illustrated in Figure 3.3b by the graph of annual mean interception ratio against annual precipitation for a number of forest sites in the maritime climate of Great Britain. Since the average rainfall rate and the potential evaporation during rainfall are surprisingly uniform the annual ratio is fairly uniform at about 0.30–0.35 (Calder, 1990).

Because the greatest interception loss occurs at the beginning of a storm, when the vegetation surfaces are dry and the interception storage capacity is large, it will be apparent that rainfall frequency, i.e. the frequency of re-wetting, is likely to be of considerably greater significance than either the duration or amount of rainfall (Rutter, 1975).

Interception loss will also be affected by the type of precipitation including the size distribution of drops and particularly by the contrast between rain and snow, which will be discussed more fully at a later stage. Another important factor, which also merits a separate discussion below, is the variation of interception loss with the type and morphology of the vegetation cover.

3.5 INTERCEPTION LOSSES FROM DIFFERENT TYPES OF VEGETATION

On the basis of the preceding discussions it would be expected that interception losses between sites will vary in response to differences in vegetation and precipitation characteristics. The main vegetational effects relate to differences in interception storage capacity from one vegetation type to another and in aerodynamic roughness and its implications for the aerodynamic resistance and the rate of evaporation from the wetted vegetation surface. In broad terms interception losses will be greater for denser vegetation, for taller vegetation and for wetter climates. The most important precipitation characteristics are duration, frequency and intensity, as well as the precipitation type (liquid or solid), which are discussed separately in Section 3.7.

Because of the complexity of the interception process and the interrelationships between the vegetational and meteorological factors which determine the magnitude of interception losses, it is often difficult to make well-founded comparisons between published data on interception loss. It is clear, however, that in most cases, interception losses are greater from trees than from grasses or agricultural crops, although the reasons for this may vary with meteorological conditions. In the uplands of Britain, for example, where long-duration, low-intensity rain is common and vegetation surfaces are wet for considerable periods of time, there are much greater evaporation losses from trees. The reason is the much increased evaporation rate in wetted conditions (i.e. interception) due to the greater aerodynamic roughness of the trees rather than their slightly higher interception storage capacity (Calder, 1979). Interception rates for conifers in upland Britain are typically 30–35% of gross rainfall due to the maritime climate of long-duration low intensity rainfall. These losses are amongst the highest in the world and it is for this reason that a substantial body of research into forest interception has been conducted in Britain.

In other climate conditions, such as where rainfall was frequent but short-lived and with rapid drying between storms, the role of interception storage capacity in determining differences in interception loss between trees and grass may be much more important. Even then, however, since grasses take a longer time to dry out than forests, the latter should have higher interception losses.

The values of interception loss for different vegetation covers which are quoted in this section must be interpreted, as far as possible, in the light of both the completeness of the measured data, where this is known, and also of weather conditions. For example, in some cases measurements were made of stemflow, in others an arbitrary allowance was made for this component and in still other cases it appears to have been ignored completely. Again the data presented in the literature are not always accompanied by an adequate analysis of meteorological conditions, particularly concerning the amount, duration, frequency, intensity and type of precipitation, all of which need to be known to permit a meaningful interpretation of the data.

3.5.1 WOODLANDS

Despite the fact that, in most cases, the leaf density is greater in deciduous than in coniferous forest, the bulk of the experimental evidence shows that interception losses are greater from the latter. Published values from a wide range of European and N American data indicate that coniferous forests intercept an average of 25–35% of the annual precipitation compared with 15–25% for broad-leaved forests (Calder, 1990; Van Stan et al, 2015). For mixed evergreen forest in South Island, New Zealand, interception losses averaged about 30% (Pearce et al., 1982).

The contrast between broadleaf and coniferous forest is illustrated in Figure 3.4.

Both types of woodland show a reduction in relative importance of interception loss as rainfall amounts increase but, over the complete range of rainfall totals, interception loss is markedly greater from the conifer forests. One of the reasons for this contrast may be that, while water droplets remain clinging to separate conifer needles, they tend to run together on the broadleaves and so drop or flow onto twigs and branches. Conifers typically have canopy storage capacities of about 1–2 mm, whilst broadleaf species tend to be below 1 mm (Harding et al., 1992). It is also likely that the open texture of coniferous leaves allows freer circulation of air and consequently more rapid evaporation of the retained moisture. Interception losses will depend upon a number of factors including tree age and forest structure. Teklehaimanot et al. (1991) showed the importance of tree spacing. Interception losses decreased from 33% to 9% of rainfall when the spacing of Sitka spruce was increased from 2 to 8 metres. The greater variability between the broad leaved tree sites in Figure 3.4 is probably due to them containing a mixture of species, of different ages and with variable understorey vegetation; in contrast the conifer sites were generally even–aged single species plantations.

A new research topic is the interception loss from individual trees and hedgerows. This measures the rainfall upwind, downwind and underneath the vegetation and deduces the interception loss taking into account of the rainshadow downwind. David et al (2006) studied a Mediterranean oak savannah. They installed transects of ground-level raingauges extending radially outward from the trunk of an isolated evergreen holm oak (Quercus ilex) well beyond the limits of its canopy. They found that since rain drops often do not fall

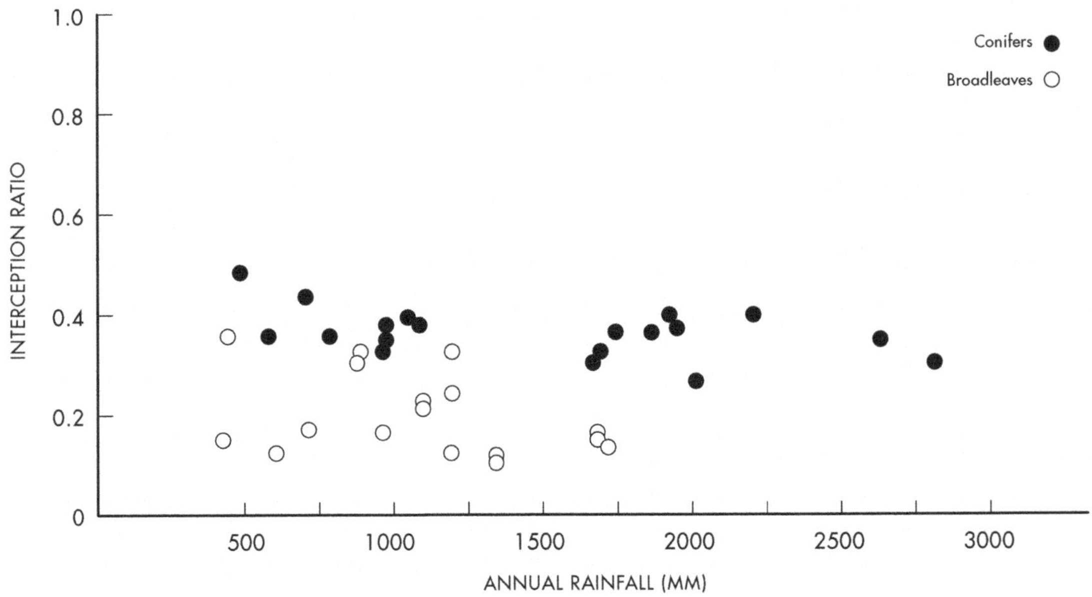

Fig 3.4: Annual interception ratios for conifers and broad-leaved trees (IH, 1998).

vertically, but have a horizontal component due to the wind, the amount of rain actually reaching the ground was enhanced on the upwind side of the tree and reduced on the downwind side. When the overall effect of the upwind and downwind impacts is taken into account the interception loss of 22% of the gross rain per unit of crown projected area, reduced to 9% when related to the total ground area affected by the presence of the tree (David et al, 2006). In a similar study Herbst et al (2006) studied two hedgerows in southern England on flat land surrounded by large fields. Interception loss averaged 57% when calculated for the ground area under the hedgerows and 24% if related to the total area influenced by the hedges.

Estimates of interception in tropical forests are very variable, as a result of the spatial sampling difficulties, but are generally much lower than those reported for temperate forests. This is for three main reasons:

- Most tropical rainfall occurs in short–duration, high intensity convective storms;
- The large raindrops are less effective in wetting foliage than finer drops; and
- Tropical rainforest leaves often have drip tips which concentrate throughfall which drains off as larger drops.

Estimates of interception losses for rainforest in Brazil include 9% (Lloyd et al., 1988) and 12% (Ubarana, 1996), 21% for secondary forest in Indonesia (Calder et al., 1986) and Asdak et al. (1998) reported losses of 11% for pristine unlogged forest (~580 trees ha^{-1}) and 6.2% for logged forest (~250 trees ha^{-1}) in Central Indonesia. Table 3.2 shows typical annual interception losses for different forest types.

With regard to seasonal contrasts, winter and summer interception percentages for evergreen coniferous forests appear to be about the same. Figure 3.5 shows this for a site in northern England (Law, 1958). In contrast, it would be expected that there would be a clear seasonal difference in interception losses from deciduous trees, being greatest during the period of full leaf. In a review of the published literature for a number of broadleaf species, Hall and Roberts (1990) indicated a median value of canopy capacity of 0.8 mm (in leaf) and 0.6 mm (leafless). Lull (1964) quoted figures showing that interception losses in northern hardwood and aspen–birch forests were 15 and 10% respectively when the trees were in leaf, compared to 7 and 4% with

FOREST TYPE	ANNUAL INTERCEPTION (%)
Upland conifers	30–35
Broadleaves	15–25
Tropical forests	10–15
Eucalyptus	5–15
Savannas	5–10

Table 3.2: Typical values of annual interception loss (% precipitation) for different forest types (Calder, 1990; Hall et al., 1992; IH, 1998, Pereira et al, 2009b)

leafless trees. Similarly, Carlisle et al. (1965) found that losses for oak trees were 17% during the summer, vegetated, period compared with under 10% in the leafless period. However, Reynolds and Henderson (1967) found little seasonal difference in interception loss due to leaf fall. This apparently surprising result may be indicative of a number of factors. Higher rainfall intensities in summer because of convective storms may give lower interception losses. It is also to be expected that the aerodynamic roughness of the forest must change, and is probably greater in winter when the trees are bare, encouraging evaporation rates. Thirdly, as noted in Section 3.7, snow interception in the winter period may give rise to higher interception losses.

An additional important aspect of interception loss in wooded areas is that this often occurs at two or more levels within the vegetation cover. Precipitation is first intercepted by the upper canopy; some of the throughfall is then intercepted again by undergrowth, or by a layer of ground litter. Comparatively little is known about the importance of this secondary interception, although the low wind speeds in the trunk space must mean that evaporation rates are quite low. It will tend to increase with the amount of rainfall, because during light rains little or no throughfall occurs from the crown canopy, whereas during long, heavy storms throughfall will probably fill the interception storage capacity of the undergrowth or ground litter.

3.5.2 GRASSES AND SHRUBS

Interception losses have been widely measured in forests but relatively few studies

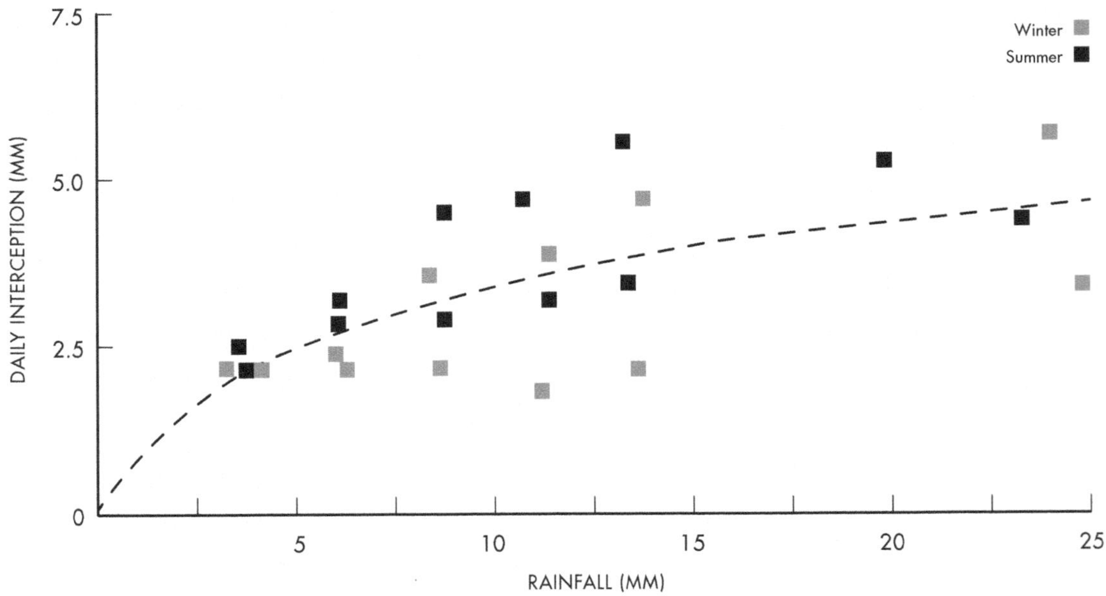

Fig 3.5: Interception loss from evergreen spruce over a two-year period showing losses in summer and winter. (Adapted from Law, 1958).

have been conducted for grasses (Dunkerley, 2000). The total leaf area of a continuous cover of mature grass or shrub may closely resemble that of a closed canopy forest, so that the interception storage capacity may also be similar to that of trees during the season of maximum development. Because of their higher aerodynamic resistance and shorter growing season, however, total annual interception loss from grasses is considerably less than from forests (including deciduous woodland). Furthermore, in areas where grass is cut for hay or silage, or is heavily grazed in the field, interception losses are much reduced. Studies have been conducted using a wide variety of techniques, including sealing the ground surface and measuring the surface runoff, or suspending grass clipping over a screen and measuring the amount of water reaching a collector underneath (Thurow et al, 1987). Values range between about 10% and 20% of gross rainfall, depending upon the climate and biomass cover . In wet temperate areas with frequent rainfall, such as maritime western Europe, interception may provide the major part of tall grass evaporation losses, although still much less in absolute amounts than forest interception losses. In well-watered areas, the rates of transpiration from grasses are similar to interception losses when the leaves are wet, since for short vegetation the aerodynamic and surface resistances are of similar size (David et al, 2005).

In contrast to the many studies of forests there is a similar lack of data on interception by herbaceous and shrubby covers (Muzylo et al, 2009). This is an important omission since they are typical of the heaths and moorlands of Europe and any land cover change, such as afforestation, in these areas occurs at the expense of heather rather than grass. Measurements by Hall (1985, 1987) and Wallace et al (1982) showed that the aerodynamic resistance for heather is lower than for grass. Therefore, during wet periods interception losses from heather are likely to be much higher than those from grass. However, in dry periods transpiration losses from heather are significantly lower than those from grass, so that in regions of moderate annual rainfall (~1500 mm) the increased interception losses are likely to counterbalance the reduction in transpiration, in high rainfall areas interception losses will dominate and in drier areas the converse will be true. Dominingo et al (1998) found significant differences in interception loss between species of semi-arid Mediterranean shrubs, caused by differences in the structure of their canopies.

3.5.3 AGRICULTURAL CROPS

There are far fewer data on the interception loss from agricultural crops than from forested areas as the presence of instrumentation may be difficult to reconcile with the need for frequent access for farming operations (tillage, sowing, spraying, harvesting etc.) not to mention the continuously changing canopy characteristics of the growing crop. In an international review of nearly 70 published interception modelling studies only 4 were of agricultural crops (Muzylo et al, 2009). The interception loss will depend not only on the rainfall characteristics, but also on the development stage of the crop including its height, leaf density and ground cover. Interception losses increase until the crops are fully developed and reach maximum values of up to 35% (Kolodziej, 2011), although such figures would have to be reduced by an appropriate amount to allow for unrecorded stemflow to represent the true interception loss (e.g. van Dijk and Bruijnzeel, 2001a,b).

3.6 MODELLING INTERCEPTION

Whilst it is much simpler conceptually to measure interception loss in the field from gross and net rainfall and stemflow, than it is to collect the climate and vegetation data necessary to model interception loss, modelling has a number of advantages. Firstly, it provides a summary of the behaviour, secondly it enables the results of one field study to be extrapolated to other areas and thirdly it can provide insights into the process.

Various approaches to modelling interception loss have been developed, although inevitably in view of the complexity of the interception process and the difficulty of establishing precise values for the major components and influencing factors (e.g. canopy and stem storage capacity, drip rates, stemflow, aerodynamic resistance, evaporation rates, etc.), most of the models suffer either from over-generalization and simplification or from exacting and extensive data and processing demands. A comprehensive overview of rainfall interception models used around the world is given by Muzylo et al, 2009.

The simplest models are those which incorporate empirical, regression–based expressions relating interception loss to gross precipitation. Horton (1919) was probably the first to propose that for storms which saturate the vegetation canopy the interception loss (I) will be equal to the sum of evaporation loss of intercepted water during rainfall and the water held on the canopy at the end of the storm (which will subsequently be evaporated):

$$I = {}_0\int^{t} E dt + S \qquad (3.2)$$

where E is the rate of evaporation of intercepted water, t is the duration of rainfall and S is the interception 'storage capacity' of the canopy, a term over which there is some confusion in the literature. Used here, it is the amount of water left on the canopy, in conditions of zero evaporation, after rainfall and drip have ceased, i.e. the minimum necessary to cover all the vegetation. As noted above, this differs from its use, for example by Herwitz (1985), to signify the 'maximum storage capacity' of the vegetation canopy.

Equation (3.2) can be elaborated, by considering separately evaporation before and after canopy saturation, to give

$$I = \underbrace{{}_0\int^{t'} E\, dt}_{\text{Wetting up}} + \underbrace{{}_{t'}\int^{t} E\, dt}_{\text{Saturated}} + S \qquad (3.3)$$

where t' is the time taken for saturation of the canopy to occur.

Although Horton (1919) recognized that Eq. (3.2) was more logical, he concluded that in practice it would often be more convenient to incorporate precipitation amounts rather than precipitation duration. Accordingly, there are many empirical models of interception loss which take the general form:

$$I = aP + b \tag{3.4}$$

where P is the gross rainfall on the vegetation canopy and a and b are empirically derived coefficients. Equation (3.4) can be used either to describe individual storm data or, if it is assumed that there is only one rainfall event per day, to describe daily interception loss as a function of daily gross rainfall.

Other similar empirical models related interception loss to factors such as the storm duration or the average evaporation rate during the storm. Useful reviews of simple models such as these were provided by Jackson (1975), Gash (1979) and Massman (1983). They stressed that, while the models are easy to use, they do not always give satisfactory quantitative results when the coefficients are derived by regression against a specific set of data and that empirical results may not be valid for similar vegetation covers at other sites.

Models that are based on more fundamental physical reasoning tend to minimize many of the weaknesses of empirical models but usually require frequent (e.g. hourly) data inputs for rainfall and throughfall rates and for meteorologically based estimates of evaporation. Probably the most rigorous of such models is that developed by Jack Rutter, which solves the vegetation water balance equation numerically.

The model was originally described by Rutter et al. (1971) and subsequently elaborated and generalized (Rutter et al., 1975) as a result of work in hardwood and coniferous forest stands. It is based on describing the water storage on the vegetation canopy and stems. Intercepted rainfall is added to this store, which is then depleted by evaporation, drip and drainage. The rates of evaporation and drip are assumed to vary with the amount of water on the canopy and, accordingly, the model is designed to calculate a running balance of rainfall, throughfall, evaporation and changes in canopy and stem storage. Evaporation from the wetted vegetation surfaces constitutes the interception loss. The rate of input of water to the vegetation canopy is:

$$(1 - p)R \tag{3.5}$$

where R is the rate of rainfall and p is the proportion of rain that falls through gaps in the canopy. The model assumes that when the depth of water stored on the vegetation canopy (C) equals or exceeds its storage capacity (S) evaporation will take place at the potential rate, E_p given by the Penman–Monteith equation (see Section 4.6.2). For a wet but unsaturated canopy (i.e. $C < S$) the evaporation is reduced proportionately so that:

$$E = E_p\,(C/S) \tag{3.6}$$

Subsequently, this relationship was given observational support by Hancock and Crowther (1979), and Shuttleworth

(1978) showed that it provides a theoretically reasonable description.

The rate of drip drainage from the canopy is assumed to be a logarithmic function of the degree of canopy saturation, so that:

$$D = D' e^{bC} \tag{3.7}$$

where *D'* and *b* are parameters which depend upon the foliage characteristics and meteorological conditions, and may be derived from observations as described by Rutter and Morton (1977). Since this equation predicts a small but continuing drip from a dry canopy other expressions have been proposed (e.g. Calder, 1977; Massman, 1980).

The storage capacity of the branch/stem system (S_t) is considered to be replenished by a constant proportion of rainfall which is diverted to that part of the branch system that drains to the trunks. When the storage capacity is completely filled potential evaporation from this store takes place at a rate linearly related to the Penman–Monteith equation. When the depth of water on the branches and trunks (C_t) is less than the capacity S_t the evaporative loss from them is further scaled down by the ratio (C_t / St). In contrast to the expression for drip from the canopy, the drainage of water from the branches and trunks in excess of the storage capacity is assumed to be immediate.

The model was tested against observed interception losses for six forest types including both deciduous broad–leaved and evergreen coniferous species (Rutter et al., 1975) and gave a very satisfactory model performance (Figure 3.6). It has also been successfully applied to other data sets (Muzylo et al, 2009).

The Rutter model, however, requires a

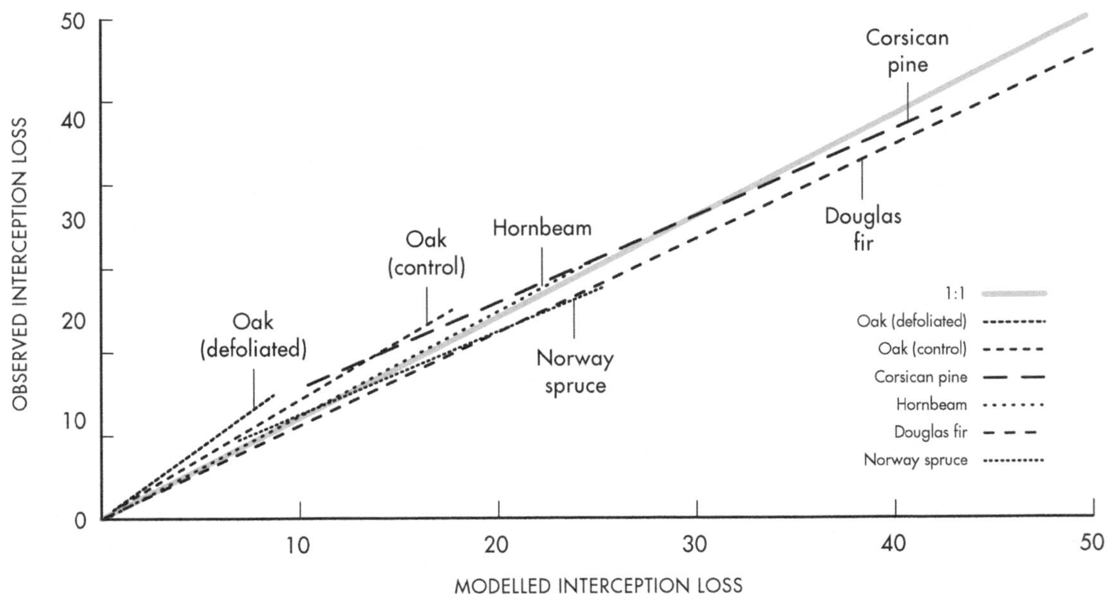

Fig 3.6: Observed and modelled interception losses (mm/month) for six forest species. (Adapted and redrawn from Rutter et al, 1975).

great deal of data and a very short time step and it is complex to use. As a consequence it is generally only used for research purposes. Furthermore, since it is based on keeping a running water balance, errors may accumulate over time. An alternative approach is to integrate the mass balance equation analytically. In this context one of the most satisfactory and widely used methods is the analytical model developed by Gash (1979). Despite a number of simplifying assumptions, it retains much of the physical reasoning of the more complex Rutter model. The Gash model calculates interception loss on a storm–by–storm basis and separately identifies the meteorological and biological controls of interception loss to give a framework within which results may be extrapolated more readily to other areas. The main simplifying assumptions are (Gash, 1979):

- The rainfall pattern may be represented by a series of discrete storms, separated by sufficiently long intervals for the canopy and trunks to dry;
- Similar meteorological conditions prevail during wetting up of the canopy and during the storms; and
- There is no drip from the canopy during wetting up, and that within about 30 minutes of the end of rainfall the canopy storage reduces to the minimum value necessary for saturation.

This model was applied to data from Thetford Forest in eastern England and produced satisfactory agreement between observed and modelled interception loss (Figure 3.7). It has also been applied in evergreen mixed forest in New Zealand (Pearce and Rowe, 1981), oak forest in the Netherlands (Dolman, 1987), tropical

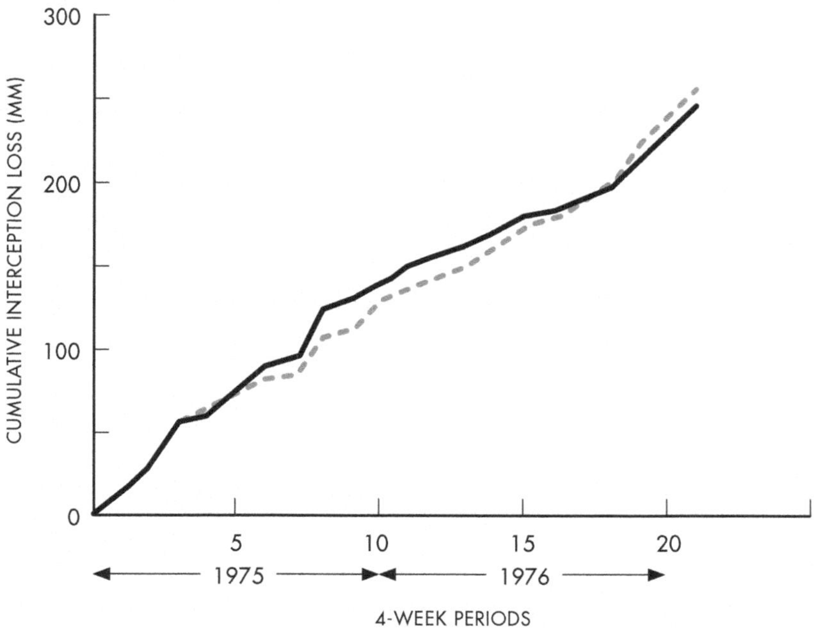

Fig 3.7: Cumulative observed and modelled interception losses for a pine forest over 21 four-week periods in 1975 and 1976 (adapted from Gash, 1979)

forests (Bruijnzeel and Wiersum, 1987; Lloyd et al., 1988: Hutjes et al., 1990), and many sites worldwide (Muzylo et al, 2009).

Dolman (1987) compared the analytical model of Gash (1979) with a more complex numerical simulation model (Mulder, 1985) which had much greater data requirements. Both models performed equally well describing losses from an oak forest, and he concluded that the Gash model was the more appropriate for practical estimates of interception loss.

Further application of the Gash model has indicated that both it and the Rutter model tended to overestimate interception loss from more open forests which are found in Mediterranean areas and many other parts of the non–temperate world (Gash et al., 1995; Valente et al, 1997). Both the Gash and Rutter models were reformulated and produced much improved results for sparse forests (Gash et al., 1995; Valente et al., 1997). Essentially this involved scaling the wet canopy evaporation by the canopy cover fraction, with wet canopy evaporation considered linearly dependent on this value (Figure 3.8). When the canopy cover reaches 100% the sparse model reduces to the original version.

The reformulated Gash sparse canopy model quickly superseded the original versions of the Rutter and Gash models and is now the most widely used interception model (Muzylo et al, 2009).

Research continues to improve both the conceptual basis and the general applicability of models of interception loss. Many of these attempts have focused upon the type of drip/drainage expression used in the Rutter model; others have attempted to simplify both the structure and the data demands of the evaporation expression.

Although the Rutter model works well in temperate forests it has been noted that it is less successful in tropical forests (Calder et al., 1986). Some authors have attributed this to the much greater spatial variability of the net rainfall and hence the larger sampling errors (e.g. Lloyd et al., 1988; Ubarana, 1996). Calder (1996) proposed a 2–layer stochastic model to describe the wetting up of a tropical forest canopy. Primary raindrop sizes are related to rainfall intensity, and the size of secondary drips from the upper foliage to lower levels is characterised by the tree species from which they fall. This model takes account of the fact that in tropical forests the canopy wetting is achieved more slowly and the maximum canopy storage capacity will be lower than for temperate forests due to the larger rain drops and higher intensities that occur in intensive convective rainstorms.

Calder and Newson (1979) proposed a very simple model for total evaporation from conifer forests, using only annual precipitation and Penman short grass potential evaporation. In essence it assumes that annual forest interception may be largely approximated as a proportion (α) of the annual precipitation P:

$$I = f\,(\alpha\, P - w\, E_t) \qquad (3.8)$$

where *f* is the proportion of the catchment with complete forest coverage, *w* is the proportion of the year when the canopy is wet (and no transpiration occurs) and E_t is the potential evaporation which approximates the forest transpiration. This very simple approach was successfully extended to other vegetation types (Hall and Harding, 1993), and to a daily interception model (Calder, 1990) enabling seasonal variations to be investigated. Its application to estimate total evaporation losses (interception and transpiration) is described in Chapter 4.

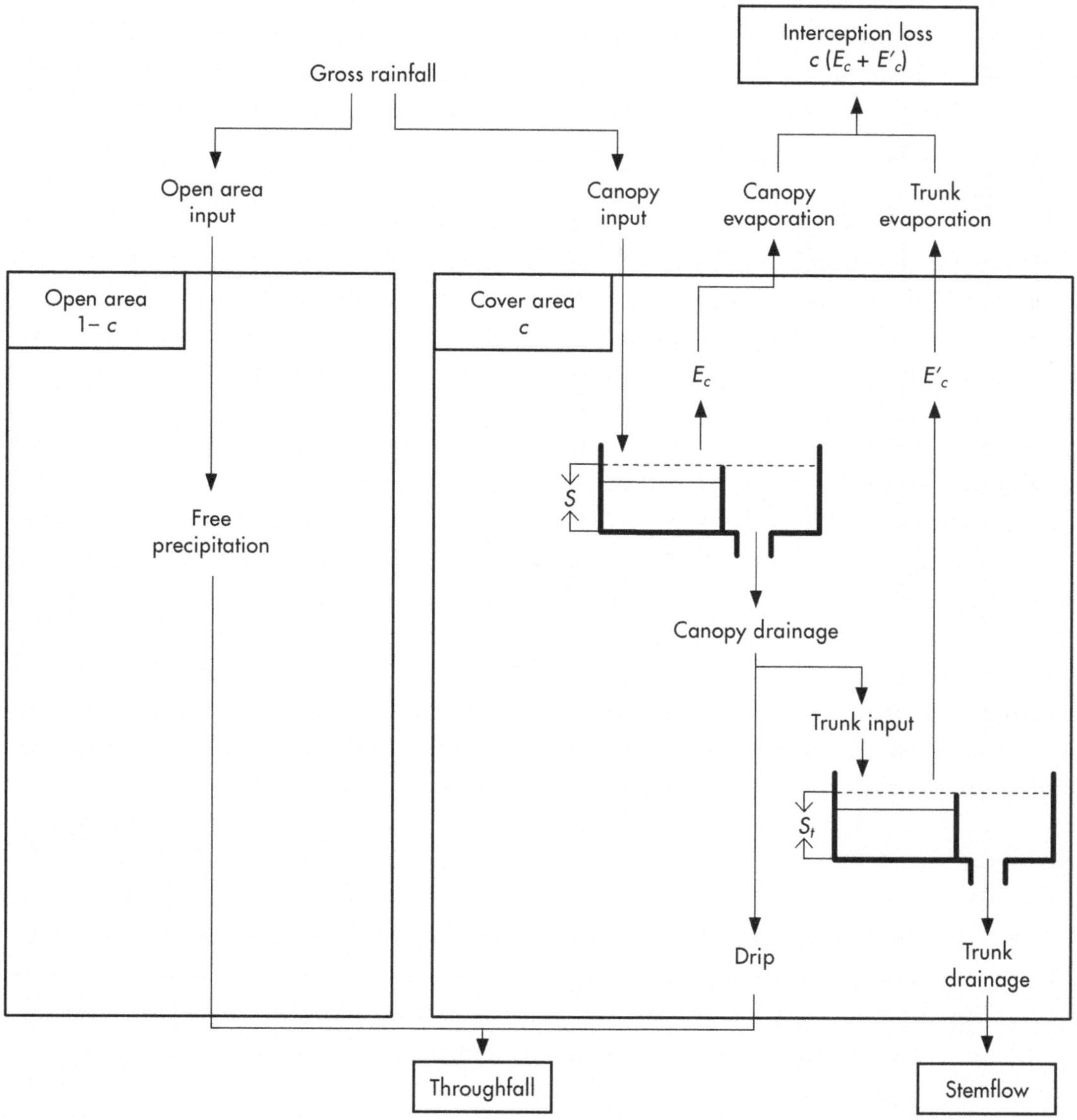

Fig 3.8: Conceptual framework of the sparse model: the same framework applies to the sparse version of both the Rutter and Gash models. (From original diagram in David et al, 2005).

3.7 INTERCEPTION OF SNOW

Due to the experimental difficulties, the evidence concerning the interception loss from snow has been frequently unsatisfactory and confusing and is largely restricted to woodland vegetation. Many of the discrepancies in the literature resulted from the difficulties of measuring snowfall, particularly in view of its tendency to drift at the edge of pronounced barriers such as forests, and in forest clearings, i.e. the very locations in which measurements are normally made.

On the one hand the water equivalent of snow stored on a forest canopy can be an order of magnitude greater than for a rain wetted canopy. Thus there is the potential for considerably larger interception evaporation over a prolonged drying period. On the other hand, it may be argued that because the accumulation of snow on a vegetation canopy will make the surface aerodynamically much smoother than a rain-covered canopy the evaporation of intercepted snow will be lower than from a rain wetted canopy. In addition, snow accumulating on vegetation surfaces is prone to large–scale mass release by rain-wash and sliding under its own weight, frequently aided by wind–induced movement of the vegetation, and also the smaller–scale release of snow particles and meltwater drip. Such release mechanisms mean that in many areas snow remains on the vegetation canopy for only a few days before falling to the ground, where due to shading from solar radiation and the wind, the opportunities for evaporation are small.

Satterlund and Haupt (1970) recorded the weight of the snow on a suspended pine tree and collected and measured the snow and water reaching a ground sheet beneath it. They concluded that only 15% of the intercepted snow was evaporated – the rest reached the ground by melting, slipping or being washed off by rain. Even when snow remains on the vegetation cover for long periods of time, the energy available for evaporation is minimal, and in some areas, such as north–west Europe, when there is a transfer of water it tends to be to the snow cover, in the form of condensation, rather than away from it, in the form of evaporation or sublimation (direct transition from solid snow to water vapour).

Later investigations (Lundberg et al., 1998) placed a much needed emphasis on the aerodynamic processes involving transport of snow through and from the forest canopy to the site of final deposition and on the energy and water fluxes above the snow cover. Gamma–ray attenuation and tree–weighing experiments in Scotland (Calder, 1990; Lundberg et al., 1998) showed that spruce canopies can hold over 20 mm water equivalent of snow and that evaporation rates from snow–covered canopies of up to 0.56 mm hr^{-1} can on occasions be as high as evaporation rates from rain–wetted canopies, especially when the snow is melting during incursions of warm maritime air. Although sublimation rates are generally lower than evaporation rates, sublimation may be important in snow interception because of the long periods of subzero temperatures during which sublimation can take place. The combination of high evaporation rates, large storage capacity and sublimation effects indicated a

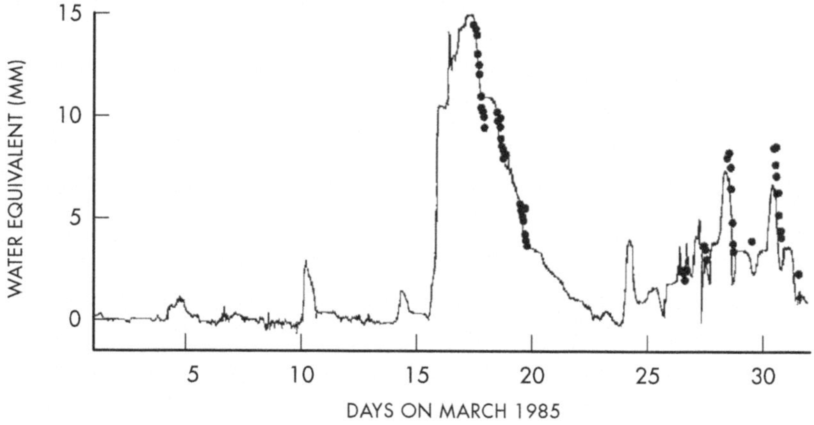

Fig 3.9 The water equivalent of snow intercepted on a forest canopy recorded by tree weighing (continuous line) and by gamma ray attenuation (dots) for three snowfall events (Lundberg et al, 1998)

potential for significant snow interception losses (Figure 3.9). Lundberg et al. (1998) suggested that the total winter losses from snow intercepted on the forest canopy might be over 200 mm. However, there are likely to be significant variations between sites, depending not only on the climate but also on the forest characteristics. For example whilst the canopy snow storage capacity on the 16 m high trees at their site was over 20 mm, Nakai et al. (1993) reported a storage capacity of only 4 mm water equivalent for 5–7 m high trees. Molotch et al (2007) used eddy correlation techniques to estimate sublimation loss rates of 0.71 mm d^{-1} for intercepted snow on a forest canopy and 0.41 mm d^{-1} from the sub-canopy snowpack in an area of 11 m tall open forest.

3.8 CLOUD WATER DEPOSITION

There are certain circumstances where vegetation may be able to gain additional water by 'stripping' fine airborne water droplets (typically ~10μm radius) from mists or low cloud which would not otherwise have fallen to the ground as precipitation at that site. Consequently it would not be measured in normal raingauges (Monteith and Unsworth, 1990). This is most likely to be true in forested areas of high relief or near to coasts, where fogs or low cloud are prevalent (Kerfoot, 1968), particularly when windspeeds are relatively high. This process is also observed, however, in agricultural crops when on calm, misty mornings the accumulation of fine mist droplets on the heads of grain crops may become so heavy that the stems bend under the load.

The main hydrological interest lies in the amount of water that is transmitted to the ground as throughfall and stemflow. Water droplets are formed on leaves, twigs and branches by impaction, and may accumulate sufficiently to fall or trickle to the ground. In the sense that this represents measurable precipitation beneath the vegetation canopy where none is recorded in the open, it could be regarded as 'negative interception'. Deposition of cloud or mist droplets onto vegetation is thought to be a sufficient addition to the normal precipitation to influence plant distributions along the west central coast of North America and Chile.

Early studies frequently emphasized that this would be essentially an edge effect and that its importance decreases markedly away from the borders of, say, a forested plot or area of relatively taller vegetation. The border nature of the phenomenon is also apparent from results of various experiments with fog gauges which used vertical gauzes to intercept horizontally driven cloud and fog droplets (e.g. Nagel, 1956).

A major conceptual advance was that of Shuttleworth (1977) who showed that cloud water deposition was controlled by the same physics as interception evaporation. Rather than simply a horizontal edge phenomenon, which is limited to the first 20 m at most, he argued that it would in fact be dominated by vertical exchanges between the atmosphere and the top of an extensive vegetation canopy. Furthermore, in the same way that interception losses are greater from forests due to their greater aerodynamic roughness, so cloud water deposition amounts would be greater to forest canopies than to shorter vegetation.

Cameron et al. (1997) estimated cloud water deposition rates of about 0.05 mm h^{-1} to 0.8 m high tussock grass at a coastal upland site in southern New Zealand. This amounted to about 60 mm or 4% of the annual precipitation. They suggested that replacement with short pasture would reduce cloud water deposition due to the lower aerodynamic conductance, whilst a change to forest may increase fog deposition due to enhanced turbulence. Nevertheless, the main hydrological impact overall would be enhanced forest interception loss due to the prevailing low intensity intermittent precipitation.

The entrainment of cloud water droplets by vegetation is mainly a feature of sites where fog and low cloud are frequent due to high altitude, climate or proximity to the sea.

If the temperature falls below the dew point of the air, condensation of water vapour may occur on plant leaves and other surfaces with the formation of dew. This is a very different process to the entrainment of airborne particles. Monteith and Unsworth (1990) estimated that for saturated air, a typical dewfall rate would be about 0.067 mm hr^{-1}, giving about 0.2–0.4 mm per night. The source of the water condensed as dew may be the atmosphere or the soil – the latter is not a net gain, but is usually the greater by a factor of about 1:5. Dew is therefore a very minor component of the hydrological cycle.

REVIEW PROBLEMS AND EXERCISES

3.1 Why was the significance of interception as net loss of water not initially recognised by many hydrologists?

3.2 Explain why the interception loss of forests is much greater than that from grass, although their canopy storage capacities may be very similar.

3.3 Discuss the different approaches required for measuring interception losses from grass, shrubs, open forest and close canopy woodland.

3.4 Discuss the importance of interception losses from a snow-covered forest.

3.5 Define the following terms: interception loss, net rainfall, throughfall, stemflow, canopy storage capacity.

3.6 Explain how evaporation losses can be greater than the net available radiant energy; what is the additional source of energy?

4

EVAPORATION

'Evaporation plays a key role in the global water cycle …. [yet] … our understanding of evaporation in the global climate system still shows significant gaps'

DOLMAN AND GASH, 2011

4.1 INTRODUCTION AND DEFINITIONS

The term **evaporation** is used by physicists to describe the process by which any liquid is changed into a gas. For hydrologists this expression is used for the loss of water from a wet surface through its conversion into its gaseous state, **water vapour**, and its transfer away from the surface into the atmosphere. Evaporation may occur from open water (including rivers, lakes and oceans), bare soil or vegetation. In addition to the evaporation of intercepted water held upon plant surfaces, which is discussed in Chapter 3, there is also direct water uptake by plants termed **transpiration**. This component of evaporation comprises soil water taken up by plant roots which moves up the plant and thence into the atmosphere, principally though the leaves. Due to the extraction of water at depth by plant roots, transpiration may continue for long after the drying out of water intercepted on vegetation foliage and held in the upper soil layer.

Although by definition evaporation includes *all* of the processes by which liquid water becomes a vapour, many textbooks still prefer to use the somewhat clumsy term **evapotranspiration** for total evaporation to emphasise the combined processes of evaporation from soil and water surfaces plus transpiration from plants. Similarly, agriculturalists use the term **consumptive use** in order to emphasise that the uptake of water by vegetation in the production of plant material represents an important 'use', rather than a mere 'loss'.

At the global scale, evaporation and precipitation are the two principal elements of the hydrological cycle. Evaporation returns to the atmosphere the same amount of water as the solid or liquid precipitation that reaches the Earth's surface. Over the entire land surfaces approximately 60% of the precipitation is returned to the atmosphere as evaporation, making it the largest single component of the terrestrial hydrological cycle (see Figure 1.4). At the global scale, land surfaces provide only a small part of the evaporated water, the bulk coming from the extensive water bodies of the seas and oceans; this has a direct impact upon the large–scale transfer of water vapour from the oceans to the continents, and hence the distribution of precipitation over land areas.

Evaporation is *also* very important in controlling the Earth's energy budget, accounting for over 75% of the net radiation reaching the Earth's surface. Part of the radiation warms the atmosphere in contact with the ground by conduction and convection and is termed **sensible heat** since its effect may be measured or sensed by a change in temperature; the energy used or liberated in evaporation or condensation is termed **latent heat** (latent = hidden) since this involves a change in state without a change in temperature. The latent heat of vaporization, λ, is 2.47 x MJ kg^{-1} at 10°C, and this is a very important factor in hydrological and energy budgets. It requires about six times more energy to change a unit volume of liquid water into water vapour than to heat that water from 0°C to 100 °C.

It is important to emphasise that there is no fundamental difference in the physics of evaporation from water, soil or plants (Shuttleworth, 2012). The only difference is in the nature of the *controls* of those surfaces. Thus, in this chapter, the evaporation process is described first, and then any important distinctions between different types of surfaces are discussed afterwards.

4.2 THE PROCESS OF EVAPORATION

The physics of the evaporation process relate primarily to two aspects:

a. The provision of sufficient energy at the evaporating surface for the latent heat of vaporization, and
b. The operation of diffusion processes in the air above the evaporating surface to provide a means for removing the water vapour produced by evaporation.

In much simplified terms the evaporation process may be described as follows. The molecules in any mass of water are in constant motion, whether a large lake or a thin film on a soil grain. Adding heat causes the water molecules to become increasingly energized and to move more rapidly, the result being an increase in the distance between liquid molecules and an associated weakening of the forces between them. At higher temperatures, therefore, more water molecules near the surface will be able to escape from the surface into the lower layers of the overlying air. In fact, all water surfaces are giving off water vapour to a greater or lesser extent. Similarly, however dry the atmosphere seems to be, it always contains some water vapour; the water molecules in the lower air layers are also in continual motion, and some of these will penetrate into the underlying mass of water. The partial pressure (or concentration) exerted by water vapour molecules in the atmosphere is termed the **water vapour pressure.** Partial pressures are generally used instead of concentrations so that changes in atmospheric pressure are taken into account. Water vapour pressures vary greatly, but are typically in the range 0.1 to 4 kPa, compared with the total atmospheric pressure of about 100 kPa (Oke, 1987; Trenberth, 1992).

The rate of evaporation at any given time will depend upon the balance between the rate of **vaporisation** of water molecules into the atmosphere and the **condensation** rate of molecules from the atmosphere (Figure 4.1). The former is determined by temperature and the latter by the vapour pressure above the surface. If more molecules are entering the air from the water surface than are returning, then evaporation is taking place; conversely if more molecules are returning to the water surface than are leaving it, condensation is said to be taking place.

In absolutely calm conditions, the net movement of water vapour molecules from an evaporating surface into the overlying air will progressively increase the water content of the lowest layers of the overlying air. This cannot continue indefinitely and eventually the vapour pressure increases until the rates of condensation and vaporisation are equal, and evaporation ceases. The air is then said to be saturated. The vapour pressure exerted at saturation is called the **saturated (or saturation) vapour pressure, SVP**. Normally, however, diffusion processes resulting from turbulence or convection mix the lowest layers with the overlying air, thereby effectively reducing the water vapour content and permitting further evaporation to take place. Warm air can hold more moisture than cold air and the SVP increases approximately logarithmically with the air temperature having, for example, a value of 1.228 kPa at 10°C and 3.169 kPa at 25°C.

Aristotle was the first to record that *both* the sun's heat and the wind are important in controlling evaporation, and it can be seen that for a given water surface the rate of evaporation will be controlled by a number of meteorological variables: the input of energy, the humidity of the air and the rate of movement of the air enabling the water vapour produced to move away from the evaporating surface.

Evaporation is an invisible process and difficult to measure, which is probably why during much of the history of hydrology it was poorly quantified and often estimated simply as the residual from other components of the water balance that were easier to measure. This meant that all the estimation errors in those components accumulated in the estimated evaporation. Symons in 1867 described evaporation as "...the most desperate art of the desperate science of meteorology". But over the last 30 years there has been a huge growth in our ability to measure, understand theoretically and to model evaporation (Gash and Shuttleworth, 2007). It is very important to understand that while most of hydrology relies on the equation of conservation of mass, evaporation is *also* constrained by the law of conservation of energy. This sets evaporation apart as a key component of the surface energy balance, and not just a residual in the water balance.

Evaporation studies have centred on **thermodynamic** and **aerodynamic** approaches, or a combination of the two. The former deals with the energy balance of the evaporating surface (providing the necessary latent heat), whilst the latter deals with the vapour flux away from the evaporating surface.

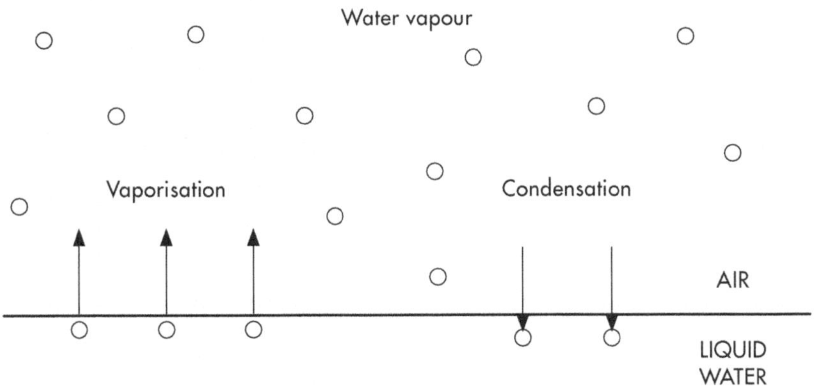

Fig 4.1: Evaporation is the net balance between the rate of vaporisation of water molecules into the atmosphere, and the condensation rate of molecules from the atmosphere.

4.2.1 THERMODYNAMIC FACTORS

The thermodynamic or energy balance approach to evaporation is concerned with estimating the latent energy available for water to change in state from a liquid to a gas. If the total amount of energy used in evaporation could be estimated, then by knowing the coefficient of latent heat it would be relatively simple to calculate the depth of water evaporated. This approach involves two main steps: a) determining the 'available energy' at the evaporating surface, b) apportioning this energy into latent and sensible heat transfers.

Of the available **net radiation**, R_n (incoming minus outgoing), some is used to heat the ground surface, G (part of which then goes to heat the overlying air, H), some is used for latent heat of evaporation (λE) and finally, a negligible part is used in plant growth.

An energy balance equation can, therefore, be written in the following terms:

$$R_n = H + \lambda E + G \quad (4.1)$$

where R_n may be determined using a net radiometer, and G from soil temperature measurements, but the convective sensible heat transfers (H) between the air and the water surface cannot be easily measured directly. The evaporation, E (mm), is multiplied by the coefficient of latent heat, λ, to convert it into units of energy. The second step of the energy balance approach is thus to determine the amount used in evaporation, E.

$$\lambda E = (R_n - H - G) \quad (4.2)$$

Bowen (1926) proposed that the ratio of the sensible and latent heat fluxes ($H / \lambda E$), now termed the **Bowen ratio**, β, may be determined from measurements of air temperature and vapour pressure at two levels, i.e.:

$$\beta = \gamma\,(T_s - T_a) / (e_s - e_a) \quad (4.3)$$

where γ is the so-called psychrometric 'constant', which actually varies with atmospheric pressure, and to a lesser extent with temperature (Shuttleworth, 2012). T_s is the mean surface temperature, T_a is mean air temperature, e_s is saturation vapour pressure at the temperature, T_s, of the evaporating surface, and e_a is actual vapour pressure of the air at a given height (commonly 1 or 2 metres). This approach makes the assumption that the turbulent transfer coefficients of heat and water vapour by eddy diffusion are equal. Then the evaporation rate is given by:

$$E = (R_n - G) / \lambda(1 + \beta) \quad (4.4)$$

Values of β are low for areas where most radiation is used for evaporation, and high where water is limited and sensible heat transfer predominates. Thus, typical *average* values of β increase from 0.1 for tropical oceans and 0.1–0.3 for tropical wet jungles, to 0.4–0.8 for temperate forests and grassland and 2–6 for semi–arid areas and as high as 10 or more for deserts (Oke, 1987; Dolman and Gash, 2010).

This approach demands accurate measurements of radiation, soil heat flux and vertical profiles of temperature and humidity. Although such measurements are relatively easily achieved in a research situation

SURFACE	CONDITION	ALBEDO
Water	Zenith angle, small to large	0.05–0.15
Snow	Old to fresh	0.30–0.90
Bare soil	Dark/wet to light/dry	0.05–0.35
Grass		0.20–0.30
Crops		0.15–0.25
Forest (deciduous)		0.15–0.20
Forest (coniferous)		0.05–0.15

Table 4.1: Typical mean values of albedo for selected natural surfaces (based on data compiled by Lee, 1980; Brutsaert, 1982; Oke, 1987).

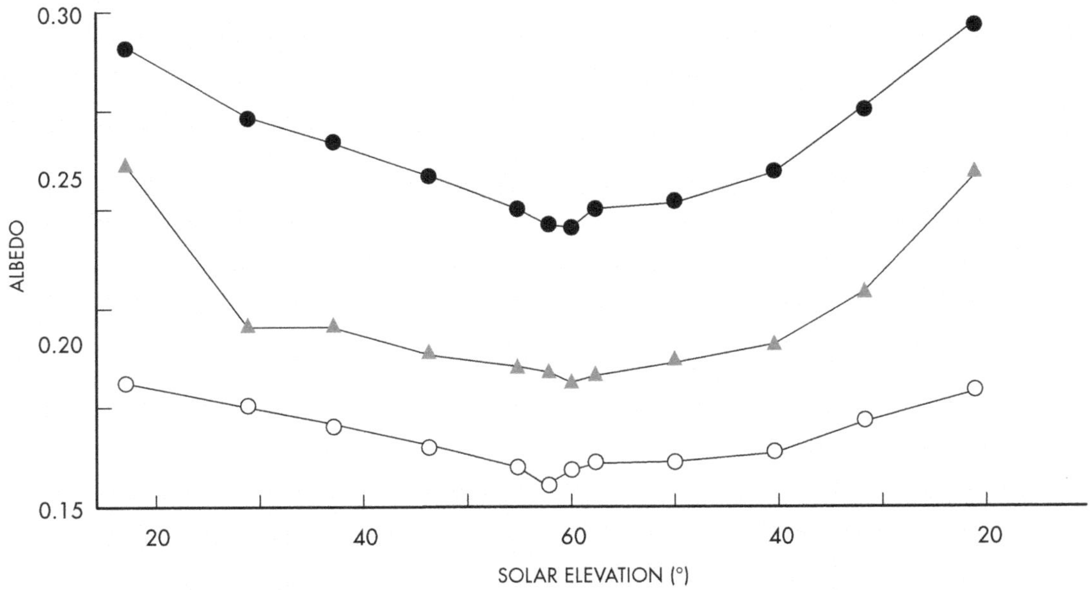

Fig 4.2: The albedo varies with sun angle and hence the time of day, season and latitude. Shown here for short grass (solid circles), bare soil (open circles) and kale (triangles) in mid-summer (Redrawn and simplified from Monteith and Szeicz, 1961).

they impose substantial limitations for the application of the energy–balance approach to the routine calculation of evaporation.

The energy balance at the evaporating surface is materially influenced by the **albedo** of the surface, the proportion of incoming radiation which is reflected back into the atmosphere. Typical values are shown in Table 4.1 for grass, agricultural crops and trees. It should be noted that the actual values change over time; the albedo varies with sun angle and hence the time of day, season and latitude, as illustrated in Figure 4.2. In general for a given sun angle and colour of vegetation, the albedo will be smaller for tall than for short vegetation since there is more opportunity for absorption by multiple reflections within deeper canopies. The overall albedo of the Earth's surface (land and oceans) is about 0.15, and the much higher planetary albedo of around 0.3 is due to the presence of clouds.

4.2.2 AERODYNAMIC FACTORS

The aerodynamic (or vapour flow) approach deals with the upward diffusion of water vapour from the evaporating surface, and is concerned with the 'drying power' of the air, comprising its humidity and the rate at which water vapour can diffuse away from the evaporating surface into the atmosphere.

Generally speaking, the evaporation of water from a given surface is greatest in warm, dry conditions and least in cold, humid conditions, because when the air is warm, the saturation vapour pressure (e_s) of water is high, and when the air is dry, the actual vapour pressure (e_a) of the water in the air is low. In other words, in warm, dry conditions the saturation deficit ($e_s - e_a$) is large and conversely in cold, moist conditions it is small. There is thus an underlying relationship between the size of the saturation deficit and the rate of evaporation.

Clearly, the stronger the wind, the more vigorous and the more effective will be the turbulent action in the air; and the greater the temperature difference between the surface and overlying air, the greater will be the effect of convection.

The aerodynamic approach to the evaporation from a wet surface was first expressed in quantitative terms by Dalton (1802b) who suggested that if other factors remain constant, evaporation is proportional to the windspeed and the **vapour pressure deficit**, i.e. the difference between saturation vapour pressure at the temperature of the water surface and the actual vapour pressure of the overlying air. Dalton's law, although never expressed by the author in mathematical terms, has provided the starting point of much of the subsequent work on evaporation:

$$E = f(u)\,(e_s - e_a) \qquad (4.5)$$

where $f(u)$ is a function of wind speed, e_s is the saturation vapour pressure at the surface, e_a is the actual vapour pressure of the free air at a reference height (commonly at 1 or 2 m).

In the lowest few millimetres of the atmosphere the air moves in straight lines or along smooth, regular curves in one direction, termed **laminar** motion. Above this layer, friction between the air and the ground surface induces eddies and whirls and air follows irregular, tortuous, fluctuating paths, termed **turbulent** motion, typified by cross–currents and gusts of wind with intervening brief lulls. This is the planetary boundary layer where the depth and strength of turbulence are largely dependent upon the roughness of the ground surface and the strength of the wind. The principal mechanism for transferring air properties, or **entities**, such as water vapour, momentum, heat and CO_2 through the atmosphere away from, or towards, the surface is a mixture of convective and mechanically generated turbulence. The greater the intensity of turbulence, the more effectively are water vapour molecules dispersed or diffused upwards into the atmosphere.

The process by which these properties are transported through a fluid is known as **diffusion**. In the laminar boundary layer close to the surface vertical transfers take the form of molecular diffusion, whilst at a larger scale with wind blowing over a natural surface turbulent diffusion is of primary importance, and is the subject of the following discussion.

The vertical flux of any entity (s) over a vertical distance (z) is proportional to the concentration gradient (ds/dz):

$$\text{Flux} = -K\, ds / dz \qquad (4.6)$$

The proportionality factor, K, is termed the turbulent transfer coefficient (or sometimes the turbulent eddy diffusivity). It is not a constant, but varies with the size of the eddies and the distance above the surface. Typical values range from about 10^{-5} $m^2\ s^{-1}$ close to a surface, up to 10^{+2} $m^2\ s^{-1}$ well above the surface (Grace, 1983). Since its value is more dependent upon the characteristics of the turbulent motion than the particular entity being transported, micrometeorologists often use the greatly simplifying assumption, called the 'Similarity Principle', namely that the different entities are transported with equal facility, and so their K values are equal.

The turbulent mixing in the boundary layer acts to homogenise the air, since it transfers atmospheric properties (such as heat and water vapour) from places of high concentration to those of low concentration, and so acts to equalise them at all heights. Thus, water vapour added to the bottom of the boundary layer by evaporation will be dispersed and diffused upwards, under a gradient from highest moisture content of the air at the ground surface and lowest at the top of the turbulent layer. For a constant intensity of mixing, an increase in the rate of evaporation will be reflected in an increased moisture gradient in the boundary layer. Similarly, with a constant rate of evaporation, variation of moisture gradient will reflect changes in the intensity of mixing. Accordingly, it should be possible to determine the rate of evaporation from any surface by reference to the moisture gradient and the intensity of turbulent mixing. Measurements are needed of the moisture content and the windspeed at a minimum of two known heights within the turbulent layer.

The vertical flux of an entity, T, between heights a and b may also be considered in relation to a resistance to flow:

$$\text{Flux} = (T_a - T_b) / r_a \qquad (4.7)$$

where r_a is the **aerodynamic** or **boundary layer** resistance ($r_a = {}_a\!\int^b (1/K)\, dz$) which is a measure of the resistance encountered by water vapour moving from the outer surface of the vegetation cover into the surrounding atmosphere.

The degree of turbulence is closely related to wind velocity and surface roughness, although this frictional turbulence may be enhanced by convective turbulence where there is a suitable gradient in mean air temperature away from the evaporating surface.

The amount of convective mixing depends upon the **stability** of the atmosphere, which is controlled by the vertical temperature gradient of the atmosphere (see Section 2.1.1). Air stability varies diurnally; normally, the environmental lapse rate increases during the daytime, reaching a maximum in the early afternoon. At night, in contrast, the ground surface cools and there may be an inversion of the air temperature profile. This decreases the buoyancy of the air, damps down convectional activity and effectively suppresses the turbulent motion of the air.

4.3 ESTIMATION OF EVAPORATION

Despite the crucial importance of evaporation in the hydrological cycle, it is inherently difficult to measure and quantify, and it remains the most difficult component of the water balance to determine with any accuracy. Unlike some other hydrological variables, such as runoff or rainfall, which can be measured directly, evaporation is commonly estimated indirectly. Several direct methods have been developed, and are outlined below, but they are not yet widely used outside of the research community.

There are two broad approaches to the measurement of evaporation. The simplest and earliest estimates involved measuring the *loss of liquid water at the surface*, whilst more recent and complex methods determine the *vertical flux of water vapour* (or latent heat) through the air.

4.3.1 LIQUID WATER LOSS

There are several traditional methods for measuring liquid water loss. Each has its limitations and disadvantages. An **evaporation pan** is a tank of water open to the air. By making regular measurements of changes in water levels and by correcting for rainfall and for any water added or removed (to maintain the water level in a certain range) the water loss due to evaporation may be derived (WMO, 2008). Due to their simplicity of manufacture and operation, evaporation pans have been widely used, and there are over 20 different designs worldwide. Some pans, including the most widely used type, the US Weather Service 'Class A' pan, are raised above the ground whilst others are sunk into the ground with their rims a few centimetres above the surface. Unfortunately due to their small surface area (typically 1–3 m^2) all designs of pans exaggerate lake evaporation. Furthermore, this is in a non–consistent manner; not only are they subject to problems of radiation on the sides of a raised pan and variable heat flow from the ground for a sunken pan, they also suffer from **advection**, the horizontal transfer of heat by moving air. Consequently it is necessary to apply an empirical pan coefficient to reduce the values to match those of the surrounding lake or vegetation. The annual value of the 'Class A' pan coefficient for transfer to lakes is often assumed to be around 0.7, but this varies enormously between years and between sites. The variation is greater over shorter periods, such as individual months (Oroud, 1998). Pans provide an even poorer indication of the evaporation from a land surface than for open water bodies, due to differences including albedo, surface roughness, and thermal storage. Nevertheless, they are still widely used, particularly to determine the irrigation water requirement in areas where meteorological data are not available. Doorenbos and Pruitt (1977) provided guidance for the selection of pan coefficients to estimate crop water transpiration – values range from 0.4 to 0.85 depending upon site conditions. Subsequently, several authors related the changing values of this coefficient to local meteorological data including daily mean relative humidity, wind speed and upwind fetch distance of low growing vegetation (e.g. Grismer, et al, 2002).

Another technique is to construct a water balance for a hydrologically isolated block of soil and vegetation termed a **lysimeter**. The simplest design is a drum packed with excavated soil, but this may differ substantially in behaviour from undisturbed soil and vegetation. A better situation is where a permeable soil overlies a naturally impermeable layer such as a heavy clay, allowing a wall or membrane to be installed in a trench cut down to the clay base (e.g. Calder, 1976). Evaporation may then be calculated from a mass balance of the difference between precipitation and drainage from the base of the block, together with measurements of changes in soil water storage. This approach may provide accurate measurements but it is difficult and expensive to install and maintain, may be subject to unmeasured leaks and it relies on evaporation being derived as the residual between precipitation/irrigation and drainage, each subject to measurement errors. Nevertheless, the water balance approach using carefully sited and operated lysimeters, and sometimes extended to whole catchment studies, has provided some of the most useful and validated evidence of evaporation rates and their changes due to alterations in vegetation or land use.

There are other more specialised techniques for the measurement of transpiration. The stomatal conductance of individual leaves may be measured using a porometer, although this is notoriously labour-intensive. The velocity of sap movement within individual plant stems may be measured using the heat balance and heat pulse techniques (Swanson, 1994; Smith and Allen, 1996), and the transpiration of whole plants may be obtained by deuterium tracing (Calder et al., 1992). All these methods suffer from the problem of extrapolating from individual trees to a whole forest stand.

4.3.2 WATER VAPOUR FLOW

The most fundamental approach is to measure the vertical transfer of water vapour away from the evaporating surface within the turbulent boundary layer. If the measurements are made close to the surface then the measured upward vapour flow rate is a good approximation to the surface exchange rate.

There are two basic approaches: the **profile** or **flux gradient** methods which are based on measurements of vertical gradients, and the **eddy covariance** or **eddy correlation** method which measures water vapour *flux* directly.

a. The **profile measurement method** relies on the assumption that over an extensive homogenous surface turbulent eddies will transfer momentum, heat and water vapour with equal facility, and so over a given height range their transfer coefficients may be assumed to be equal. This avoids the need to measure coefficients which are difficult to determine.

The Bowen ratio or energy balance profile method measures the components of the overall surface energy budget and apportions the available energy, R_n, between sensible and latent heat according to the Bowen ratio β. Since the rate of evaporation is determined on the basis of very small humidity and windspeed differences over a narrow height range within the boundary layer the frequency and the accuracy of the instrumental observations must be very high. It requires accurate measurements of net radiation, air temperature and humidity at two heights (to calculate β from Eq. 4.3), plus soil heat flux, *G*. Several factors must be considered when determining the heights of the two sets of sensors. A wider separation will increase the difference in temperature and humidity at the two levels to give better resolution, but they must not be placed so far apart that they sample air from different environments. Particular care must be taken over rough surfaces, such as a forest canopy, due to the greater turbulent mixing, as the vertical gradients will be much smaller than over a smooth surface, requiring measurements over a larger height range, perhaps equivalent to the height of the forest.

b. The **eddy correlation (EC)** approach is the most direct measure of water vapour flux with minimum theoretical assumptions. The measurement of fluxes of water vapour (or sensible heat) in turbulent air requires the instantaneous and simultaneous measurement of vertical velocity and vapour density (or air temperature). Thus, for water vapour to be transferred upwards by turbulent air motion it is necessary that on average the upward moving air is moister than the corresponding down currents.

Such instruments must sense virtually every variation in the vertical wind velocity and the entity under study (heat or humidity), and be able to process and integrate very large amounts of data to derive estimates of the turbulent fluxes of sensible heat and water vapour. The UK Institute of Hydrology's *Hydra* (Shuttleworth et al., 1988) was the first compact portable eddy correlation system used in field studies around the world. This comprised an ultrasonic anemometer (measuring wind speed by the time of flight of sonic pulses between pairs of transducers) together with humidity and temperature sensors linked to a microprocessor to enable analysis of measurements and calculation of fluxes at very fine temporal resolution directly in the field. Subsequent developments in eddy correlation include consideration of fully three-dimensional air movements to provide measurements over rough terrain, and consideration of fluxes of water vapour, CO_2 and other gases. The stage has been reached where eddy correlation systems are sufficiently robust to be run continuously and there are many hundreds of EC systems being used around the world. Some early instruments did not work satisfactorily during rainfall as raindrops passing though the sonic path distorted the readings and water lay on the transducer. Subsequent developments included software to remove the 'spikes' caused by raindrops, and the use of special coatings to make water run off the sensors have overcome this. Mizutani et al (1997) and Grelle et al (1997) showed that sonic anemometers can work in the rain, and this was confirmed in later studies including Gash et al (1999) and van der Tol et al (2003). A five year record of eddy flux data collected at a site in S England is shown in Figure 4.3. This shows the clear annual cycle of evaporation, with a minimum in September –March, when there is an excess of rainfall, and recharge to the underlying Chalk aquifer takes place (there is negligible runoff at this site).

Fluxnet (http://fluxnet.ornl.gov/) is a global network of several hundred micrometeorological towers that use eddy correlation to measure CO_2, water vapour and energy fluxes (Baldocchi et al, 2001). Typical instrumentation at each field site includes a three-dimensional sonic anemometer, to measure wind velocities and virtual temperature, and a fast responding sensor to measure CO_2 and water vapour. Sampling rates between 10 and 20 Hz are needed ensure complete sampling of the high frequency portion of the flux co-spectrum, and the duration must be long enough to capture low frequency contributions to flux covariances, typically 30–60 minutes. Although the primary motive for much of this work is to study and quantify greenhouse gases, particularly the carbon budget, they measure evaporation as a by-product and these data are being used by hydrologists (van Dijk et al, 2015).

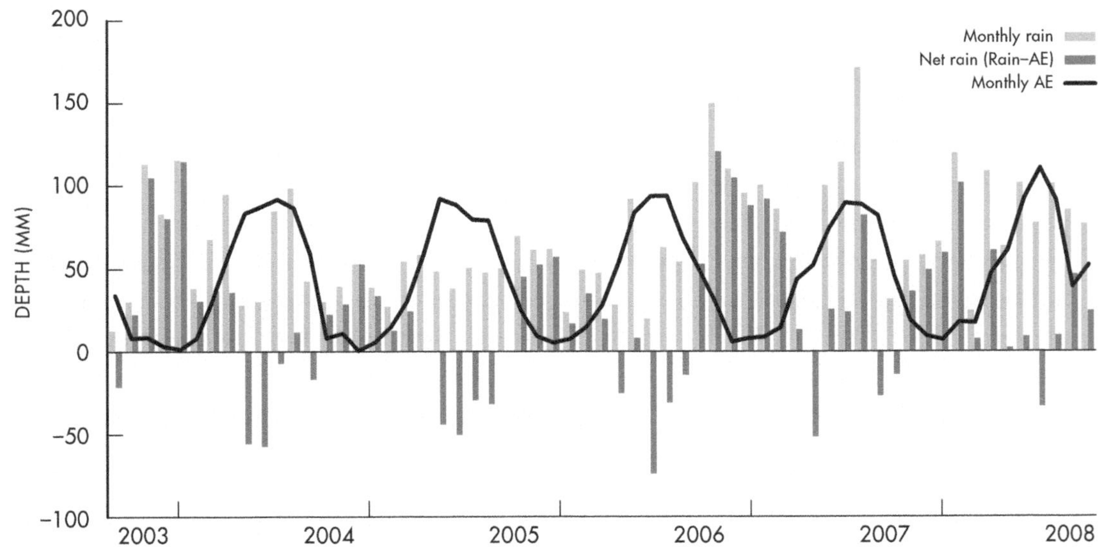

Fig 4.3: Monthly Actual Evaporation (AE) by eddy covariance, and recorded Rain (R), together with computed net rain (R minus AE). Positive values indicate net recharge. (Data for 2003–2008 at Sheepdrove Farm, supplied by CEH).

The Integrated Carbon Observation System (ICOS) is a pan-European initiative to provide harmonized and high precision scientific data on carbon cycle and greenhouse gas budget (www.icos-ri.eu). It began in 2015 and will potentially become an enormous resource for hydrologists.

Scintillometers provide average evaporation estimates over large areas by transmitting a beam of electromagnetic radiation and measuring the intensity of fluctuations in the received signal due to air density differences altering the refractive index between the transmitter and receiver (De Bruin and Evans, 2010). Depending on the wavelength used this will be mainly due to heat movement (visible and near-infrared) or moisture fluctuations (microwaves). This approach can be used to obtain line averaged sensible heat fluxes over several kilometres of heterogeneous terrain (e.g. Ward et al, 2014), which can then be compared with measurements of net radiation to derive the latent heat flux.

Despite great advances in technology much of the essential instrumentation for such measurements lacks standardization, can be demanding in terms of maintenance and is expensive. As a consequence attention for routine estimates of evaporation is still often focused on empirical or semi–empirical models of the evaporating system, especially total evaporation from a vegetation–covered surface, in which the model input relies on readily available, routine meteorological measurements.

Measurement representativeness

All of these approaches to the measurement of evaporation from water, soil or vegetated surfaces, together with the theoretical approaches discussed in the remainder of this chapter, suffer to a greater or lesser extent from the overriding problem of point sample representativeness, and most of them suffer also from the distorting effects of advection which is inversely proportional to the size of sample evaporating surface. Often, it is difficult to assess the extent of the upwind fetch sensed by these micrometeorological measurements, in terms of a valid representation of areal evaporation at a field, catchment or regional scale. If the surface is not homogeneous, the measured

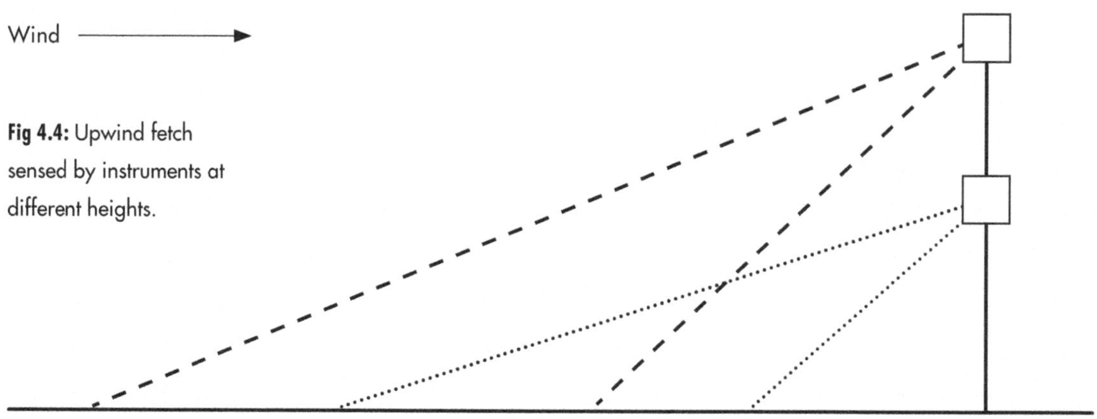

Fig 4.4: Upwind fetch sensed by instruments at different heights.

value will depend on which part of the upwind surface had the strongest influence on the sensor, and thus on the size and location of its 'footprint' (Figure 4.4). Gash (1986) derived a simple method of estimating the upwind area sampled by micrometeorological measurements as a function of measurement height and aerodynamic roughness. Subsequent work on estimating the flux footprint is reviewed by Schmid (2002).

4.4 EVAPORATION FROM DIFFERENT SURFACES

As noted earlier, the physics of evaporation is the same regardless of the evaporating surface, but different surfaces impose different controls on the process. These differences are discussed below for three broad surface types: open water, bare soil and vegetation.

4.4.1 OPEN WATER

An open water surface is the simplest situation, with little limitation on evaporation. The supply of water is by definition, so plentiful that it exerts no limiting influence on the rate of water loss. Accordingly, evaporation rates from lakes may be very high – as much as 2,000 mm y^{-1} in some arid areas (Van der Leeden et al., 1990), and this must be taken into account when, for example, planning a new reservoir since the additional loss of water from the open water surface will offset to some extent the increase in water supply from the reservoir.

The rate of open water evaporation is determined by a number of factors, both meteorological (energy, humidity, wind) and physical (especially the size of water body), in accordance with the factors outlined earlier controlling the evaporation process.

Evaporation rates broadly follow variations in radiation, modified by considerations of heat storage in the water body. Evaporation may increase as the air temperature rises, and is able to hold a larger amount of water vapour below the saturation level. Air movement is necessary to remove the lowest moist air layers in contact with the water surface and to mix them with the upper drier layers, so that the rate of evaporation is almost always influenced to some extent by turbulent air movement. In fact, the relationship between windspeed and evaporation holds good only up to a certain speed, above which further increases lead to no further increase in evaporation since evaporation is then limited by the energy and humidity conditions.

For a given water body the rate of evaporation under identical meteorological conditions may vary as a result of physical differences, the most important of these is the water surface area.

Size of water surface

As air moves from the land across a large lake there will initially be high evaporation and then as the humidity of the air increases and its temperature decreases, the rate of evaporation will decrease. The larger the lake, the greater will be the total reduction in the depth of water evaporated, although of course the total volume of water evaporated may well increase with the size of the water surface.

For an extensive water surface, such as the oceans, the humidity of the air will be largely independent of the distance it has travelled, except in coastal areas. Consequently evaporation will be uniform over extensive areas and will be closely related to the amount of heat energy available. At the other extreme, small water surfaces such as evaporation pans exert little influence on the temperature or humidity of the overlying air. The small amount of water vapour which leaves the surface, even with high rates of evaporation, is quickly diffused so that a continuous high rate of evaporation is maintained. This enhanced local evaporation has been termed the 'oasis' effect. Consequently there is an inverse relationship between the evaporation rate and the size of the evaporating surface. This difference is greatest if the humidity of the incoming air is low.

Water depth

The effect of water depth upon the seasonal distribution of evaporation may be quite considerable, due to the thermal capacity of water and to the mixing of waters. The seasonal temperature regime of a *shallow* lake will normally approximate closely to the seasonal air temperature regime, so that maximum rates of evaporation will occur during the summer and minimum rates during the winter. For large, deep lakes, however, there may be a time lag of several months between net radiation and evaporation. During the spring and summer, heat entering the surface is absorbed in the first few metres of water and the warm surface waters are mixed downward by turbulence; so evaporation is lower than for a shallow water body. In winter, however, the water may be warmer than the air and heat stored in the lake may be released for evaporation to take place at higher rates than would be expected from meteorological measurements alone (Blanken et al, 2000). The heat storage is crucial for short term estimates, and requires regular measurements of the temperature profile of the water. The net result is that water temperatures are lower than air temperatures during the summer and higher in the winter; thus evaporation for a large water body may be higher in winter than in the summer (Figure 4.5).

The situation is further complicated by the fact that water has its maximum density at 4°C, ie if it gets any colder or warmer, it will rise. Ice floats on top of lakes, preventing evaporation, and lakes stay liquid underneath, allowing fish and other life to survive. This can have an important seasonal effect, but over the course of a year there will be little net change in heat storage overall and the total annual evaporation is little altered.

Finch and Gash (2002) overcame the general lack of water temperature measurements using a finite difference scheme to calculate a running balance of lake energy storage. Evaporation could then be calculated from standard land-based daily observations of sunshine, relative humidity wind run and air temperature.

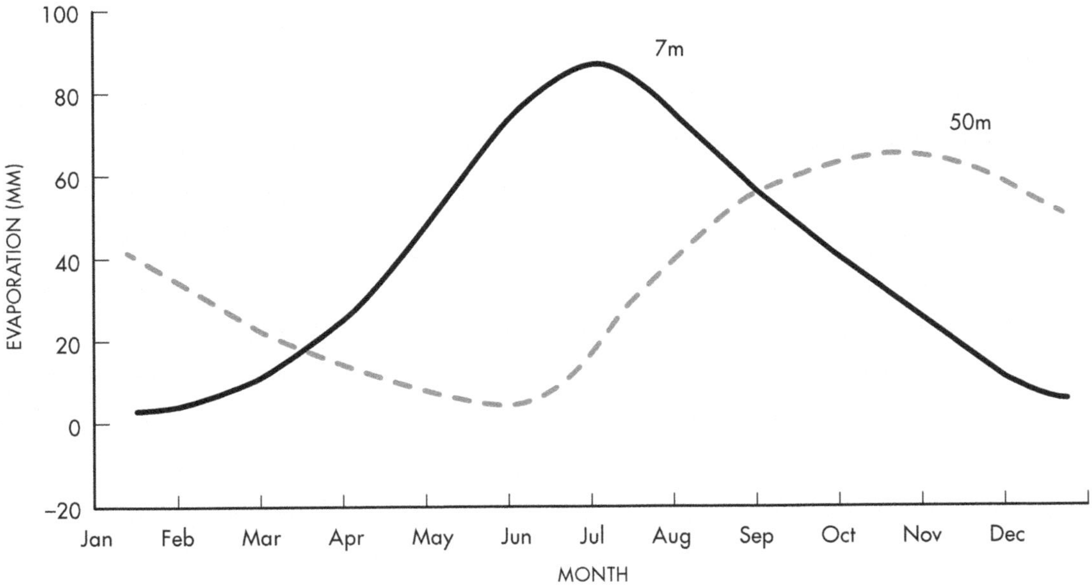

Fig 4.5: Simulated monthly average lake evaporation for different water body depths. Changes in the heat stored in the lake can cause a substantial lag between solar radiation and lake evaporation. Based on data in Finch and Calver, 2008.

Salinity

If a substance is dissolved in a liquid, the motion of the liquid molecules is restricted and the vapour pressure of the liquid is lowered ('Raoult's law'). Thus, salinity may affect evaporation rates since the SVP over saline water is lower than over pure water. Evaporation decreases by about 1% for each 1% increase in salinity. Accordingly, evaporation from sea water (average salinity of about 35 parts per thousand, or gram kg^{-1}), is about 2–3% less than evaporation from fresh water. This effect is normally small enough to be discounted when comparing evaporation rates from different 'fresh' water bodies, but it may be important for very saline waters and brackish lakes, e.g. Dead Sea (Calder and Neal, 1984).

Evaporation from snow

It is generally accepted that evaporation rates from snow-covered land are usually very low. This may be contrasted with the situation already noted for snow interception (see Section 3.7). Lying snow generally occurs at locations and times of the year when radiation inputs and air temperatures are low; furthermore, snow cover has a high albedo and a generally smooth surface which inhibits turbulent transfer of heat and water vapour (Calder, 1986). Evaporation can occur only when the vapour pressure of the air is less than that of the snow surface; evaporation from snow will cease when the dew point rises to 0°C, and as temperatures rise above freezing, the rate of snowmelt must exceed the rate of evaporation, since approximately ten times greater energy is required to sublimate snow (change from solid to vapour) than just to melt it, the mass of water melted and available as runoff, is substantially greater than the amount evaporated.

4.4.2 EVAPORATION FROM BARE SOILS

Evaporation from a soil surface comprises the evaporation of films of water surrounding the soil grains and filling the spaces between them, and it is governed by the same meteorological factors as the loss from a free water surface. However, the rate of evaporation from soils is often less than that from a free water surface under the same meteorological conditions, because the supply of water may be limited by the amount of water in the soil and by the ability of the soil to transmit water to the surface. These factors are discussed in more detail in Chapter 6, and are only briefly described here.

In semi–arid areas, with sparse natural vegetation, soil evaporation may be may be a major component of the total evaporation (Jacobs and Verhoef, 1997; Kabat et al., 1997). In contrast, for well watered and vegetated areas, such as north western Europe, the evaporation from plants will generally provide most of the total evaporation. Exceptions include bare ploughed land and areas of immature, widely spaced crops, where evaporation from the soil surface may deplete the surface soil water and potentially affect the growth and yield of the crop.

The water content of the surface layers of bare soil exerts the most direct influence on evaporation rates. Evaporation decreases rapidly after rainfall as the surface water content falls until, with a dry soil surface, it is zero. In the absence of plant roots, the upward movement of soil water to the surface from the wet layers beneath is controlled by capillarity and will, therefore, tend to vary with soil properties. In fine–textured soil, with small intergranular pores, capillary movement may be effective over large vertical distances, such as a metre or more, whereas in coarse–textured sand it may be only a few centimetres. However, the speed of water movement tends to vary inversely with the height of capillary lift so that in neither case does the supply of water to the soil surface by capillary activity normally significantly increase total evaporation.

In areas with hot, dry, soil surfaces water vapour gradients are built up in the soil air, and water vapour may be transported up to the surface. The thickness of the evaporating zone will be greater in fine textured soils (Yamanaka et al., 1998). Soil colour will tend to affect evaporation; darker soils will absorb more heat than lighter soils and the resulting rise in surface temperature may increase the evaporation rate. The rainfall regime is also important for soil water replenishment, in just the same way as for the loss of water intercepted on vegetation foliage. On average evaporation from a soil surface will be greater if it is frequently wetted by intermittent showers than occasionally soaked by the same quantity of rain falling in a few large storms.

Bare soil evaporation can be described as a two-stage process. Immediately after rainfall, soil water is sufficient for evaporation at the potential rate, and then once soil water becomes a limiting factor, evaporation declines as a function of time (Wallace and Holwill, 1997).

Remote sensing

Evaporation remains one of the biggest unknowns in the global water balance. Remote sensing has a potentially important role in providing areal estimates of evaporation for a wide range of scales from individual fields to major river basins. Evaporation cannot be directly measured by remote sensing techniques, and current techniques combine variables that are relatively easy to measure; such as visible channels for surface albedo to estimate net radiation, infrared channels for surface temperature to estimate the vapour pressure deficit of the overlying air, together with additional ground based meteorological measurements, in order to complete the energy balance. Miralles et al (2011 a, b) adopt a pragmatic approach for estimating global evaporation from a variety of available measurements including the GOES, Meteosat and AQUA satellites. They combined satellite-derived estimates of precipitation, land surface temperature and surface soil water content, with the minimalistic Priestley-Taylor equation for potential evaporation, a simple soil model (to reflect when evaporation is below the potential rate), and a rainfall interception model.

4.4.3 EVAPORATION FROM VEGETATION COVERS

The hydrologist's greatest interest in evaporation losses concerns those from vegetated surfaces, including agricultural crops and natural vegetation. Water is essential for plant life and growth and comprises as much as 95% of the weight of a plant. Water transports nutrients within the plant, prevents the plant from overheating, and produces an internal pressure that helps the plant maintain its rigid yet flexible form. It has been emphasized earlier that the total evaporation from vegetation comprises the *sum* of the evaporation of intercepted water from the wet surface of the vegetation (see Chapter 3) and transpiration by plants. The transpiration system provides a particular example of the evaporation process in which water is evaporated from plant tissues.

Transpiration is closely linked to **photosynthesis**, which is an essential process by which plants form carbohydrates that are fundamental to the life of plants. Most plants takes in CO_2 from the air and combines it with water to produce carbohydrate, and oxygen is lost from the plant as a gaseous by-product.

The photosynthetic tissue, the mesophyll, in the leaf is protected by an outer epidermis which consolidates the leaf structure and protects the inner tissues from physical damage, from attack by micro-organisms, and also from desiccation. Numerous small pores called **stomata** (singular **stoma**) on the leaf enable the diffusion of CO_2 and O_2 during photosynthesis, but they also allow the loss of water vapour from a plant.

Figure 4.6 illustrates the complexity of transpiration for a single plant in a vegetation–covered surface, and shows that it is dependent upon a sequence of water–moving processes. Soil, plant and atmosphere form parts of a continuous flow path of varying resistances in which water moves at varying rates and undergoes both chemical and phase changes. Water at a point in the soil profile moves under the influence of a moisture gradient towards the root hairs, is there absorbed and then travels up the water-conducting vessels (**xylem**) of the plant stem, which are a low resistance hydraulic conductor, to the plant leaves from where it is finally vaporized in the stomatal cavities of the leaves before passing through the stomatal apertures to the atmosphere.

The stomatal conductance of leaves depends upon a combination of factors of which the most important are temperature, vapour pressure deficit, leaf water potential and CO_2 levels. At larger scales the balance of control shifts as the sensitivity of evaporation to canopy conductance declines as feedbacks of temperature and moisture in the air above the canopy increase in importance (Jarvis and McNaughton, 1986).

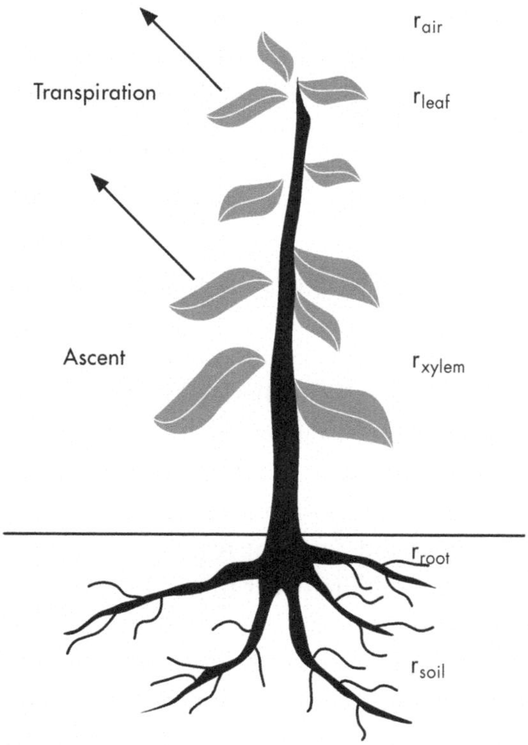

Fig 4.6: Transpiration through a single plant in a vegetation-covered surface, showing some of the main resistances (r) to water movement.

Transpiration – a necessary evil?

Very little of the water taken up by plant roots is actually retained for growth, most is lost in transpiration. This loss of water vapour occurs along the reverse pathway to the uptake of CO_2 in photosynthesis. Transpiration has been called a 'necessary evil' because the assimilation of CO_2 from the atmosphere requires intensive gas exchange, whilst the prevention of excessive water loss requires gas exchange to be kept low. Nevertheless, transpiration is an effective way to cool a plant in hot weather (due to the heat used up in latent heat of vaporisation) and the movement of water transports nutrients through the plant.

Plants can control water loss by varying the opening of the stomatal apertures. Thus, one of the consequences of water shortage in plants is a reduced uptake of CO_2, which reduces the rate of plant growth, and so necessitates the irrigation of agricultural crops.

Cell water potentials

Cells are the basic structural elements in plants and control the fundamental plant response to water as it moves through plants. The cell wall is porous and water moves across a series of cells due to water potential gradients. This alone would be too slow to replace the water lost by transpiration, if it were not for the vital role of **osmosis**. This is the flow of water across a membrane separating two solutions of different concentrations. The **water potential**, ψ, in a plant cell may be considered as:

$$\psi = \pi + P \qquad (4.8)$$

where π is osmotic potential, P is the internal cell wall pressure potential or **turgor**. At times of extreme water shortage, the **turgor pressure** force exerted outward on a plant cell by the water within the cell may be reduced to zero, causing the leaves to become limp or flaccid, and the plant condition may be said to be 'wilting'.

Leaves maintain a low water potential by accumulating a high concentration of salts in their cell sap, which results in a large negative osmotic potential. This is partially offset by the positive turgor potential. During active transpiration the leaf cells partially dehydrate, decreasing the turgor potential, and the cell sap concentrations increase, which in turn causes the osmotic potential to become more negative. The combined effect is to lower the water potential within the leaf cells. Thus, the loss of water vapour by transpiration from the leaves generates a reduction in the water potential, which acts as a suction force which is transmitted from cell to cell right through the system from the leaf to the root and is capable of drawing up water through even the highest trees against the force of gravity. Water potentials in plants are analogous to soil water potential discussed in detail in Chapter 6.

The flow of water from soil to leaves along a gradient of decreasing water potential, ψ, occurs against a series of resistances to movement (e.g. from soil to root, root to leaf, leaf to atmosphere). Each may be represented by the general transport equation:

$$\text{Flux} = (\psi_1 - \psi_2) / r \qquad (4.9)$$

where r is termed the resistance. For a system comprising a sequence of resistances in series, such as a whole plant, the total resistance (R) is, by analogy with Ohm's law, simply the sum of the individual resistances (i.e. $R = r_1 + r_2 + r_3 \ldots$). In plant physiology it is often convenient to replace the resistance by its reciprocal the conductance, g (i.e. $g = 1/r$), since for a number of resistances in parallel, such as individual leaves, the overall canopy conductance is the sum of the individual conductances, (i.e. $G = g_1 + g_2 + g_3 \ldots$). Then for a given difference in water potential, the flux is directly proportional to the conductance.

Considering a *steady state* situation the transpiration flow through the series of resistances will be equal; thus:

$$\text{Flux} = (\psi_s - \psi_l)/r_p = (\psi_l - \psi_a)\ r_g \qquad (4.10a)$$

$$= (\psi_s - \psi_a) / (r_p + r_g) \qquad (4.10b)$$

where the water potential is ψ_s at the soil–root interface, ψ_l in the leaf, and ψ_a in the air; and the resistances are in the plant, r_p, and the gas phase, r_g. Although this is an over–simplification, since it assumes steady state flow and constant resistances (conditions that seldom occur in practice), it provides a useful model of water flow in plants. By assigning reasonable values to the potentials at different points in the system, the relative importance of the different resistances in controlling the transpiration flux can be illustrated. Thus, if the soil water is at saturation (i.e. $\psi_s = 0$), the leaf water potential, $\psi_l = -2$ MPa, and the water potential of the air, $\psi_a = -100$ MPa (relative humidity ~50%), substituting these values in equation 4.10a indicates r_g/r_p ~50. This indicates that the resistance of the gas phase is very much larger than that of the plant, and consequently will have the controlling influence on transpiration.

Most physicists studying the transpiration process ascribed 'primary' control to the stomata, although in some conditions it is the root resistance that is the most important link in the transpiration flow. Some studies have shown that the stomatal opening is controlled by the leaf water status, whilst in other cases the root water status appears to be more important (Kramer and Boyer, 1995). There is also evidence that plants adjust their stomata to optimize CO_2 uptake and hence carbon assimilation (Ball et al, 1987).

The factors affecting transpiration from vegetation covers are numerous and will vary in importance depending on conditions. Discussion here will concentrate on two main vegetational influences on evaporation, i.e. stomatal control and root water uptake.

Stomatal control

On average stomata occupy about 1% of the total leaf area for a wide range of plant species (Monteith and Unsworth, 2013). They generally occur only on the underside (*abaxial*) surface of leaves. They may be completely absent from the upper (*adaxial*), leaf surface, especially in broad leaved trees. Each stoma consists of a pair of elongated guard cells surrounding the stomatal aperture. The cells in the walls of the aperture are usually impregnated with water as the gases CO_2 and O_2 involved in photosynthesis can only be utilised in solution, in which state they travel within the plant by diffusion. As water evaporates from these walls the water vapour pressure in the stomatal cavity increases and diffusion of water vapour molecules takes place through the stomata into the external air.

Stomata combine an efficient means of gas exchange needed in photosynthesis, with a means to limit the risk of damage to the plant from losing water. The size of the stomatal aperture can be varied by changes in the turgor of the guard cells. When guard cell turgor is low the pore closes. Since stomata are primarily photosynthesis structures, they are extremely sensitive to changes in light intensity. Thus, in the majority of plants there is a diurnal pattern to stomatal movements; they open during the daytime (for photosynthesis) and close at night when it is dark to avoid unnecessary water loss when photosynthesis would not be taking place. They are also sensitive to atmospheric CO_2 concentrations, opening if concentrations reduce. The **stomatal resistance** of a leaf is a physiological resistance imposed by the vegetation itself and is considered further in Section 4.5.1.

In general, the stomatal control of vapour diffusion is associated with the maintenance of leaf turgidity. However, the factors affecting stomatal resistance and the trigger mechanisms are varied, and include the pumping of potassium ions into the guard cells, the effect of carbon dioxide concentration and the role of abscisic acid. The effect of light appears to differ little among species, and on sunny days the stomatal resistance on exposed leaves decreases rapidly at sunrise, remains at a minimum value all day, if the water supply to the leaf is adequate, and increases again at sunset (Federer, 1975).

Physiological factors may be very important causes of interspecies variation in r_l. The low rate of transpiration from heather, for example, is caused by its large stomatal resistance of ~50–170 s m^{-1} (Wallace et al., 1984; Miranda et al., 1984). This results from the stomata occurring only on the abaxial side of the leaves in a groove lined with fine hairs, so that free movement of water vapour is impaired (Hall, 1987).

As transpiration occurs, the leaf water potential declines, causing increased inflow of water from regions of higher potential in the stems, roots and soil. If the soil is sufficiently moist, minimum values of resistance (i.e. maximum conductance) may be maintained throughout the day, and evaporation is controlled by meteorological factors. However, when evaporation from the leaves exceeds the rate of water supply to them through the soil–stem system, a critical leaf water potential is reached, at which stomata begin to close, thereby limiting further water loss. This critical value may depend on plant type and history, and on leaf location, i.e. leaves in the upper canopy may have lower values, so that leaves in the lower canopy experience earlier stomatal closure, implying a preferential water supply to the more exposed leaves in the upper canopy. The stomatal resistance increases above the minimum value and water shortage reduces actual evaporation below the potential rate via the mechanism of stomatal control. Exactly *when* this occurs is still not clear but, as Federer (1975) observed, "it is the classic question of availability of soil water stated in terms of stomatal control". It illustrates clearly the difficulty of distinguishing between meteorological, plant and soil influences on evaporation.

The divergence of evidence appears to be wide, ranging from work that showed stomatal control of evaporation in wet soil conditions to other studies in which evaporation appeared not to be reduced below the potential rate until *very* low values of soil water content had been attained (see review by Turner, 1986).

Given that the soil–plant–atmosphere system is characterized by great variability of water content in space and time, it is perhaps not too surprising that different vegetation types, growing in different soil and atmospheric conditions, will limit evaporation at different values of soil water status. Indeed, it might be argued that much more remarkable is the similarity of transpiration from different forest species in Europe noted by Roberts (1999) and explained in terms of a possible similarity both of $r_{l\text{–}min}$ and also of the relationship between r_l and atmospheric humidity. Roberts further suggested that similar $r_{l\text{–}min}$ values in European forests might result from tree species being genetically adapted to similar soil water conditions.

There is also a sense in which the integration of the complex heterogeneity of, say, water potentials and surface resistances of individual leaves imposes a certain homogeneity on the system as a whole. Thus evaporation from a vegetation canopy is the sum of the evaporation losses from the much larger area of all the individual leaf surfaces. The leaf area index (LAI) is the surface area of all the leaves in a unit area of land, and if $\bar{r}_l$ is the mean surface resistance of the leaves then the overall surface resistance of the canopy r_c (also called the **canopy resistance** or the **bulk surface resistance**) may be regarded as:

RESISTANCE	DESCRIPTION
Aerodynamic resistance, r_a	Varies with wind speed and vegetation height
Stomatal (or surface) resistance for individual leaves, r_l	Obtained from porometry measurements on individual leaves
Canopy (or bulk) resistance, r_c	Obtained by scaling up from r_l, using the Leaf Area Index
Surface resistance, r_s	Obtained from back calculation of Penman-Monteith eq. and meteorological readings. Includes non-physiological resistances (e.g. soil and leaf litter). Thus $r_s = r_c + r_{soil}$.

Table 4.2: Nomenclature and symbols for the different resistance terms for water vapour movement used in this text.

VEGETATION	INDIVIDUAL LEAF, g_{max}	BULK VEGETATION, G_{max}
Temperate grass	8	17
Cereals	11	32.5
Conifers	5.7	21.2
Deciduous trees	4.6	20.7
Tropical rainforest	6.1	13
Eucalyptus	5.3	17

Table 4.3: Average maximum values for stomatal conductance (s mm^{-1}). (Values from Kelliher et al., 1995).

$$r_c = \bar{r}_l / L \qquad (4.11)$$

This has been described as the "overall physiological control of transpiration" (Lee, 1980). Since except for very sparse vegetation cover the LAI is usually greater than unity, the bulk stomatal resistance is generally much lower than the leaf stomatal resistance. It should be noted, that whilst this is a commonly employed technique (e.g. Szeicz et al., 1973) it represents a gross simplification. It makes no allowance for canopy structure and microclimatological interaction between the individual leaves, such as the reduction in radiation to the lower canopy. The relation of mean r_l to r_c is major gap in knowledge.

Since L is difficult and laborious to calculate the two measures of stomatal control at leaf and canopy scales are generally derived independently (Kelliher et al., 1995); r_l may be obtained from porometry on individual leaves, whilst r_c for canopies is commonly obtained by inverting the Penman–Monteith equation (see Section 4.6.2).

There is scope for considerable inconsistency in the different terms used in the literature, and the symbols used in various publications. Table 4.2 attempts to clarify this. In some cases r_s is used for the canopy

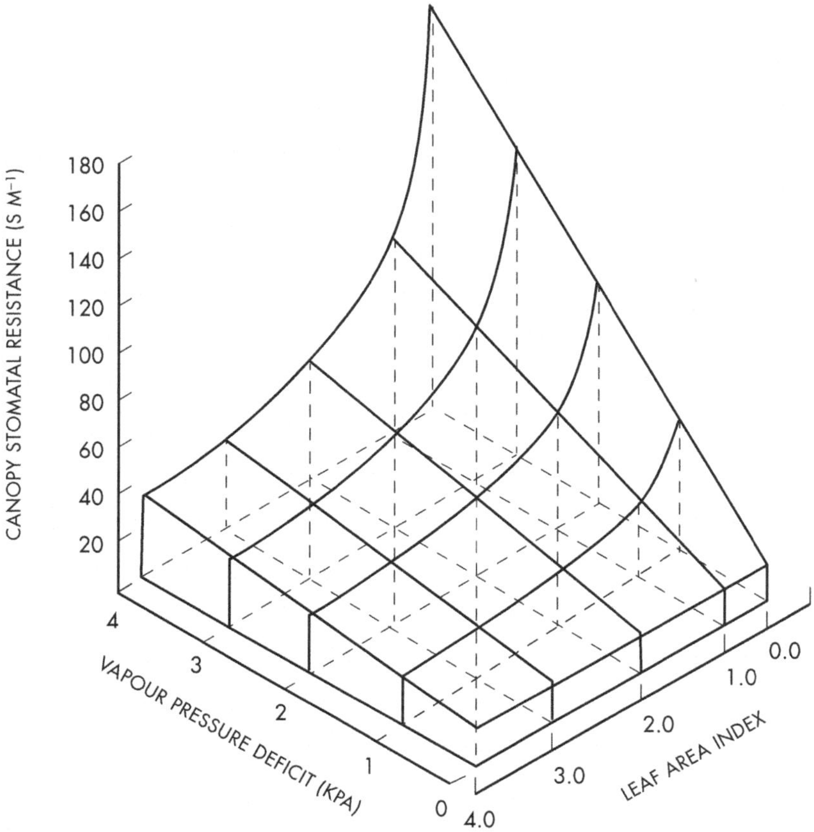

Fig 4.7: The canopy resistance of a crop varies with the leaf area index and vapour pressure deficit (Smith et al, 1985)

resistance, whilst in others it is used for the leaf stomatal resistance. In practice there is generally little difference in the numerical values of canopy resistance and surface resistance.

The stomatal resistance continually varies over time due to factors especially changing light intensity. Smith et al. (1985) found that r_c for well–irrigated wheat varied significantly with the leaf area index and vapour pressure deficit, as shown in Figure 4.7. Similarly, Stewart (1988) derived semi–empirical relations between r_c and meteorological factors for a pine forest.

For many purposes it is the *minimum* resistance or *maximum* conductance (the inverse of resistance), rather than the average value that is of most significance when comparing transpiration of different vegetation. Table 4.3 compares average values of maximum stomatal conductance for major world vegetation types. An alternative means of describing the behaviour of stomata is the theory that their primary function is to optimise CO_2 uptake (Ball et al, 1986) and the need to predict both evaporation and CO_2 flux, means that this approach is now the one adopted in most global climate models (Gash and Shuttleworth, 2007).

Root control

The preceding review of stomatal control emphasised the necessity of adequate soil–plant water supply to the leaves to maintain leaf water potential above critical levels. Indeed, traditional treatments of the subject tended to treat the roots as inert, uniform sinks for soil water, although subsequently this passivity has been questioned (Turner, 1986; Monteith, 1985). It is now better appreciated that under some situations it is the roots rather than leaves which may act as the primary sensors of water stress. It appears that a water shortage to the roots may cause changes in root metabolism, such as an increase in abscisic acid production and a disturbance in nitrate metabolism, which can send biochemical signals to the plant leaves leading to physiological changes, including a decrease in stomatal conductance, and rate of photosynthesis, *regardless* of the current leaf water potential (Kramer and Boyer, 1995). There is still discussion of the relative roles in controlling stomatal conductance of roots sensing soil water stress, and leaf water status (Comstock, 2002; Jones, 2007).

The uptake of water by plants during dry weather leads to a reduction in soil water around each root and a consequent fall in soil conductivity, which will considerably increase the resistance to liquid flow. The extent and efficiency of the root system should help to determine the total amount of water available to the plant cover and the rapidity with which, in drying conditions, the rate of actual evaporation falls below the potential rate (see Section 4.6.1).

Considerations of rooting depth suggest that a forest will transpire more than a pasture because trees, on the whole, are more deeply rooted than grass. The relationship between root density and water uptake is, however, not as simple as this. Field experiments (e.g. McGowan et al., 1984) have shown that the capture of soil water by crop roots is not solely dependent upon root development and soil water potential, but upon the *differences* of potential that can

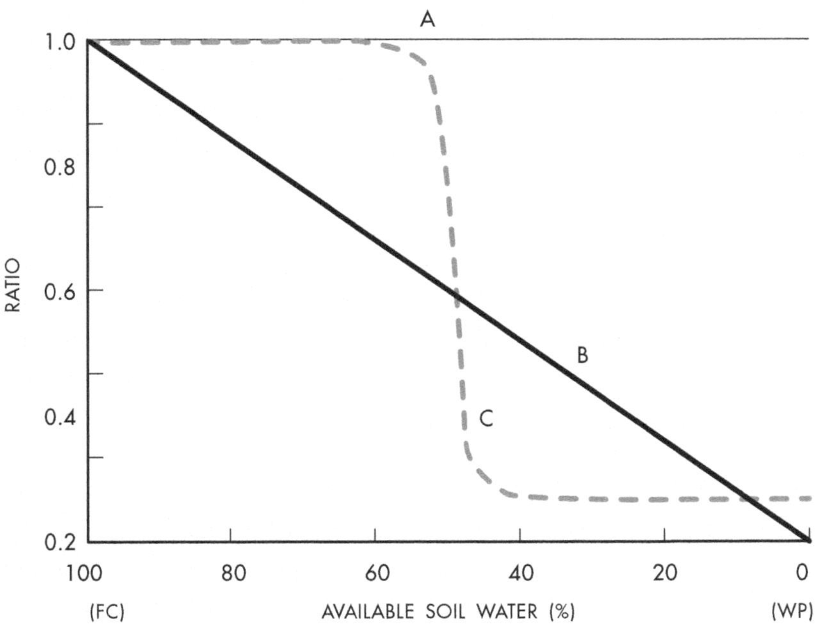

Fig 4.8: Several possible patterns of changing ratio of actual transpiration to unstressed transpiration for vegetation as soil water availability is reduced from field capacity (FC) to wilting point (WP). See text for details.

develop between soil and root. Their evolutionary ecological survival strategies are also different. Trees are long-lived and must survive the worst drought, whilst grasses grow quickly and set seeds; then they can die or be dormant. This also this affects stomatal resistance, being larger for trees than for shorter vegetation. Measurements during and after the great British drought of 1975–6 on three consecutive crops of winter wheat grown in the same field showed that the crop with the largest root system, indeed an unusually large root system, grown in the second drought year, 1976, was the least efficient in extracting soil water and so failed to dry the soil as thoroughly as the crops in 1975 and 1977. Plant water potential data showed that this restricted use of the available soil water was associated with "... failure to make any significant osmotic adjustment, leading to premature loss of leaf turgor and stomatal closure" (McGowan et al., 1984).

4.4.4 SOIL FACTORS

Numerous attempts have been made to relate plant transpiration to soil water status. It will be clear, however, that the influence of soil water conditions on evaporation from vegetated surfaces is so closely linked to the vegetational 'control' exerted via the stomatal and root systems that it is very difficult to separate them. Consequently, correlations between soil water potential and evaporation have often been explored through the medium of drying curves showing the ratio of transpiration to unstressed transpiration for vegetation in conditions of reducing water availability (Figure 4.8). These represent a 'black–box' substitute for a detailed understanding of the complex soil–plant–water relationships and the necessary associated sophisticated experimental investigation of aerodynamic and physiological resistances to water movement through the soil–plant–atmosphere continuum.

Despite their obvious limitations drying curves have enabled significant advances to be made in the understanding, measurement and modelling of evaporation from vegetated covers, and have formed the basis of many of the current operational techniques for estimating actual evaporation losses (see Section 4.6.3).

There is general agreement on the normal range in soil water contents – from the upper limit of water held at **field capacity** after initial rapid drainage ceases to the residual water held at the **wilting point** (but see also Section 6.3). The availability to plants over this range is still subject to much debate, and Figure 4.8 shows a variety of possible relationships. At one extreme (line A) is the view that transpiration continues at a maximum rate until the soil water available to the plant falls below a critical level (the wilting point) when water uptake by the roots stops. This insensitivity to a wide range of soil water may be expected to be particularly relevant in situations where roots extend widely throughout the soil. At the other extreme many would argue that evaporation decreases throughout the range of soil moisture drying (line B), long before the wilting stage is reached, because the soil hydraulic conductivity will decrease with decreasing soil water content (see Section 6.4).

Between these two extremes there are several intermediate models, one of the best known of which is that of the **root constant** proposed by Penman (1949) (line C). This may be described as the maximum soil moisture deficit below field capacity (see Section 6.3.4) that can be built up without checking transpiration. This is based on the premise that because water, on the whole, moves relatively slowly through the soil, the readily available moisture is effectively restricted to that in the root zone, and so be largely dependent upon the plant type (Penman, 1963). The value for grass was estimated by Penman at between 70 and 120 mm, compared to as much as 250–300 mm or more for some trees. In practice, the Penman-type root constant, even for the same species growing on the same soil, is not a true constant but can vary significantly from year to year.

These simple models must now be viewed in the light of our better understanding of the physiology of plant–water relations and assessed accordingly.

4.5 THE COMPONENTS OF EVAPORATION FROM VEGETATION COVERS

The individual components of total evaporation from a vegetation–covered surface, i.e. bare soil evaporation, transpiration and the evaporation of water intercepted by the vegetation surfaces, have been described and discussed in this and the preceding chapter. Throughout, the relationships of these principal evaporation components have been implied without directly addressing the question "Which component is the most important?" The answer clearly depends upon local conditions. Thus, in areas where the vegetation cover is dense and continuous the component of soil evaporation is the least important. More generally, it depends upon the availability of moisture for the evaporation process and specifically on whether evaporation is taking place from a wetted or a non–wetted vegetation cover.

First, it is necessary to consider the resistances that discourage that vapour flow.

4.5.1 RESISTANCES TO WATER VAPOUR FLUX

There are two principal resistances to evaporation from vegetation: the aerodynamic (controlling the loss of intercepted water) and the stomatal (controlling transpiration).

The aerodynamic roughness of the vegetation impacts on the role of turbulence and diffusion processes in evaporation. It commonly varies between 10 and 100 s m^{-1}, and depends solely on the *physical* properties of the vegetation cover. Vegetation height will clearly be important since the coefficient of turbulent exchange increases by a factor of over 2 with a change in vegetation height from a short cut surface at about 2 or 3 cm to 10 cm, and doubles again to a vegetation height of 90 cm (Rijtema, 1968). Similarly, the aerodynamic resistance for trees is an order of magnitude less than for grass, because trees are not only taller but also present a relatively rougher surface to the wind and so are more efficient in generating the forced eddy convection which, in most meteorological conditions, is the dominant mechanism of vertical water vapour transport (Calder, 1979).

Water vapour diffuses through the leaf stomata into the atmosphere partly in response to meteorological variables (available energy, vapour pressure deficit and wind speed), and is partly dependent on the vegetation type (stomatal conductance and the structural characteristics of the canopy

LAND COVER	AERODYNAMIC RESISTANCE, r_a	STOMATAL RESISTANCE, r_l	CANOPY RESISTANCE, r_c
Open water	125	0	0
Grass	50–70	100–400	40–70
Arable	30–60	100–500	50–100
Heather	20–80	200–600	60–100
forest	5–10	200–700	80–150

Table 4.4: Typical values of aerodynamic resistance and stomatal resistance (s m–1). (Data from Szeicz et al., 1969; Kelliher et al, 1975; Miranda et al., 1984; Oke, 1987; Hall, 1987).

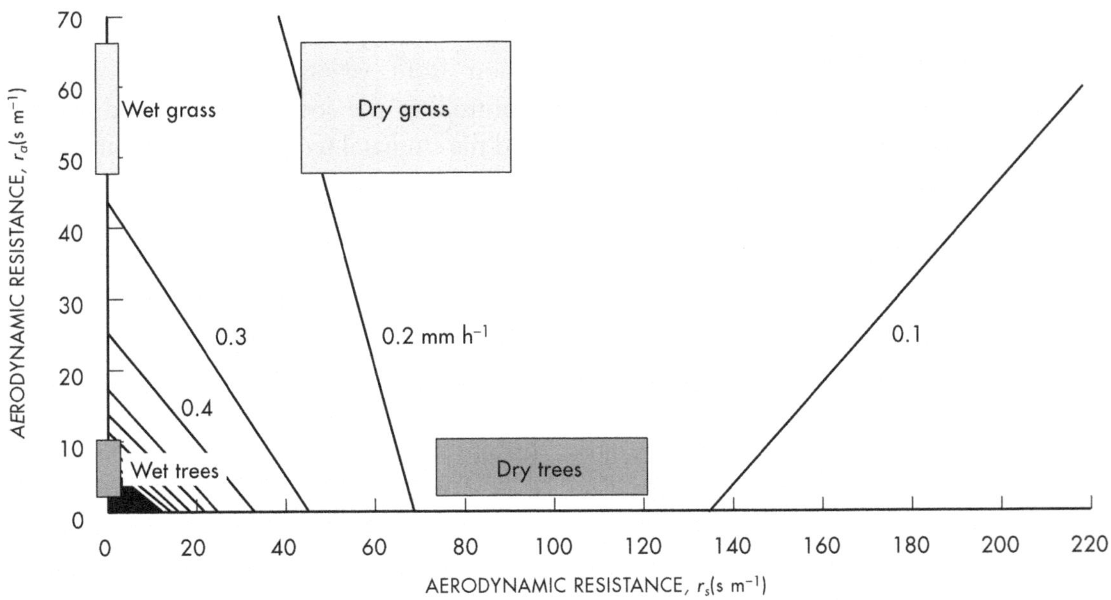

Fig 4.9: Computed evaporation (mm h^{-1}) for grass and coniferous forest, as functions of aerodynamic resistance, r_a and surface resistance, r_s, for typical cool summer daytime conditions in Britain (redrawn after Calder, 1979).

which influence the aerodynamic conductance). The stomatal or surface resistance is of major importance in the evaporation process because it is usually an order of magnitude greater than aerodynamic resistance, commonly varying between 100 and 1000 s m^{-1} (Lee, 1980).

Table 4.4 shows typical values of aerodynamic and stomatal resistances and illustrates the larger aerodynamic resistance (smaller conductance) over smooth surfaces (water, short crops) than over taller, rougher vegetation. Conversely the bulk stomatal resistance is larger for forests than for shorter vegetation.

Figure 4.9 illustrates the transpiration and interception components of evaporation in terms of aerodynamic and surface resistances of two different vegetation covers. It shows calculated evaporation rates for typical meteorological conditions on a cool summer day in Britain (i.e. net radiation = 200 W m^{-2}; vapour pressure deficit = 0.5 kP; air temperature = 10°C). Evaporation rates from grass and coniferous trees are similar in dry conditions (i.e. transpiration), but when the foliage is wet the evaporation (i.e. interception) rates differ substantially, increasing by a factor of 5–15 for the trees, but by only 1.5 for grass (Calder, 1979).

Eddy diffusion depends on windspeed and surface roughness, and r_a will be inversely related to windspeed, although this does not take the form predicted by classical eddy–diffusion theory since large–scale turbulence in the form of intermittent energetic gusts are a major mechanism in water vapour transport and weaken the classically predicted relationship between r_a and mean windspeed (Hall (1987).

4.5.2 INTERCEPTION AND TRANSPIRATION

It is now generally accepted that for tall and *wet* vegetation the evaporation of intercepted water will normally take place at a rate much higher than transpiration by means of stomatal diffusion, i.e. that the interception component will dominate the evaporation total. Furthermore, it seems likely that the evaporation of intercepted water will be relatively more important where surface wetting by precipitation occurs predominantly at night rather than during the day (Pearce et al., 1980). Even in high–rainfall areas, however, vegetation surfaces are normally dry more often than they are wet, and the relative importance of interception and transpiration components over a period such as a year will reflect the relative duration of wet and dry vegetation conditions as well as the interaction of other associated soil, vegetational and atmospheric conditions. For example, Shuttleworth (1988) reported that, in an area of Amazonian rainforest, average evaporation over two years was within 5% of potential evaporation. However, in wet months average evaporation exceeded potential estimates by about 10% and fell below such estimates by at least this proportion in dry months.

Given the complexity of the processes involved, it is not surprising that the evidence varies widely even within geographically restricted areas. Table 4.5 demonstrates the importance of climate and vegetation type on the balance between the interception and transpiration components of total annual evaporation. For coniferous forested areas in Britain, for

SITE	P	I	T	I / T
UK CONIFERS:				
Scotland (Balquhidder)	2500	710	280	2.5
Mid-Wales (Plynlimon)	1820	520	310	1.7
N England (Stocks)	980	370	340	1.1
S England (Thetford)	600	210	350	0.6
Heather	2500	350	170	2.1
Grass	2500	200	160	1.3
Rainforest (Amazonia)	2640	330	990	0.3
Eucalyptus (mining groundwater)	700	80	1000	0.1

Table 4.5: Comparison annual precipitation (P), interception (I) and transpiration (T) depths (mm) for areas of differing climates and vegetation characteristics. Annual totals rounded to nearest 10 mm. (Based on data in Calder, 1976; Calder, Hall and Adlard, 1992; Hall, 1987; Hall and Harding, 1993; Law, 1958; Shuttleworth, 1988).

example, interception (*I*) is much greater than transpiration (*T*) for high annual rainfall areas such as Plynlimon in mid–Wales, whereas for low rainfall areas such as Thetford in eastern England, transpiration is the larger component. The difference is not just one of amount of precipitation. Comparison with the figures for the Amazon rainforest shows the importance of the rainfall regime, with rain falling in relatively few very intense convective storms, compared to the long duration low intensity rainfall typical of UK conditions. Differences due to vegetation type are shown for the very wet Balquhidder site in southern Scotland, where interception losses of heather are double those of grass, but still much lower than those from forests. Finally figures are presented for a Eucalyptus forest in India where due to the highly seasonal rainfall regime the interception losses are quite modest, but total evaporation is greater than the rainfall because the Eucalyptus species there has little stomatal control and at that particular site over the study period the trees were effectively 'mining' water as their roots extended to deeper layers.

Figure 4.10 shows the energy budget components measured for a young Douglas fir forest in British Columbia (McNaughton and Black, 1973) and indicate that although evaporation accounted for a substantial proportion of the energy available, the sensible heat flux from the comparatively dry canopy was numerically very similar.

Figure 4.11 shows comparable data for a hardwood forest in Canada whose canopy was initially wet and then dried progressively through the day. The graphs confirm the high rates of evaporation from the wet canopy which exceeded the net radiation supply at times, indicating that the wetted canopy acts as an important sink for advected energy from either the overflowing air or the canopy space itself (Singh and Szeicz, 1979).

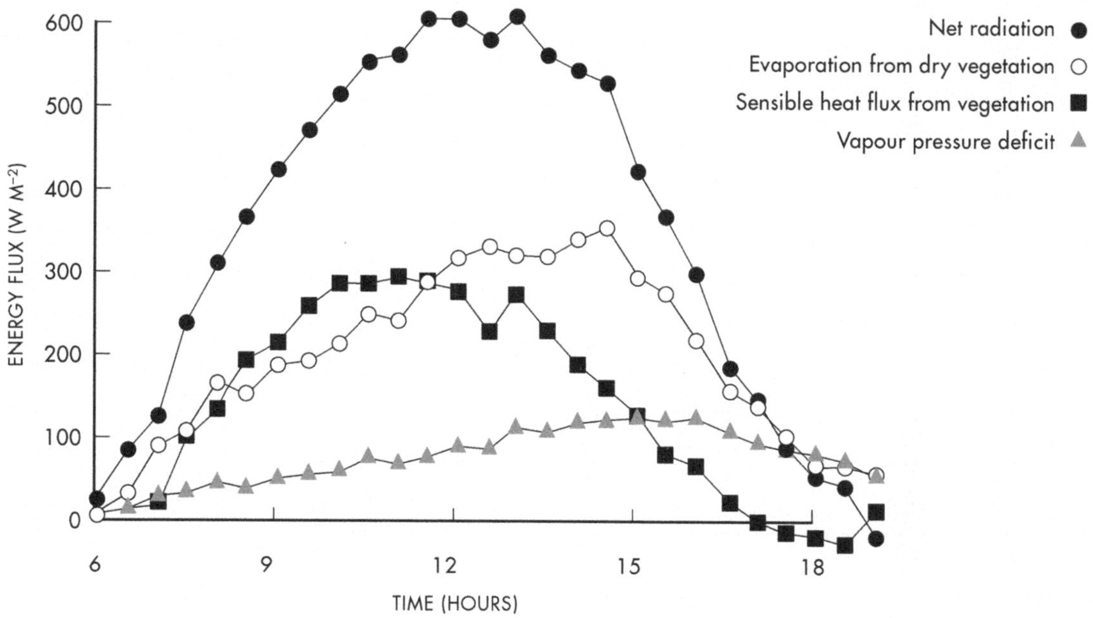

Fig 4.10: The energy budget components for a young Douglas fir forest at Blaney in British Columbia (based on an original diagram by McNaughton and Black, 1973).

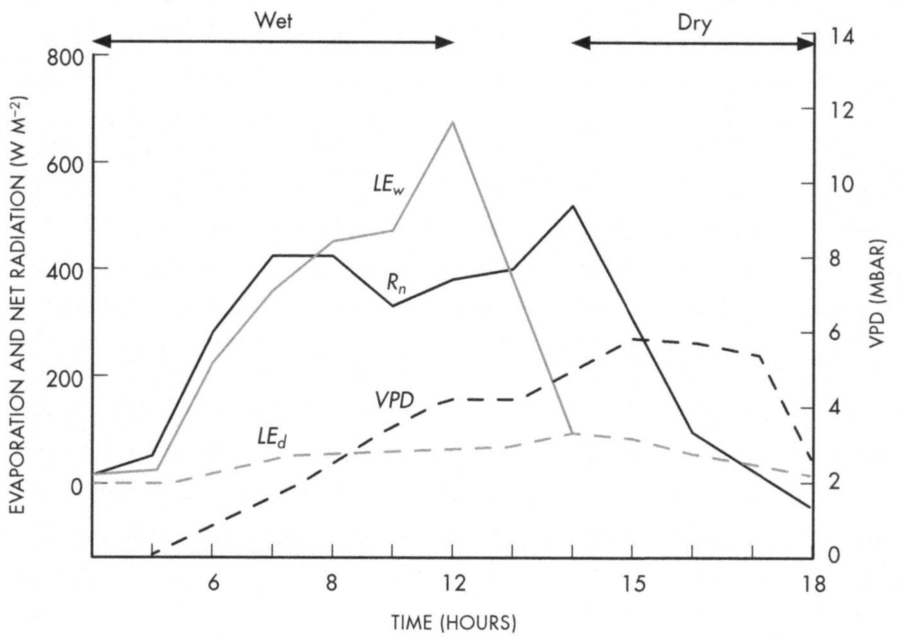

Fig 4.11: The energy budget components for a hardwood forest in Canada whose canopy was initially wet (LE_W) and then dried progressively (LE_d) through the day (based on an original diagram by Singh and Szeicz, 1979).

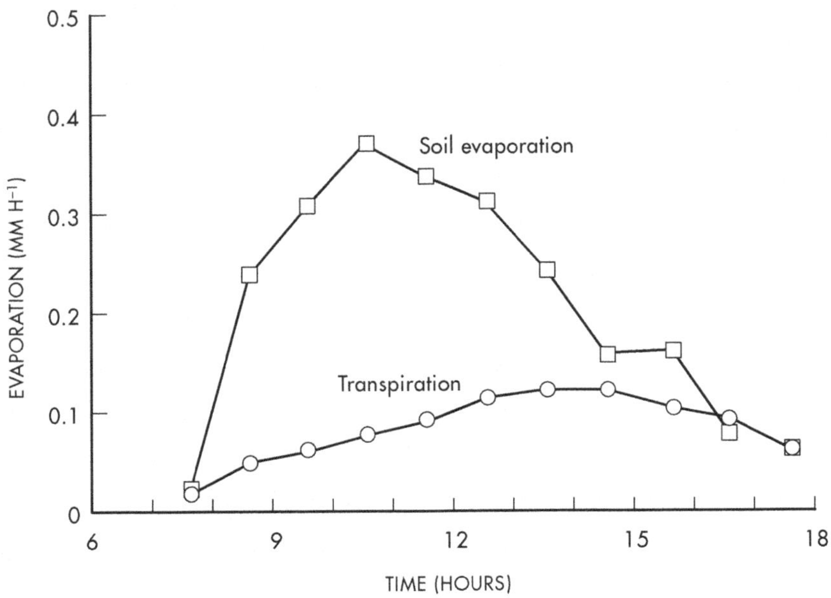

Fig 4.12: Following rainfall, direct soil evaporation under sparse vegetation may account for double the water loss from transpiration (from Wallace et al, 1989).

Mixed vegetation

Except for special situations such as agricultural crops and forest plantations, vegetated surfaces usually comprise a mixture of species. One such instance is the case of forests having an understorey of shorter vegetation. In areas with an open forest and a vigorously growing understorey the non–tree component of transpiration may account for over 50% of the total forest transpiration over the growing season (Black and Kelliher, 1989). The relative contributions to total forest transpiration from overstorey and understorey vegetation will change with environmental conditions. Thus, Roberts et al. (1980) working in a Scots pine forest with a bracken understorey, found that the bracken contributed about 25% of transpiration in normal conditions; this rose to 60% in very warm and dry conditions when the bracken was less affected than the pine by the large vapour pressure deficit.

Sparse vegetation

Where the vegetation cover is not continuous or where it has a relatively small plant mass, evaporation comprises the sum of two components: evaporation from the vegetation and from the soil. Micrometeorological measurements can provide a single integrated flux, but not the individual contributions which are necessary to create accurate models for situations, such as row crops or semi-arid regions, which have large bare soil areas. Wallace and Holwill (1997) used a Bowen ratio system very close to the ground surface to obtain short fetch area average evaporation from just the bare soil patches in an area of 'patterned' woodland and bare soil in Niger. At their site, bare soil evaporation accounted for 28% of the annual rainfall, and a higher percentage in dry years. Wallace et al. (1989) reported on a study of the detailed processes of evaporation from sparse dryland crops typical of those grown in many low rainfall areas of

Africa and the Middle East. In these vegetation conditions, direct soil evaporation may on occasions, e.g. after rainfall, account for double the water loss resulting from transpiration (see Figure 4.12). Shuttleworth and Wallace (1985) provided a framework through which measurements (or modelled values) of crop mass, stomatal and soil resistances and soil heat flux can be combined to calculate the partition of the available energy.

4.6 MODELLING EVAPORATION

Although the driving forces for evaporation are well known and relatively easy to measure (i.e. radiation, wind speed and humidity), the actual rates of evaporation are difficult to either measure or to estimate due to the complex interactions of meteorological 'demand' and surface factors controlling 'supply'. Consequently, for the routine estimation of evaporation, especially from vegetation covered surfaces, attention has focused on empirical or semi–empirical models, in which data requirements are limited to readily available routine meteorological measurements.

4.6.1 POTENTIAL EVAPORATION AND ACTUAL EVAPORATION

The most important simplification has undoubtedly been the development of the concept of *potential* evaporation, E_p, determined by atmospheric demand, as distinct from the evaporation that *actually* takes place. The calculated potential rate may then be reduced to yield an estimate of the actual evaporation according to the soil water content and a model, whereby inadequate soil water supply reduces evaporation below the atmospheric demand. The potential evaporation concept assumes that water is not limiting and is at all times sufficient to supply the requirements of the transpiring vegetation cover. It was defined by Thornthwaite (1944) as "the water loss which will occur if at no time there is a deficiency of water in the soil for use of vegetation". Subsequent modifications to this concept have limited its application to the evaporation that takes place from a continuous and unbroken green vegetation cover; e.g. Penman (1956) defined potential evaporation as the "evaporation from an extended surface of short green crop, actively growing, completely shading the ground, of uniform height and not short of water".

It is worth examining this definition more closely. The use of the term 'extended' means of sufficient extent to minimize advectional influences, 'short' excludes effects of vegetation height, 'uniform' excludes the effects of shape and roughness, and the role of soil water movement is eliminated by the term 'never short of water'. Thus defined, potential evaporation may be regarded as a climatological parameter, only affected by vegetation differences through albedo and stomatal control.

Clearly, such a restrictively defined concept is likely to be of only limited practical significance, as these conditions are unlikely to be fulfilled except, perhaps, for a very large surface of close-mown grass in a humid environment. In less constrained, and more realistic, circumstances theoretical argument and experimental evidence indicate that the shape and height of the vegetation cover and the supply of large-scale advective energy affect the transpiration rate in such a way that, even in the humid conditions, the actual rate of evaporation under optimum water supply *can* exceed potential evaporation. Furthermore, it is shown in Chapter 3 that the evaporation of water intercepted on the foliage of tall vegetation may greatly exceed the rate of transpiration from the same vegetation cover. Potential values of evaporation are often not achieved because of the resistances imposed on water vapour movement through the soil-plant-atmosphere continuum.

Indeed, such are the ambiguities surrounding the various notions of a climatologically determined maximum rate of evaporation from a vegetated surface, that it may be argued that the simple concept of potential evaporation, as originally advanced by Thornthwaite and Penman, has outlived its usefulness. Whilst acknowledging its usefulness in providing an upper limit to evaporation loss in a given environment, Lhomme (1997) outlined some of the inherent difficulties. First there is the need to provide a very precise definition of the 'ideal' surface conditions, and it is for this reason that many authorities now prefer the term 'reference' evaporation (e.g. Shuttleworth, 2012; Allen et al., 1998). Secondly, if the surface is sufficiently extensive to prevent any local advection, as required in the definition, then there is likely to be a feedback relationship between the evaporation process and the meteorological variable. Thus, air moving across an extensive surface adequately supplied with water, will become progressively more humid and cooler (due to latent heat of vaporisation) until it reaches an equilibrium with the surface. In practice, however, potential evaporation is often calculated under conditions of water shortage at the surface, resulting in a higher air temperature (as more radiation goes to sensible heat) and a lower air humidity.

There is a range of conditions of surface wetness and/or adequate soil water supply that will permit actual evaporation to take place at a potential rate determined largely by atmospheric variables; otherwise actual evaporation will take place at a lower rate and may be greatly influenced by plant and soil conditions.

4.6.2 MODELS OF POTENTIAL EVAPORATION

Whilst many evaporation models have been developed, only a small minority are used on a routine basis. Probably the most widely used are: a) the purely empirical model developed by Warren Thornthwaite, especially where data are limited, and b) the physically based model of Howard Penman, especially in the form as subsequently modified and developed by John Monteith and co-workers and now known as the 'Penman–Monteith' model.

The Thornthwaite model

Thornthwaite (1948) presented an empirical formula for estimating potential 'evapotranspiration' based on lysimeter and catchment observations in the central and eastern United States. Monthly potential evaporation (cm) is calculated as an exponential function of air temperature:

$$E_p = 1.6\,(10\,t/I)^a \tag{4.12}$$

where t is the mean monthly temperature (°C), I is the annual heat index (defined below) and a is a cubic function of I. The annual heat index is the sum of twelve individual monthly indices, i:

$$i = (t / 5)^{1.514} \tag{4.13}$$

where t is the mean temperature of the month.

The only data requirements are mean air temperature and hours of daylight (to adjust for the unequal day lengths in different months). It makes no allowance for variations in windspeed or air humidity. The method therefore appears to be extremely empirical and has been criticized for that reason. However, Thornthwaite's choice of an empirical model was quite deliberate, as also was his choice of mean air temperature as the main variable influencing potential evaporation. He justified the selection of mean air temperature on the grounds that there is a fixed relationship between that part of the net radiation which is used for heating and that part which is used for evaporation when conditions are suited to evaporation at the potential rate, i.e. when the soil is continuously moist (Thornthwaite, 1954). This means, in effect, that although the Thornthwaite model is empirical, it nevertheless estimates potential evaporation by an indirect reference to the radiation balance at the evaporating surface.

The Thornthwaite model was probably more widely adopted for operational use than any other, due to its modest data requirements and to its relative ease of use. The formula seems to work well in the temperate, continental climate of North America where it was derived, and where temperature and radiation are strongly. In other circumstances, however, it has been less successful. It may give rise to severe underestimation of evaporation rates in dry climates (Monteith, 1985; Pereira and Paes de Camargo, 1989).

Over the course of a year, air temperature lags behind radiation so that temperature is not a good indicator of the energy available for evaporation. Over short periods the Thornthwaite estimates may differ significantly from observed values using lysimeters, but the accuracy will increase as longer periods are considered. The method may also give poor results in areas like the British Isles, where advection effects resulting from frequent air mass changes lead to frequent rapid changes in mean air temperature and humidity, without corresponding changes in radiation. Even so, during the summer months, when E_p is of greatest significance, more stable air mass conditions tend to prevail, thereby strengthening the relationship between temperature, the radiation balance and evaporation.

It is unfortunate that the enthusiasm with which the Thornthwaite model has been adopted in so many different conditions has often been allowed to obscure his own caution that the chief obstacle to the development of a rational evaporation model "is the lack of understanding of why [E_p] corresponding to a given temperature is not everywhere the same" (Thornthwaite, 1948).

The Penman–Monteith model

Although imposing greater data demands than the Thornthwaite approach, the Penman-Monteith evaporation model has become firmly established, both as a routine method of quantifying evaporation and also as the basis for most of the significant conceptual advances in evaporation research that have taken place in recent decades.

Penman (1948) devised a model for potential evaporation which combined the turbulent transfer and the energy-balance approaches. This was later restated by Penman in slightly modified forms (1952, 1954, 1956, 1963). Basically, there are three equations. The first equation is a measure of the drying power of the air. This increases with a large saturation deficit, indicating that the air is dry, and with high windspeeds. The first equation is, therefore, derived from the basic pattern of the turbulent transfer approach and takes the form:

$$E_a = 0.35(e_a - e_d)\ (1 + u/100)\ \text{mm/d} \quad (4.14)$$

where e_a is the saturation vapour pressure of water at the mean air temperature, e_d is the saturation vapour pressure of water at the dew point temperature, or the actual vapour pressure at the mean air temperature, and u is the windspeed (miles d^{-1}) at a height of two metres above the ground surface.

The second equation provides an estimate of the net radiation (H – Penman's notation instead of the more usually used R_n) available for evaporation and heating at the evaporating surface and takes the form:

$$H = A - B\ \text{mm/d} \quad (4.15)$$

where A is the net short–wave incoming radiation and B is the net long–wave outgoing radiation, as estimated in the following expressions:

$$A = (1 - r)\ R_a\ (0.18 + 0.55(n/N)\ )\ \text{mm/d} \quad (4.16)$$

$$B = \sigma T_a^4\ (0.56 - 0.90 \sqrt{e_d})\ (0.10 + 0.90\ (n/N))\ \text{mm/d} \quad (4.17)$$

where r is the albedo of the evaporating surface, R_a is the theoretical radiation intensity at the evaporating surface in the absence of an atmosphere, expressed in evaporation units, n/N is the ratio of actual/possible hours of bright sunshine; σT_a^4 is the theoretical back radiation which would leave the area in the absence of an atmosphere, T_a being the mean air temperature in kelvin and σ is the Stefan–Boltzman constant, and e_d is as in Eq. 4.14.

Penman assumed that the heat flux into and out of the soil, which usually represents about 2% of the total incoming energy is small enough to be ignored, so he simply apportioned the net radiation between heating the air and evaporation. The proportion of it used in evaporation is then estimated by combining equations (4.14) and (4.15) to give:

$$E = ((\Delta/\gamma)H + E_a)\ /\ ((\Delta/\gamma) + 1)\ \text{mm/d} \quad (4.18)$$

where Δ is the slope of the saturation

vapour pressure curve for water at the mean air temperature. This relationship assumes equality of the coefficients of water vapour and convective heat transfer, a requirement that is well met in windy and unstable conditions but which may be less certain in conditions of strong radiation and light winds.

The Penman equation combines two terms, namely the aerodynamic or evaporativity term E_a which describes the drying power of the air, and the energy term, H, which is an estimate of the available net radiation. The relative importance of these terms in total evaporation depends upon the dimensionless ratio Δ/γ. The value of this weighting factor varies with air temperature, for example being 1.3 at 10°C, and 2.3 at 20°C respectively (Penman, 1963). Thus, during the warm summer months, when totals of evaporation are significantly high, the net radiation term is given more weight than the evaporativity term. In humid areas, H is usually greater than E_a, and therefore H tends to be the dominant term in the equation. On the other hand, in conditions of zero net radiation, the evaporation will be determined solely by the aerodynamic term (Calder, 1990).

In its basic form, the Penman (1948) equation describes the evaporation from an open water surface, E_T. This may then be related to a vegetated surface, E_T, by a coefficient, f; i.e. $E_T = f\ E_T$, where f for grass varies from 0.8 in summer to 0.6 in winter. Subsequently, it was shown that the potential evaporation for vegetation could be obtained directly by incorporating the reflection coefficient for an extended short green crop (r = 0.25) instead of that for an extended sheet of open water (r = 0.05).

If all or part of the available energy represented in Eq. 4.15 is measured directly then even short–period estimates of potential evaporation may be tolerably accurate. Several techniques may be employed. First, measured incoming radiation may be substituted for Eq. 4.16. A second option is to replace the whole of the expression for H, i.e. Eq. 4.15, by measured net radiation. Third, in addition to net radiation, measurements may be made of the heat flow through the soil. In this way the complete energy balance is measured and errors caused by the neglect of heat storage may be obviated. The choice of options has unfortunately introduced a certain degree of uncertainty into the exact definition of the Penman equation (Calder, 1990). For example, Robinson (1999) illustrated the introduction of spurious trends into the long term record of Penman short grass potential evaporation at the Eskdalemuir Observatory in Scotland, due to different versions of the Penman equation being used for different time periods.

Much of the discussion of evaporation from vegetated surfaces earlier in this chapter focused on the important roles of the aerodynamic resistance and surface resistance in accounting for variations of evaporation between vegetation covers that differ in terms of wetness, structure and physiology. Neither resistance appears to be incorporated in the Penman model of the evaporation process, although interestingly Thom and Oliver (1977) showed that the wind function in Eq. 4.14 implicitly incorporates some reasonable assumptions about both. The explicit incorporation of these resistances was subsequently presented in its most familiar form by

Monteith (1965), whose 'big leaf' model treated the vegetation canopy as a single extensive isothermal leaf so that:

$$\lambda E = \Delta H + \rho c(e_a - e_d)/r_a\} / \{\Delta + \gamma [1 + (r_s/r_a)]\} \quad (4.19)$$

where ρ is the density of the air, c is the specific heat of the air and r_s is the surface resistance which combines the canopy resistance, r_c and any additional resistances, such as contributions from the soil surface; all other terms are as previously defined.

This modification permits incorporation of the important aerodynamic and physiological influences of the vegetation cover on evaporation as well as the largely meteorological influences on which the original Penman model focused. The Penman–Monteith equation is one dimensional and assumes a bulk canopy conductance equal to the parallel sum of the individual leaf stomatal conductances. The big leaf approach gained preference over more complex multi-layer canopy models once it was realized that detailed representation of within-canopy exchanges is less important than adequate representation of the major controls of stomatal resistance and bulk aerodynamic transfer between the canopy and the overlying air (Shuttleworth, 2012).

The Penman–Monteith model has played a valuable role in the development of conceptual understanding of the complex evaporation process from wetted and unwetted vegetation surfaces. Much useful research has been conducted on the dependence of r_a and r_s on environmental variables using the Penman–Monteith equation. Work in Thetford Forest, for example, showed how surface resistance values were highly dependent on the atmospheric humidity deficit, and less so on the input of solar radiation (Stewart, 1988). Other similar work has been conducted on bracken (Roberts et al., 1980) and heathland vegetation (Wallace et al., 1984; Hall, 1987).

Jensen et al. (1990) compared 20 equations with data from weighing lysimeters at 11 locations, and concluded that the Penman–Monteith equation performs better than the other models tested. Uptake of the model outside of the research community for evaporation estimates was initially slowed by the difficulty of obtaining adequate measurement of the vegetational factors, particularly r_s (surface resistance) which is a complex function of many climatological and biological factors including radiation, saturation deficit, soil water status and biomass characteristics. This was admitted by Monteith (1985, 1995) who noted that the model had been used mainly as a diagnostic tool for estimating canopy resistance, when the transpiration rate and the other variables are known, rather than as a predictive tool for estimating λE when r_s is assumed. However, following extensive testing and the derivation of values of canopy resistance for different vegetation, it has now been widely adopted and it is the internationally recognised reference standard for the estimation of crop water requirements, being recommended by organisations such the Food and Agriculture Organization (FAO), and detailed in their official guidelines (Allen et al., 1998). The reference evaporation is defined as the evaporation from a hypothetical reference crop, not short of water, with a height of 0.12 m, an albedo of 0.23 and a surface resistance of 70 s m^{-1} (Allen et al 1998).

It has generally been found to work well in a wide range of studies (Calder, 1990), although in heterogeneous vegetation canopies such as open forest with herbaceous plants or sparse vegetation with substantial bare soil (Figure 4.12), the one-dimensional approximation may lead to significant errors, and a dual source model (e.g. Shuttleworth and Wallace, 1985) is generally needed.

4.6.3 TEMPORAL AND SPATIAL VARIATIONS IN EVAPORATION

Annual estimates of potential evaporation have been computed for stations with long periods of data, including Penman (1948) short grass estimates for the Radcliffe Observatory in Oxford, 1881–1966 (Rodda et al., 1976) and Penman and Thornthwaite values for the Edgbaston Observatory in Birmingham, 1900–68 (Takhar and Rudge, 1970). These indicate the presence of long term variations with, for example, a period of low annual values at both stations centred on 1930. Subsequently, Burt and Shahgedanova (1998) used air temperature data to estimate Thornthwaite values for Oxford back to 1815, but found a relatively poor agreement when comparing recent years with Penman estimates at a nearby site. The Thornthwaite figures lagged behind the Penman values by approximately one month, which in fact was to be expected due to the thermal lag between solar radiation and air temperature (see Section 4.6.2).

Long term data sets are extremely valuable, but great care must be taken to ensure their consistency, both in site and in data collection, and maintain a detailed **metadata** record of the instrumentation used and data processing procedures adopted. Wallace (1997), for example, provided a detailed account of the numerous meteorological instrument and siting changes at the Radcliffe Observatory in central Oxford that are essential for a proper interpretation of any apparent trends. Similarly, Crane and Hudson (1997) showed that siting changes can be important even in a rural environment, and Robinson (1999) studied two long-term records, one in S Scotland the other in mid-Wales, and in both cases identified a number of general problems including changes over time in the meteorological observations used, and in the version of the Penman formula adopted, potentially creating spurious time trends in the potential evaporation estimates.

Knowledge of the spatial variation of potential and actual evaporation from vegetated surfaces is also of value for a variety of climatological, water balance and resource evaluation purposes. Figure 4.13 emphasizes the overriding control of available energy in determining the pattern of potential evaporation across Europe. The comparatively regular pattern of isolines, increasing southwards, is broken only by the mountains of central and southern Europe, where relatively lower evaporation rates prevail. The pattern of isolines shows a broad latitudinal decrease from south to north, and in some mountainous areas, reflecting the lower availability of net radiation.

Operational use of potential evaporation models

One important prognostic use of the Penman–Monteith model has been its incorporation in the UK Meteorological

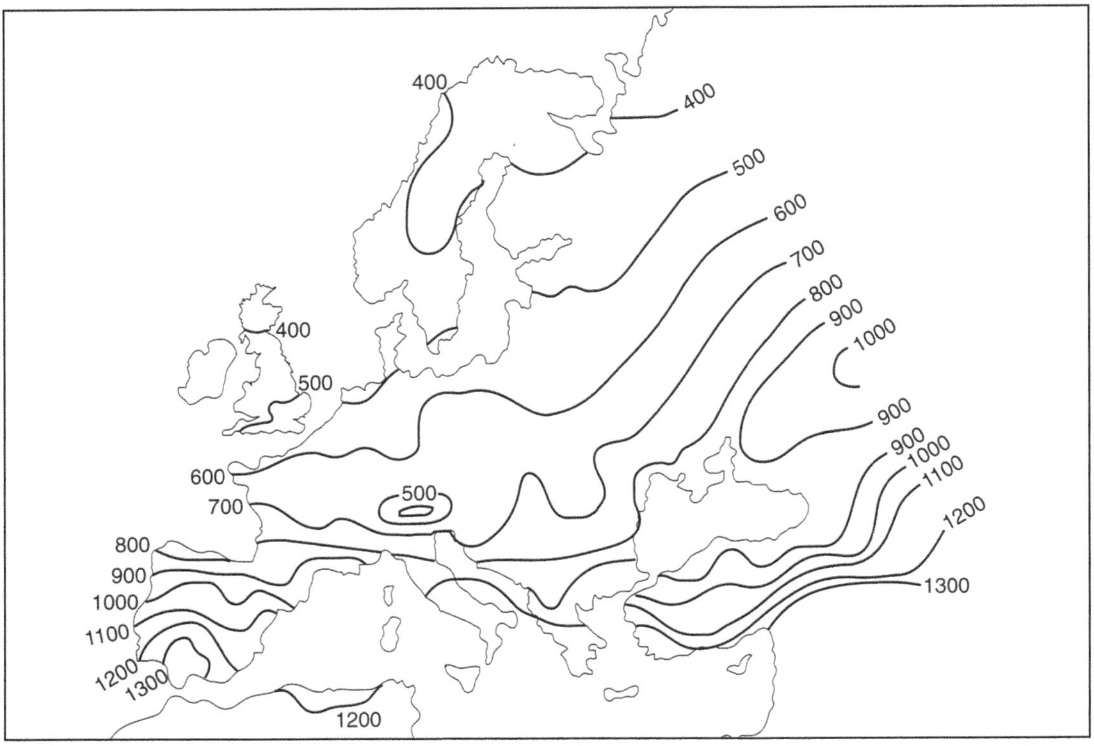

Fig 4.13: The spatial variation of mean annual Penman-Monteith potential evaporation across Europe, 1961–90. Adapted from a map produced by Prof Nigel Arnell, Reading University, based on the gridded climatology from the Climate Research Unit, University of East Anglia.

Office Rainfall and Evaporation Calculation System (MORECS). This became operational in the late 1970s and was revised in the early 1980s, since when it has remained basically unchanged to the present day though in the mid-1990s further improvements were made to the treatment of soils and land use (Hough and Jones, 1997). This system uses daily meteorological data to produce weekly estimates of evaporation, soil moisture deficit and effective rainfall for 190 grid squares (40 x 40 km) (Thompson et al., 1981). It uses a modified version of the Penman–Monteith equation to calculate daily potential evaporation, with a soil moisture accounting system to reduce actual evaporation at times when soil water is limiting. It has been widely utilized as an input to conceptual hydrological models for British catchments and used to underpin water resources and risk management planning (Prudhomme and Williamson, 2013).

MORECS models soil water extraction as a two–layer water reservoir. When the soil is at field capacity, upper layer X contains 40% and layer Y 60% of the soil water. When soil drying commences, water is extracted initially from layer X and is freely available to the vegetation (i.e. $E_a = E_p$). Subsequently water is extracted from layer Y, from which it becomes increasingly difficult to extract as the moisture status is reduced, and the actual evaporation falls progressively below Ep. This is achieved by increasing the canopy resistance in relation to the decrease in moisture content in Y. Water is extracted from Y only if X is empty, and rainfall replenishes layer X first, and then Y only when X is full.

A useful description and evaluation of the calculation of evaporation and soil moisture extraction was given by Thompson (1982), and Gardner and Field (1983) compared estimates with independent field measurements of actual soil water contents. As Gardner and Field (1983) observed, the value of X_{max} is important (a) because it defines a soil moisture deficit threshold rather similar to the Penman root constant, and (b) because being 40% of the available soil water it defines the total available water in the soil, i.e. 2.5 X_{max}. Different X_{max} values are allocated to different surface covers,

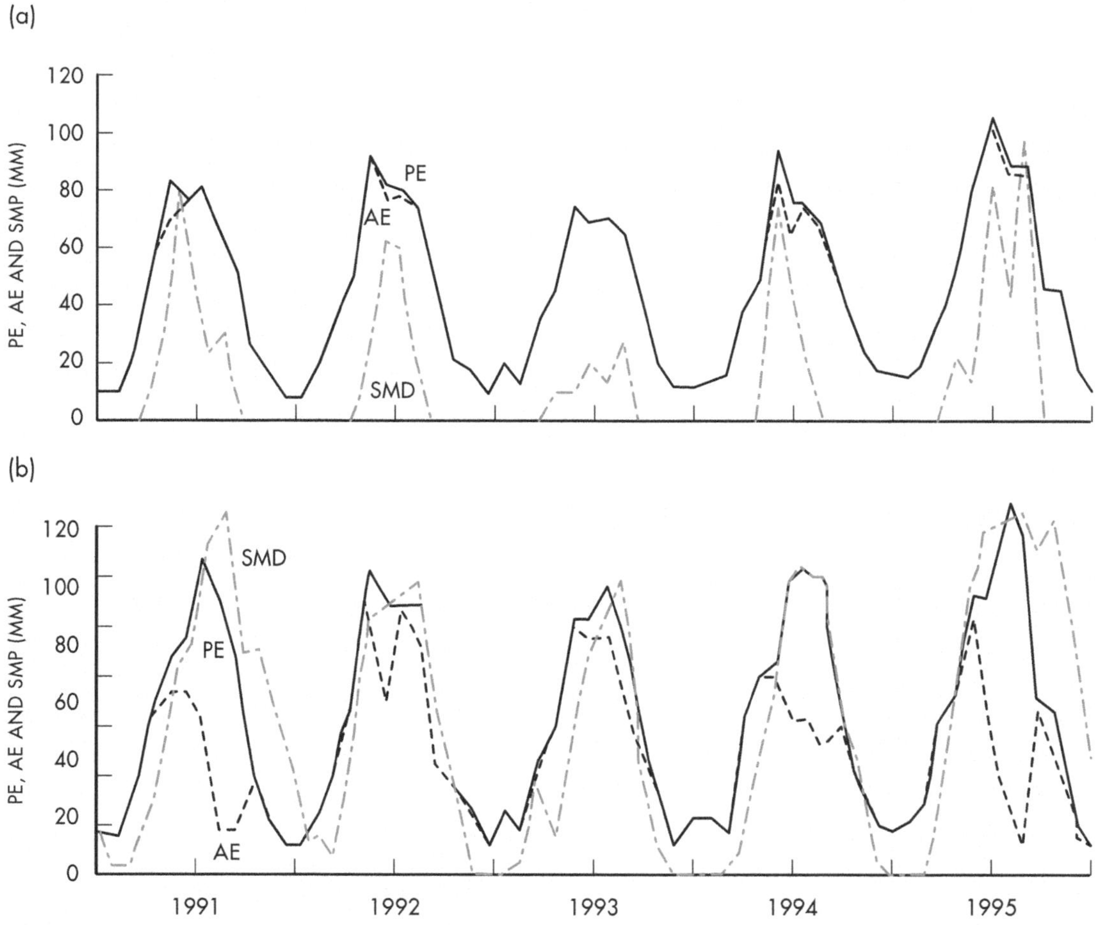

Fig 4.14: Monthly values of potential evaporation (PE) and actual evaporation (AE), and soil moisture deficit (SMD) below field capacity for: (a) wet climate of central Scotland (N Ayrshire) , and (b) drier climate of south eastern England (N Lincolnshire) (Adapted from IH/BGS, 1996, based on data supplied by the UK Meteorological Office).

ranging from bare soil to forest, and for each surface cover three values of X_{max} are used to represent soils having low, medium and high available water capacity. Thus, for grass the X_{max} values are 37.5, 50.0 and 62.5 mm, representing total available water (or maximum permitted deficits) of 94,125 and 156 mm respectively for soils of low, medium and high water availability.

Specimen MORECS results are shown in Figure 4.14. The mean monthly estimated evaporation and rainfall data for two 40 km grid squares in Britain over a 5-year period. This indicates the close correspondence between actual and potential evaporation in wetter area 'a' where soil moisture deficits are low and actual evaporation is close to the potential rate, and the marked divergence in the drier area 'b' which experiences high soil moisture deficits, and actual evaporation falls below the potential rate in summer.

Calder et al. (1983) compared the performance of a number of combinations of different potential evaporation formulae and different soil water depletion models relating E_p to E_a via the moisture status of the soil. The results were judged against over 3,000 field measurements of soil moisture at grassland sites in the UK. They concluded that the most sophisticated models did not necessarily produce the best results. Thus, the most detailed evaporation equations did not provide better soil moisture estimates than a simple climatological mean. There was little difference between different root constant functions, and a simple linear function performed as well as mathematically more complex models. They suggested that this was the result of the conservative nature of annual potential evaporation (both spatially and temporally). Similar conclusions were reached by Andersson and Harding (1991) for grass and forest sites in Norway.

4.7 PROGRESS IN UNDERSTANDING EVAPORATION PROCESSES

At the beginning of this chapter it is noted that evaporation is generally considered to be the most difficult component of the hydrological cycle to measure. From the subsequent discussions in this chapter it can be seen that there have been great advances in our knowledge of evaporation processes, and in our ability to measure or to estimate it. Nevertheless, despite this Morton (1994), in a thought-provoking paper, questioned the value of much of this work, and in particular the physical basis for evaporation equations and the treatment of advection and atmospheric feedback.

4.7.1 PHYSICAL BASIS FOR EVAPORATION EQUATIONS

The Penman–Monteith equation is internationally accepted as the preferred method for computing a "reference" evaporation

(Allen et al, 1998); yet as noted above, it contains the surface resistance term, r_s which may not be known beforehand. It is assumed that r_s will be primarily dependent upon stomatal resistance, and so will reflect changes in that resistance. Morton (1985) noted, as others have before him, that there is experimental evidence to show that under certain circumstances leaf stomata may close in response to an increase in vapour pressure deficit, thus reducing evaporation (to conserve water and leaf turgor), at a time when atmospheric demand is actually increasing. This is the opposite to the expected behaviour incorporated in the Penman–Monteith equation.

Advection

Much of the discussion of evaporation in the literature, and including this chapter, tends to be imprecise about the role of advection. Reference, for example, was made in the definition of potential evaporation to an 'extensive' evaporating surface and to the assumption that a sufficiently extensive surface would obviate advectional effects. Implicit in such terminology is the recognition that evaporation is influenced strongly by vapour pressure deficit and that, in turn, the vapour pressure deficit is linked to the available energy and to the canopy resistance. Conceptually it is possible to move in opposite directions from this position, i.e. that potential evaporation may be viewed as both a cause and an effect of actual evaporation. Thus, as water becomes limiting, the rate of actual evaporation decreases but the potential evaporation may increase due to the rise in air temperature (as the ground warms) and the reduction in water vapour pressure.

With heterogeneous surfaces and contrasting itinerant air masses, the linkage between vapour pressure deficit, available energy and canopy resistance may be broken when air with greatly different characteristics is imported from another area. In other words, when an air mass moves across the boundary between two surfaces of different wetness a horizontal gradient of saturation deficit is produced, leading to 'advectional enhancement' (McNaughton et al., 1979) if the air moves from a drier to a wetter surface and the saturation deficit increases and to advectional depression when the movement is in the opposite sense and the saturation deficit decreases. As McNaughton et al. (1979) observed, although "some significant effect of advection on local evaporation rates is the rule rather than the exception", the same methods, including the Penman–Monteith model, are normally used to estimate evaporation when advective enhancement is large as at other times. In fact the use of the Penman–Monteith model would not be unreasonable if values of canopy resistance were known to sufficient accuracy. At present they are not, although research in this area continues.

With an 'infinite' rather than an extensive surface, surface exchange rates control the behaviour of the atmosphere and give rise to an 'equilibrium' evaporation rate in which λE is nearly proportional to H. This may be stated as:

$$\lambda E_{eq} = \alpha \, (H \, \Delta) \, / \, (\Delta + \gamma) \qquad (4.20)$$

where α is an empirical constant. Equilibrium evaporation thus represents the lower limit of evaporation from a wet surface. Doubt persists, however, about the extent to which $\Delta/(\Delta + \gamma)$ does, in fact, represent an equilibrium rate. Priestley and Taylor (1972), in discussion of their own proposed simplification of the Penman model which eliminates the term involving windspeed and atmospheric humidity deficit, noted that the average rate of evaporation from vegetation freely supplied with water was given when $\alpha = 1.26$. The reason why this value is greater than unity has not been satisfactorily explained although the additional energy implied has been ascribed to the entrainment of relatively warm, dry air downwards through the upper surface of the planetary boundary layer (Monteith, 1985). The Priestley-Taylor equation has been widely used in areas not suffering from water stress or significant advection, but it has been found to produce serious underestimates in semi-arid areas (Gunston and Batchelor, 1983) and for forests with wet foliage (Shuttleworth and Calder, 1979).

The partitioning of the radiation absorbed by a surface between sensible and latent heat depends upon the availability of water. If the ground is drying out there will be a decrease in the amount of latent heat transfer to the atmosphere and a corresponding increase in the transfer of sensible heat. If there is no change in the prevailing air mass the atmosphere will become warmer and drier and consequently the potential evaporation will increase whilst the actual rate decreases. In typical British conditions (temperate maritime) actual annual evaporation tends to be higher in wet years than in dry years due largely to differences in summer evaporation. In dry summers soil evaporation is considerably less than that from a free water surface, whereas in wet summers, when re-wetting of the soil is frequent, soil evaporation will be much higher.

Miralles et al (2011b) estimated the contribution to global terrestrial evaporation as being 80% due to transpiration, 11% interception, 7% soil evaporation and 2% snow sublimation.

4.7.2 CHANGES IN EVAPORATION WITH CLIMATE CHANGE

There is uncertainty about the effects of climate change. Increasing air temperatures will increase the ability of the air to hold water, and so increase the air's drying power – the vapour pressure deficit. The impact of higher air temperatures may increase potential evaporative demand, but a reduction in summer soil water due, for example to reduced rainfall, would mean that actual evaporation would increase by a smaller amount, particularly in the driest catchments. Changes in potential evaporation are likely to be particularly critical for regions where evaporation and precipitation are currently of a similar magnitude.

Increasing levels of CO_2 are likely to increase stomatal resistance in many plants (particularly C3 plants, which includes virtually all woody plants and temperate crops and grasses), which may reduce transpiration, although plant water use efficiency (biomass produced per unit of water used) may increase.

REVIEW PROBLEMS AND EXERCISES

4.1 Define the following terms: evaporation, water vapour, latent heat, water vapour pressure, saturation water vapour pressure, and saturation deficit.

4.2 Describe the principal meteorological factors controlling evaporation from a water surface.

4.3 Explain what is meant by the Bowen ratio; what is its significance for evaporation studies?

4.4 Discuss the advantages and disadvantages of evaporation pans. Why are they still so widely used throughout the world?

4.5 Is transpiration a necessary evil?

4.6 Describe and compare the importance of stomatal and root controls on transpiration.

4.7 Define and clearly distinguish the following three sets of terms: stomatal or surface resistance, canopy or bulk surface resistance, and surface resistance.

4.8 Describe different black box models of transpiration reduction with progressive soil water depletion.

4.9 Discuss the role of climate and vegetation type on the relative importance of transpiration and interception to total evaporation.

4.10 Discuss the physical basis for a definition of potential evaporation.

4.11 Explain how factors such as rainfall regime, soil water content and vegetation type may lead to actual evaporation rates that are: a) lower than the potential evaporation rate; b) greater than the potential evaporation rate.

4.12 Discuss the effect of advection on potential evaporation estimates.

WEBSITES

Fluxnet:
http://fluxnet.ornl.gov/

Integrated Carbon Observation System (ICOS):
https://www.icos-ri.eu

MORECS:
http://www.metoffice.gov.uk/services/industry/data/specialist-datasets

5

GROUNDWATER

"Know the geology" is the first rule of groundwater studies

SCHNOOR, 1996

5.1 INTRODUCTION AND DEFINITIONS

Much of the precipitation that reaches the ground is absorbed by the surface layers of the soil. This water may subsequently be evaporated, or flow laterally close to the surface as throughflow, or else it may **percolate** downwards under gravity to the **water table**, which marks the upper surface of the **groundwater** body in which all the pore spaces are completely filled with water.

Groundwater is the Earth's largest accessible store of fresh water and, excluding ice sheets and glaciers, and has been estimated to account for about 95% of all fresh water (see Figure 1.1). Groundwater resources are abundant in many parts of the world and the primary source of drinking water worldwide, particularly in developing countries where it is often the best or only source of cheap potable water, being a generally high quality source that requires less treatment than surface water. The sheer size of this invisible store of water is even more strikingly illustrated by converting it to a precipitation equivalent. If distributed evenly it would cover the Earth's land surfaces to a depth of 67 m, compared to the average annual precipitation of about 0.75 m. Groundwater occurs almost everywhere beneath the land surface, but its distribution is quite variable. For example in the USA the total pore space occupied by water (together with gas and petroleum) ranges from 3 m under the Piedmont plateau to about 2500 m under the Mississippi Delta (Heath, 2004). The role of groundwater as a vast regulator in the hydrological cycle can be seen from its long residence time, averaging about 300 years, although with considerable spatial variation. As will be shown later, deep groundwater basins will have longer flow paths, and much greater travel times before re-emergence back at the surface, than shallow more local flow systems. Due to its long residence time, areas that currently have an arid climate, with little opportunity for water to percolate deeply, may nevertheless have significant groundwater reserves as a result of replenishment or **recharge** in former pluvial periods, hundreds or thousands of years ago. There are, for example, enormous groundwater reserves under the Sahara. The volume of freshwater groundwater resources in Africa is estimated to be about 0.66 M km^3, which is over 100-times the annual renewable freshwater resources and 20-times the freshwater stored in African lakes (MacDonald et al, 2012). Globally as many as two billion people depend directly on groundwater for drinking water, and the 40% of the world's food which is produced by irrigated agriculture depends largely on groundwater (UNEP, 2003). Groundwater supplies about 75% of the water needs in Europe and 30% in England and Wales. Fan et al (2013) compiled water table data from over 1.6 M well sites, and concluded that the water is within 10 m of the ground surface over about half of the global land area, but there are substantial areas such as the Sahara Desert, southwestern North America and the Andes where the top of the groundwater zone is more than 80 m below the ground surface. Valuable reviews of the progress in the development of groundwater hydrology are provided by Anderson (2008) and by Narasimhan (1998).

Groundwater sustains streamflow during periods of dry weather and is vital for many lakes and wetlands, maintaining aquatic ecosystems in dry periods. Many plants and aquatic animals depend upon groundwater outflows to streams, lakes and wetlands. However, due to human use the supplies are diminishing, and it has been estimated that 21 of the world's 37 largest aquifers are now severely over-exploited in locations from China and India, to France and the United States (UN, 2014).

Various criteria have been adopted to distinguish between soil water and groundwater, such as limiting consideration of the former to the surface metre or so, which contains the depth of plant roots (Price, 2004). However, **soil water** is normally defined as the subsurface water in the zone of aeration, i.e. the unsaturated soil and subsoil layers above the water table, and groundwater is defined as the subsurface water in soils and rocks that are fully saturated. This distinction between saturated and unsaturated conditions is an important one. In the saturated zone, the pore spaces are completely filled with water and the pressure of water is equal to or greater than atmospheric pressure. In the zone of aeration, the pore spaces contain both water and air and the pressure of water is less than atmospheric. However, both saturated and unsaturated conditions are part of the continuum of subsurface water which is in a perpetual state of flux. The zone of aeration is really a transition zone in which water is absorbed, held or transmitted, either downwards towards the water table or upwards towards the soil surface from where it is evaporated. At times of prolonged or very intense rainfall, part of the soil zone may become temporarily saturated although still separated by unsaturated layers from the main body of groundwater below. Such transient, perched water tables may result when an impeding layer in the soil slows the drainage of infiltrating water, or when the surface layers of the soil are so slowly permeable as to result in saturated conditions. These often short-term and possibly localized areas of saturation within the zone of aeration may be very important in generating lateral flows to stream channels (see Section 7.4).

5.2 GEOLOGICAL BACKGROUND

Layers of rock or unconsolidated deposits that contain sufficient saturated material to yield significant quantities of water are known as **aquifers** and less permeable formations that transmit water more slowly than the adjacent aquifers are commonly known as **aquitards**. The terms are deliberately imprecise and indicate relative rather than *absolute* properties. Thus a silt bed would be an aquitard in a stratigraphic sequence of alternate sand and silt layers but, if interlayered with less permeable clay beds, the silt would be an (albeit poor) aquifer. Most major aquifers, for example sandstones and limestones as well as unconsolidated materials, comprise **sedimentary** deposits formed from the erosion and deposition of other rocks. In contrast, igneous and metamorphic rocks, formed under conditions of high temperatures and pressures, generally have no interconnected pore spaces and are only water bearing where they are weathered or have fractures.

The lower limit of groundwater occurs at a depth where interstices are so few and so small that further downward movement is virtually impossible. This groundwater boundary is frequently formed by a stratum of very dense rock, such as clay, slate or granite, or by the upper surface of the parent rock where the groundwater body occurs within a surface deposit of weathered material. Alternatively, the compression of layers or **strata** with depth, which results from the increasing weight of the overlying rocks, means that a depth is eventually reached beyond which the interstices have been so reduced in both size and number that further water movement is effectively prevented. The depth at which this occurs will depend on the nature of the water-bearing rock, and so be shallower in a dense granite than in a deep porous sandstone. Nevertheless, the number of interstices tends generally to decrease with depth, and below about 10 km all rocks may, for practical purposes, be considered to be impermeable (Price, 2004).

William Smith is widely considered as the '*Father of British Geology*' and produced the first national geological map in 1815 of Britain. He was the first person to conclusively demonstrate that the same sequence of strata (with the same rocks and same fossils) occurred in two adjacent valleys, the Cam and Wellow Brooks in northern Somerset, and was able to map the layers across the wider landscape. This led to an appreciation of the truly three-dimensional nature of rock strata that ultimately underpins our understanding of aquifers and groundwater flow.

5.3 CONFINED AND UNCONFINED AQUIFERS

The diagrammatic section in Figure 5.1 shows the four main zones into which subsurface water has been traditionally classified. Precipitation enters the soil zone at the ground surface and moves downwards to the water table which marks the upper surface of the **zone of saturation**. Immediately above the water table is the **capillary fringe** in which all but the largest pores are full of water (see Section 6.3); between this and the soil zone is the **intermediate zone**, where the movement of water is mainly downwards. These zones vary between different parts of a river basin. On the valley flanks, water drains from the soil zone proper into the intermediate zone, and may or may not eventually reach the zone of saturation perhaps several hundred metres below. In the floodplain areas, however, the capillary fringe often extends into the soil zone or even to the ground surface itself, depending on the depth of the water table and the height of the capillary fringe. Although convenient as an introduction, this classification tends to obscure the fact that subsurface water is an essentially dynamic system. As well as varying spatially within a river basin, these zones may also vary over time, as when seasonal fluctuations of the water table bring the capillary fringe up into the soil zone.

The upper boundary of the zone of saturation varies according to whether the groundwater is **confined** or **unconfined** (see Figure 5.2), which relates to the permeability of the overlying rock strata. In the case of unconfined groundwater, the upper boundary of the saturated layer, the water table, is the level where the pore water pressure is equal to atmospheric pressure.

The water table tends to follow the contours of the overlying ground surface, although in a more subdued form. The locations of the discharge zones determine the location of the lows and intervening highs in the water table. Assuming a similar amount of infiltration from rainfall over both high and low ground, the amplitude of relief of the water table depends largely upon the texture of the material comprising the zone of saturation. In the case of very open-textured rock, such as well-jointed limestone, groundwater will tend to move through the interstices at such a rate that it rapidly attains a near-hydrostatic level, thus forming a more or less horizontal surface. On the other hand, with a fine-textured rock, groundwater movement will be so slow that water will still be draining towards the valleys from beneath the

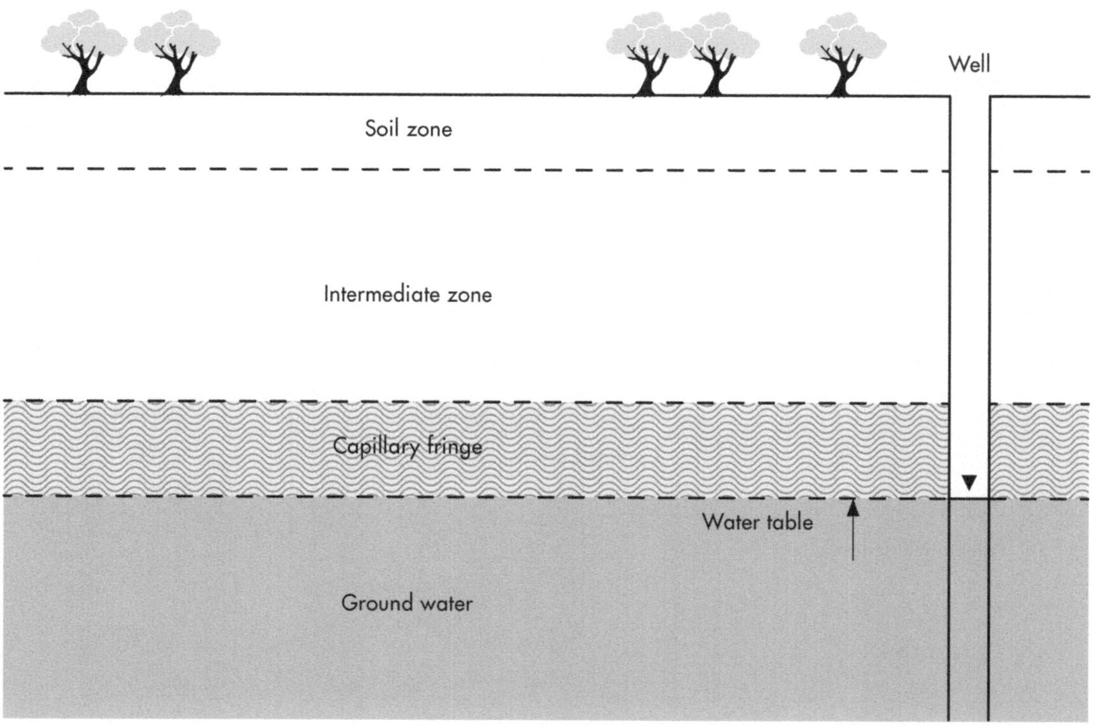

Fig 5.1: Schematic diagram showing the main zones into which subsurface water is usually classified: comprising the unsaturated and saturated zones, together with the capillary fringe.

higher ground when additional infiltration from subsequent precipitation occurs, so that its height is built up under the latter areas. This tendency is magnified by the fact that precipitation normally increases with relief.

Perched groundwater represents a special case of unconfined groundwater where the underlying impermeable or semi-permeable bed is not continuous over a very large area and is situated at some height above the main groundwater body. Perched groundwater commonly occurs where an impermeable bed either exists at a shallow depth or intersects the side of a valley. In many areas the first unconfined groundwater encountered in drilling a borehole is of this perched type. As indicated earlier, water percolating through the zone of aeration after heavy rainfall may also be regarded as a temporary perched water body (see also the discussions of throughflow in Chapters 6 and 7).

The upper boundary of a **confined groundwater** body is formed by an overlying less permeable bed (see Figure 5.2). The distinction between unconfined and confined groundwater is often made because of hydraulic differences between the flow of water under pressure and the flow of free, unconfined groundwater. Hydrologically, however, the two form part of a single, unified system as all rock formations have some permeability, however small. Most confined aquifers have an unconfined area through which recharge to the groundwater occurs by means of infiltration and percolation, and in which a water table, as defined above,

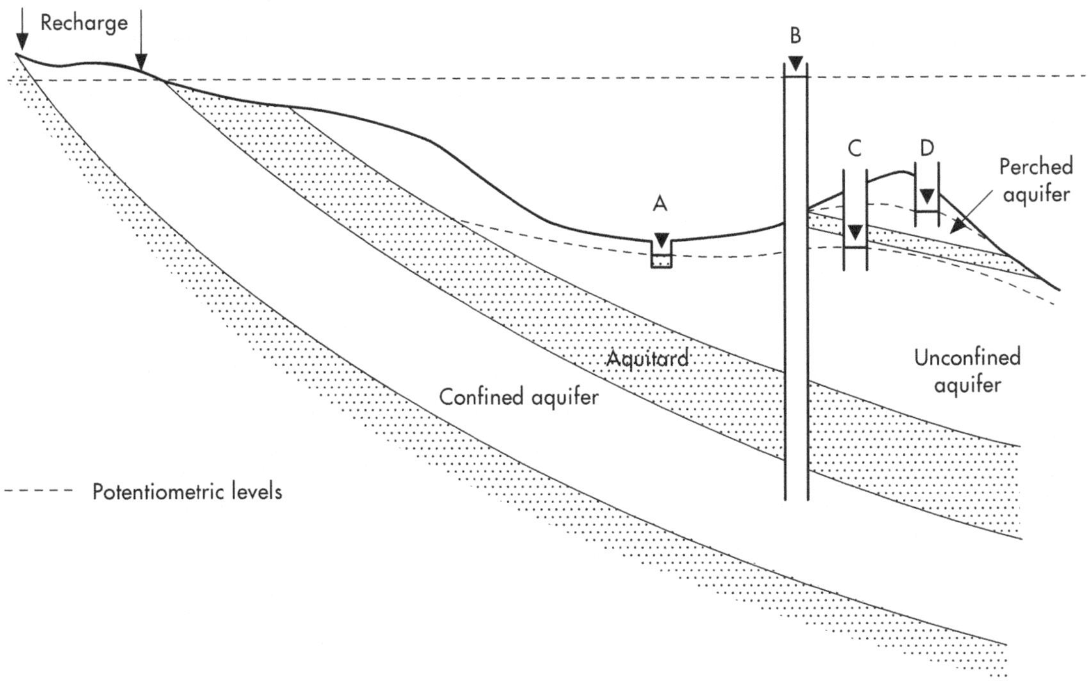

Fig 5.2: The relationship between the unconfined, confined and perched aquifers. The potentiometric levels measured in piezometers, A (unconfined), B (confined), C (locally confined) and D (perched) may be different.

represents the upper surface of the zone of saturation. Furthermore, the confining impermeable bed rarely forms an absolute barrier to groundwater movement so that there is normally some interchange, and therefore a degree of hydraulic continuity, between the confined groundwater below the confining bed and the unconfined groundwater above it. Indeed, attention has already been drawn to the relative sense in which terms such as aquifer and aquitard must be used and to the fact that a rock forming an aquitard in one situation may form an aquifer in another.

Since the water table in the unconfined groundwater area, through which recharge takes place, is situated at a higher elevation than the confined area of the aquifer, it follows that the groundwater in the latter area is under a pressure equivalent to the difference in hydrostatic level between the two. If the pressure is released locally, as by sinking a well into the confined aquifer, the water level will theoretically rise in the well to the height of the hydrostatic head, i.e. the height of the water table in the recharge area minus the height equivalent of any energy losses resulting from friction between the moving groundwater and the solid matrix of the aquifer between the point of recharge and the point of withdrawal. The imaginary surface to which water rises in wells tapping confined aquifers is called the **potentiometric surface** (Lohman, 1972; Freeze and Cherry, 1979). This term has replaced earlier names, such as piezometric surface, and can be applied to both confined and unconfined aquifers. In practice, the elevation of the potentiometric surface is measured, not in a well, but in a **piezometer**, which is a tube having an unperforated casing except for a short length at the base. In the case of an unconfined groundwater when there is no flow, the water table and the potentiometric surface occur at the same level. In flowing unconfined groundwater, however, the elevations of the water table and the potentiometric surface will differ, hence the curvature of streamlines in local groundwater systems.

The term **artesian** has been used in different ways to describe either the confined aquifer itself, or a well that penetrates a confined aquifer, or any well producing freely flowing water up to the ground surface. As will be seen later (Section 5.6) this does not have to be a confined aquifer if flow is upward, and a tube well provides a short-circuit enabling water to reach above the aquifer top. Some of the classic and best-known free-flowing 'artesian' conditions are found in areas of gently folded sedimentary strata such as the type area in the province of Artois in northern France, the London Basin in England, or the great artesian basins of east-central Australia and the Great Plains of the USA. Early wells in these last two basins initially encountered water with sufficient pressure head to gush more than 45 m above the ground surface, although the pressure subsequently diminished rather rapidly (Davis and De Wiest, 1966). Artesian conditions have also been found in fissured and fractured crystalline rocks, particularly where they are overlain by relatively impervious superficial deposits. Natural artesian springs may also result from faulting in areas of folded sedimentary rocks. Artesian conditions do not always require an overlying confining bed and may occur in steep areas as a result of topographic controls.

Photo 5.1: Flowing artesian well in southern Jordan close to the banks of the Red Sea. The pressure is provided by water infiltrating into the hills bounding the rift valley. The aquifer there is unconfined, but it becomes confined as water flows from the recharge zone to deeper levels. (Photo source: John Bromley).

Categorising groundwater as 'unconfined', 'confined' and 'perched' tends to overemphasize differences between the three types which may be difficult to recognise in practice, even in simple hydrogeological conditions. In areas of complex hydro-geology the terms become almost meaningless. However, the categories have been widely adopted in the literature and are used in this chapter as a convenient framework for the ensuing discussions of groundwater storage and groundwater movement.

5.4 GROUNDWATER STORAGE

Aquifers serve both as reservoirs for groundwater storage and as pipelines for groundwater movement. Because groundwater moves so slowly and has such a long residence time in the aquifer, the storage function is often more obvious. The age of water in some aquifers in England and Libya, for example, has been estimated at more than 20,000–30,000 years (Downing et al., 1977; Wright et al., 1982) and in central Australia some groundwater may be 1.4 million years old (Habermehl, 1985). Clearly the accurate determination of groundwater age will be important for assessing both the resource potential of the groundwater body and also its vulnerability to pollution. Dating methods are often based on the use of dissolved species derived either from the atmosphere or from beneath the ground surface, e.g. chlorofluorocarbons (CFCs) and noble gases such as helium and argon.

This Section considers the main features of groundwater storage, particularly the aquifer characteristics which affect it, such as porosity and specific yield and retention, and the mechanisms of storage change in both unconfined and confined aquifers.

5.4.1 POROSITY

The amount of groundwater stored in a saturated material depends upon its **porosity**. This is normally expressed as the proportion of the total volume of a rock or soil which is represented by its **interstices**, or voids. While most interstices are small intergranular spaces some are cavernous. A knowledge of the nature of these interstices is clearly essential to an understanding of the storage and movement of groundwater, and several methods have been proposed to classify them. The most frequently used classification is based upon their mode of origin, and considers original and secondary interstices (Todd and Mays, 2004; Heath, 2004). **Original interstices**, as the name implies, were created at the time of origin of the rock in which they occur; thus in sedimentary rocks they coincide with the intergranular spaces, while in igneous rocks, where they normally result from the cooling of molten magma, they may range in size from minute intercrystalline spaces to large caverns. **Secondary interstices** result from the subsequent actions of geological, climatic or biotic

MATERIAL	POROSITY	HYDRAULIC CONDUCTIVITY (M/D)
UNCONSOLIDATED MATERIAL		
Clay	0.40 – 0.70	10^{-8} – 10^{-4}
Silt	0.35 – 0.45	10^{-3} – 10^{-1}
Sand	0.25 – 0.45	10^{-1} – 10^{2}
Gravel	0.25 – 0.40	10^{2} – 10^{4}
CONSOLIDATED ROCKS		
SEDIMENTARY		
Limestone	0.01 – 0.20	10^{-3} – 10^{1}
Karst limestone	0.05 – 0.50	10^{-1} – 10^{3}
Sandstone	0.05 – 0.30	10^{-4} – 10^{-1}
Chalk (matrix)	0.20 – 0.50	10^{-1} – 10^{2}
IGNEOUS AND METAMORPHIC		
Shale and Slate	0.1 – 0.10	10^{-7} – 10^{-4}
Crystalline rocks	0.0 – 0.05	10^{-8} – 10^{-5}
Fractured crystalline rocks	0.0 – 0.10	10^{-4} – 10^{-1}

Table 5.1: Typical range of key hydrogeological properties for selected aquifer types (based on data from various sources). Porosity can vary by over 3 orders of magnitude and hydraulic conductivity by 13 orders of magnitude. Conductivity and porosity are not directly correlated. e.g. clay has a very high porosity but very low permeability.

factors upon the original rock. Faults and joints, enlarged perhaps by weathering and solution, are the most common. Such interstices are often found in old, hard, crystalline rocks which have virtually no intergranular porosity, and so play a significant role in the storage and movement of groundwater over large areas of Africa, northern North America, northern Europe and India, for example. A problem with this type of 'genetic' classification of the interstices is that the original intergranular spaces are often later modified by processes including cementation and solution. A very similar, but perhaps more appropriate, classification is, therefore, that between the primary porosity due to intergranular spaces in the soil or rock matrix (Figure 5.3 a, b, c) and secondary porosity due to processes such as solution along joints and bedding planes or to jointing and fracturing (Figure 5.3 d).

Confusion sometimes arises, in the case of a well-jointed rock, for example, between the porosity of the solid rock matrix (which may be very low) and the overall porosity of the whole stratum or formation that it comprises (which may be relatively high). It is important to realize that all interstices are involved in the concept of porosity, so that joints, bedding planes and fractures, including those greatly enlarged by solution and weathering, must be included as part of the total interstitial volume. This has important implications for the size of sample used in measuring porosity and hydraulic conductivity (see Sections 5.5.2 and 5.6), since the larger the sample, the more likely it is to include a large interstice, such as a joint or a fracture. Sometimes porous media contain voids that are not interconnected to other voids and which are therefore hydrologically inert. Such voids are not part of the **effective porosity**, which can be defined as the porosity that plays an active part in the storage and movement of water, and therefore are not considered further.

In analyses of aquifer systems it is common to assume that the aquifer is **homogeneous** and **isotropic**; that is that certain properties, such as porosity, have the same values in different parts of the aquifer (homogeneity) and in different directions from the same point (isotropy). The very nature of primary and secondary geological processes means, however, that even apparently uniform deposits may have a preferred orientation of particles or fractures (anisotropy), and the stratification in most sediments often imparts a marked **heterogeneity**. Since rocks are generally formed in layers or strata, their

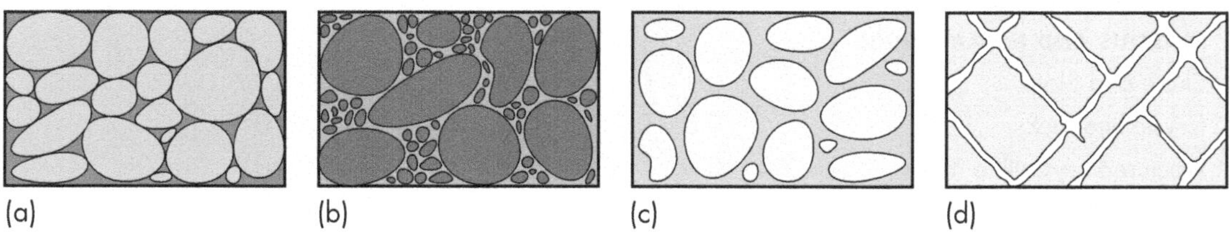

Fig 5.3: Types of interstices between: a) well-sorted particles, b) poorly sorted particles, c) grains that have been partially cemented, and d) rocks with fractures and joints.

properties such as hydraulic conductivity may vary with the direction of measurement, and such materials are termed **anisotropic**, whereas if the hydraulic conductivity is the same in all directions the layer would be called **isotropic**.

The porosity of a medium will depend upon a number of factors, including the shape, arrangement and degree of sorting of the constituent particles, and the extent to which modifications arising from solution, cementation, compaction and faulting have occurred. Poorly sorted material (with a large range of particle sizes) will have a low porosity since the interstices between the larger fragments will be filled with smaller particles and the porosity correspondingly reduced in comparison with material composed of uniformly sized grains. The combined effect of these various factors is illustrated in Table 5.1, which shows typical ranges of porosity for a number of different types of material. In general, rocks such as sandstone, shale and limestone have lower porosities than soils and other unconsolidated deposits. Initially it may seem strange that clay, which so often forms a barrier to water movement, has a very high porosity, while good aquifers, such as sandstone, have low to medium porosities. Further consideration, however, reveals that although porosity determines how much water a saturated medium can hold, by no means all of this water will be readily available for movement in the hydrological cycle. The proportion of the groundwater that is potentially 'mobile' will depend partly on how well the interstices are interconnected and partly on the size of the interstices and, therefore, by implication on the forces by which the water is retained in them.

5.4.2 SPECIFIC YIELD AND SPECIFIC RETENTION

Porosity determines the theoretical maximum amount of water that an aquifer can hold when it is fully saturated. In practice, however, only a part of this porosity is readily available to sustain groundwater discharge at seepages, springs or wells. This part is the **specific yield** which may be defined as the volume of water that can freely drain from a saturated rock or soil under the influence of gravity, and it is normally expressed as a percentage of the total volume of the aquifer (not just the pore space). It can be measured by a variety of methods, but well pumping tests generally give the most reliable results (Todd and Mays, 2004). The remaining volume of water (also usually expressed as a percentage of the total aquifer volume), which is retained by surface tension forces as films around individual grains and in capillary openings, is known as the **specific retention**. This term is analogous to 'field capacity' which is used when referring to soil water (Section 6.3.4), and is similarly imprecise in the sense that there is no fixed water content at which gravity drainage ceases. An extreme example of the difference between the porosity and specific yield of an aquifer is given by chalk. This is a very fine-grained limestone in which the matrix pore sizes are typically less than 1 mm and only a very small part of the pore water can drain freely under gravity (Price et al., 1976). At a site in southern England, Wellings (1984) found that the porosity of the chalk was about 30%, but the specific yield was only about 1%. A much larger proportion of the water was available to plants, and in the upper layers water tensions (see Section 6.3.1) in excess of 1,000 kPa were recorded.

The relationship between typical values of porosity, specific yield and retention for different types of unconsolidated material is shown in Figure 5.4. This indicates that, as the texture of the material becomes coarser, and by implication the importance of the larger interstices increases, both the specific retention and the total porosity decrease. Although clay has a high total porosity, the available water in terms of the specific yield is very small.

5.4.3 STORAGE CHANGE

Porosity, specific retention and specific yield control the ability of an aquifer to store and retain water, but the amount of water actually present at any one time reflects the changes of storage which are, in turn, determined by the changing balance between recharge to and discharge from the aquifer. If recharge during a given time interval exactly equals discharge, the amount of water in storage will remain constant; if recharge exceeds discharge, storage will increase while, if discharge is greater than recharge, storage will decrease. This may be conveniently expressed in the form of a simple water-balance equation:

$$\Delta S = Q_r - Q_d \quad (5.1)$$

where ΔS is the change in groundwater storage, Q_r is the recharge to groundwater and Q_d is the discharge from groundwater. A knowledge of the main components of the groundwater balance equation is normally an essential prerequisite of successful attempts to develop a groundwater resource.

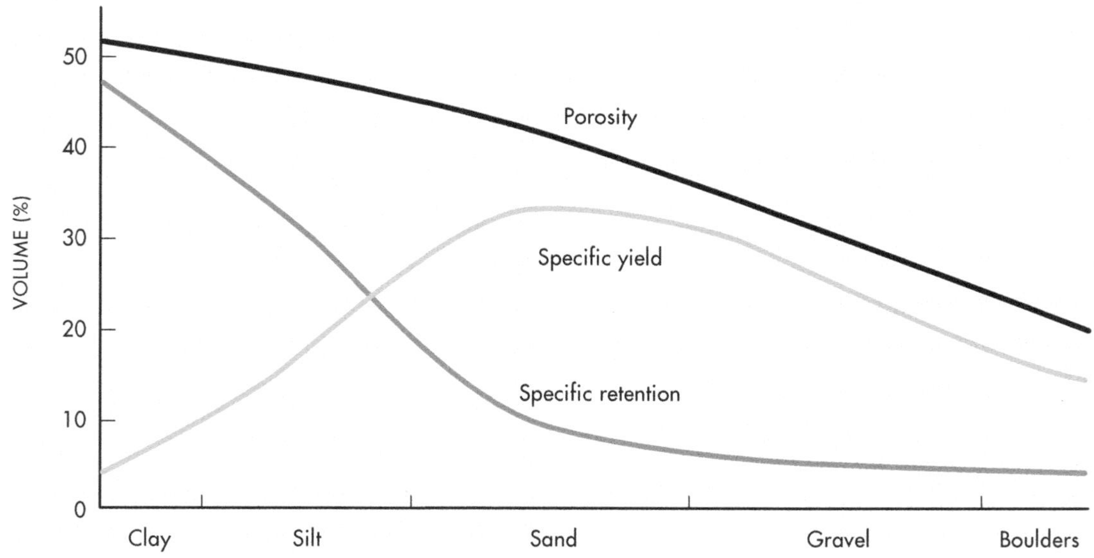

Fig 5.4: Typical values of porosity, specific yield and specific retention for different types of unconsolidated material. Individual site value may vary differ significantly from these averages. (Based on data from many sources).

The main components of groundwater *recharge* are:

- infiltration of precipitation at the ground surface which may result in water draining below the root zone and thus becoming potentially available for groundwater recharge;
- seepage through the bed and banks of surface water bodies such as lakes and rivers, especially in arid or semi-arid conditions, and even the oceans;
- groundwater leakage and inflow from adjacent aquitards and aquifers; and
- artificial recharge from irrigation, injection wells and leakage from water supply pipelines and sewers.

The main components of groundwater *discharge* are:

- evaporation, particularly in low-lying areas where the water table is close to the ground surface;
- natural discharge by spring flow and more diffuse seepage into surface water bodies;
- groundwater leakage and outflow through aquitards and into adjacent aquifers; and
- artificial abstraction.

The amount of water added to an aquifer by a given amount of recharge (or removed from an aquifer by a given amount of discharge) can be expressed as the **storativity**, or **coefficient of storage**, of the aquifer. Formally, this is defined as the volume of water that an aquifer takes into, or releases from, storage per unit surface area of aquifer per unit change in head. This is simply illustrated for a prism of unconfined aquifer in Figure 5.5a. As the water table falls by 0.5 m over the prism's cross-sectional area of 10 m^2 groundwater drains from 5 m^3 of rock. If the amount of water draining out is 50 litres (0.05 m^3) then the value of the dimensionless coefficient of storage is 0.05/5 = 0.01. In unconfined conditions the coefficient of storage corresponds to the specific yield (Section 5.4.2), provided that gravity drainage is complete, and it normally ranges from about 0.01 to 0.3 (Heath, 2004). In confined conditions (Figure 5.5b) no dewatering of the aquifer occurs as the potentiometric surface declines. Instead, the volume of water which is released as the potentiometric surface falls is a consequence of the slight compression of the granular structure of the aquifer and a very small expansion of the water in the aquifer. The storage coefficients of confined aquifers tend to be significantly smaller than those for unconfined aquifers, falling within the range 0.00005 to 0.005 (Todd and Mays, 2004). In other words, the potentiometric change associated with a given volume of recharge or discharge in a confined aquifer is much larger than that associated with the same volume of recharge or discharge in an unconfined aquifer.

Clearly, there are significant differences between the mechanism of storage changes in confined and unconfined conditions, and some of the more important of these differences will now be discussed.

Storage changes in unconfined aquifers

Storage changes in unconfined conditions are relatively uncomplicated and are usually reflected directly in variations of groundwater level. When recharge exceeds discharge, water table levels will rise, and when discharge exceeds recharge, they will fall. Recharge to and discharge from the same groundwater system usually occur simultaneously, so that groundwater level fluctuations reflect the net change of storage resulting from the interaction of these two components. The study and interpretation of water table fluctuations thus forms an integral part of the study of groundwater storage.

The precise hydrological linkage between a possible recharge event (e.g. percolating rainfall) and the consequent rise in water table level depends on conditions in the zone of aeration. Particularly important are the water content and hydraulic conductivity and the size and distribution of interstices. Apparently similar infiltration/percolation events at the ground surface can result in very different water table responses. For example, the rapid response of water table levels to precipitation, when the soil water content is high, almost certainly results from **translatory flow** (Hewlett and Hibbert, 1967). This is a displacement process by which water reaching the water table during rainfall is not the 'new' rainfall but previously stored rainfall that has been displaced downwards by successive bouts of infiltration (see also Section 7.4). Translatory flow means that water tables may respond rapidly to precipitation even in low-permeability materials where rates of downward percolation are low. It is

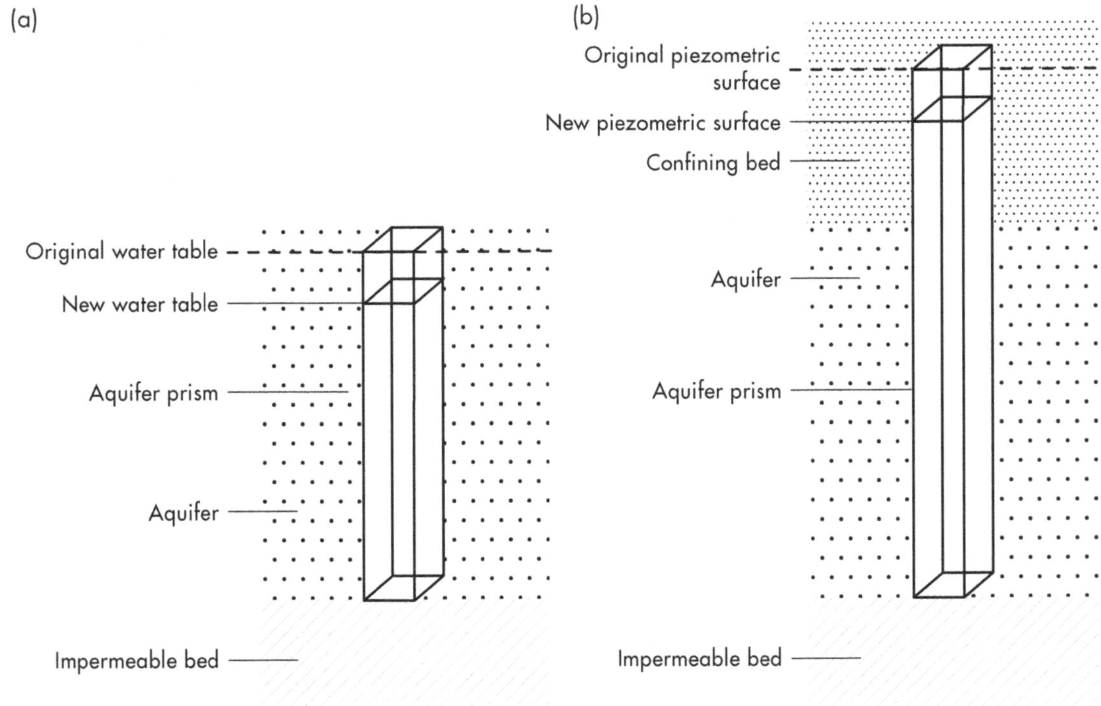

Fig 5.5: Definition diagram showing some of the basic terms in the calculation of storage change in (a) unconfined and (b) confined aquifers.See text for explanation. (Redrawn and simplified from an original diagram by Todd, 1980).

not effective in drier conditions, when the necessary continuity of interstitial water, from ground surface to water table, is broken. Tracer studies have shown that percolation through the zone of aeration is often spatially varied and that water may move preferentially along easier flowpaths such as cracks, fissures and decayed root channels (see Section 6.5.2 'Macropores'). These may be important in enabling pollutants such as fertilisers, pesticides and bacteria to bypass the filtering and purifying medium of the soil and be transmitted directly to the groundwater.

Seasonal fluctuations of water table level, reflecting the annual cycle in storage and in water availability, are normally of considerable hydrological significance. In many areas, including Western Europe, high water levels occur during the winter months and low levels during the summer months, so that the hydrological year is regarded as beginning in October and ending in September. Annual water table peaks and troughs as illustrated in a chalk aquifer in southern England (Figure 5.6) are fairly typical. In some areas the normal climatically determined groundwater regime may be modified significantly by artificial abstractions and pumping.

Also of hydrological significance are long-term or **secular** fluctuations of water table level. These are mainly associated with secular variations of rainfall (see Section 2.5.3) but may also result from changing patterns of groundwater abstraction for industrial, agricultural and domestic purposes. Excessive abstraction over a prolonged period results in a gradual lowering of the potentiometric surface over a

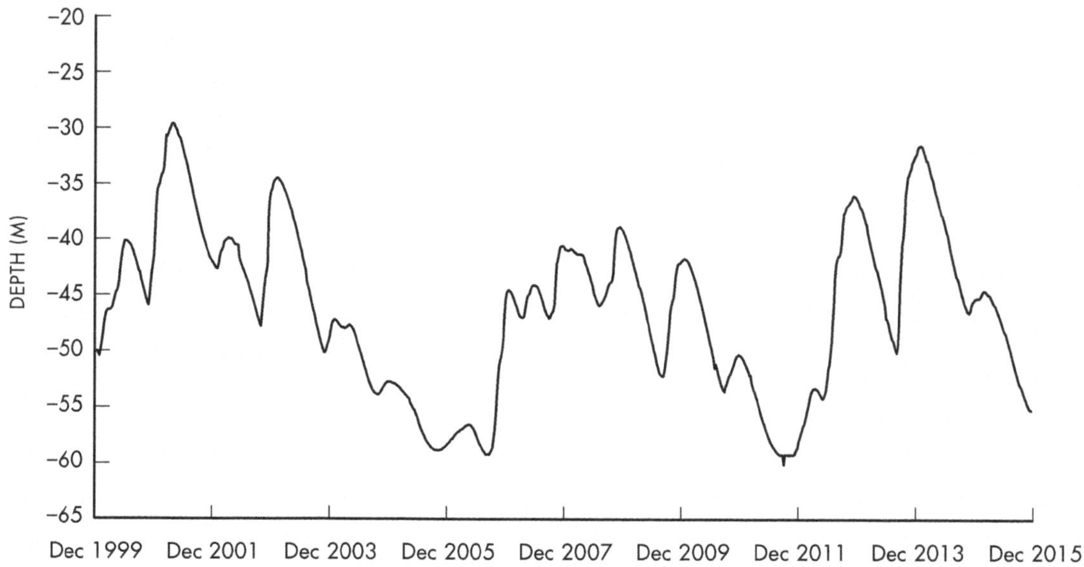

Fig 5.6: Groundwater level in the unconfined Chalk aquifer at Stonor Park, near Henley on Thames, Oxfordshire (2000–2015). Flow in the nearby Assendon Spring is initiated when water levels in this well rise to within 35.5 m of the ground surface. Data supplied by British Geological Survey.

wide area, and this has already affected many major groundwater basins, especially in low-rainfall areas. Even in Europe significant groundwater decline has been observed around Kharkov (Ukraine), Lille (France) and in the Ruhr basin (Germany) (Arnold and Willems, 1996). Groundwater abstraction from the Chalk aquifer under London caused a steady decline in groundwater levels beginning in the 1820s until the 1960s. Thereafter, levels began to recover due to a decline in abstractions for industry, and from increased leakage from the mains water supply network (Lerner and Barrett, 1996). Subsequently, as a result of collaborative effort between the Environment Agency, Thames Water and Transport for London this rise which was beginning to cause property damage has been stopped (Figure 5.7a). A similar pattern has been observed elsewhere including the sandstone aquifers under Birmingham (Figure 5.7b) and Liverpool. Elsewhere over-abstraction of groundwater has caused widespread streamflow reductions and in common with many coastal areas throughout the world, groundwater abstraction has resulted in the incursion of saline groundwater from adjacent aquifers or by direct seepage from the oceans into coastal aquifers (Section 5.6.4).

Short-term fluctuations of water table level, usually on a much smaller scale, are of less importance in interpreting storage changes but may provide useful hydrogeological information in particular circumstances. In valley bottom areas, discharge of groundwater by evaporation during the hottest part of the day may exceed the rate at which groundwater inflow from surrounding higher areas takes place and, as a result, the water table falls. During the evening and at night, the evaporation rate is much reduced, and so that the water table level will recover. This regular diurnal rhythm may be maintained for much of the summer, although it may be interrupted by periods of rainfall and reduced evaporation. If each daily drawdown exceeds the subsequent recovery, the water table level will gradually decline until it reaches a depth at which the capillary rise of water will be unable to satisfy evaporation demands at the soil surface.

The relationship between storage and the level of the potentiometric surface is complicated by the fact that the latter responds to factors other than storage changes. This is most apparent in confined groundwater conditions, when variations of the potentiometric surface may result, for example, from changes of loading at the ground surface such as atmospheric pressure or from earthquake shocks. In some circumstances, unconfined groundwater may also be affected, although the water table variations involved are likely to be very small.

Storage changes in confined aquifers

In unconfined conditions, aquifer compressibility is virtually negligible. By contrast, in confined conditions, the compressibility and elasticity of the aquifer greatly complicates the relationships between changes in potentiometric level and changes of groundwater storage.

In a confined aquifer, the total stress at a given depth is made up partly of **effective stress** (i.e. the pressure at the points of contact between individual solid particles, which is due to the weight of the overlying

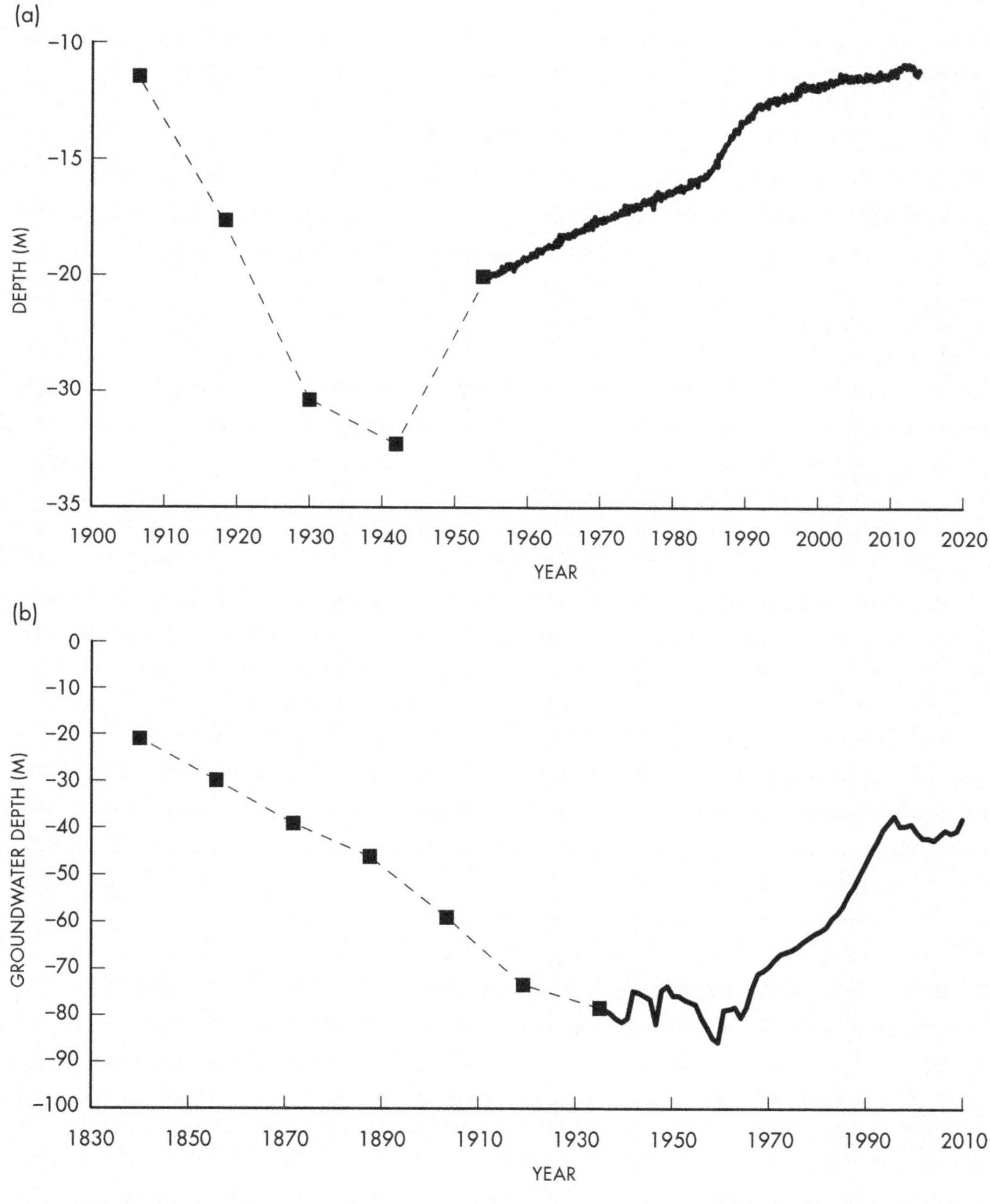

Fig 5.7: Pattern of groundwater decline (depth below ground level) due to abstraction, and subsequent recovery in observation boreholes under two British cities: a) London, Trafalgar Square (1845 – 2014), b) Birmingham, Constitution Hill (1900 – 2012; site closed). Data from British Geological Survey and Environment Agency.

deposits) and partly of **neutral stress** (which is due to the pressure of the groundwater contained in the pore space of the aquifer). The neutral stress acts equally on all sides of the solid particles of the aquifer matrix and has little effect upon the overall porosity of the aquifer. However, when groundwater is removed from the aquifer the neutral stress (pore water pressure) is reduced, causing the aquifer matrix to compress slightly and thereby reducing porosity. Conversely, when recharge to the aquifer occurs, the neutral stress is increased, causing slight compression of the pore water and a reduction in the effective stress which is accompanied by aquifer expansion and an increase in porosity. In other words, aquifer compression causes an increase in the grain-to-grain pressures within the matrix and aquifer expansion results in a decrease in the grain-to-grain pressure. If the confined aquifer is **perfectly elastic** and the water levels in the recharge and discharge areas do not change, the original potentiometric head will ultimately be restored when discharge is followed by an equivalent amount of recharge. However, not all aquifers are perfectly elastic and especially in those having a high clay content, there may well be some permanent compaction and loss of porosity following a period of severely reduced potentiometric head.

The foregoing discussion and the earlier definition of the coefficient of storage both indicate that, for a given variation of potentiometric level, the change of groundwater storage in a confined aquifer will be small compared with that in an unconfined aquifer. Furthermore, any variation of loading on a confined aquifer may result in a fluctuation of the potentiometric surface. Such variations of loading may result from several causes, including barometric changes, variable tidal and gravitational loading and, in certain circumstances, from the occurrence of an earthquake or a nuclear explosion; in each case they may provide valuable information about the elastic and storage properties of the aquifers concerned.

With a continuing excess of discharge over recharge to a confined aquifer, the ever-increasing intergranular pressures and the resulting compression of the aquifer may ultimately result in the subsidence of the overlying ground surface (Konikow and Kendy, 2005; Daito and Galloway, 2015). Such subsidence is normally inelastic and permanent. Localised ground surface subsidence of 10 m or more has been reported in countries including Mexico, Japan and the USA. Subsidence of 2–4 m has occurred at Osaka and Tokyo in Japan, Houston-Galveston in Texas and the Santa Clara valley, California (Domenico and Schwartz, 1998) and less severe but still significant subsidence in many other areas, including Venice (Teatini et al, 2007), and parts of northern and eastern China including Shanghai (Xue et al, 2005). In parts of Beijing the ground surface is dropping by up to 10 cm per year as a result of excessive groundwater extraction (Chen et al, 2016). The amount of subsidence at any location depends upon the subsurface lithology, the thickness of the compressible materials and their storage characteristics, as well as upon the magnitude and duration of the decline in head. Almost all the main areas suffering from subsidence due to groundwater extraction are underlain by deposits of young, poorly consolidated material of high porosity, and much of the subsidence occurs due to compression of the clayey aquitards (Poland, 1984). Unsurprisingly, therefore, the relationship between head loss and subsidence is complex, with the main examples appearing to fall into two well-defined categories in which the ratio of subsidence to head loss is either less than 1:10 or more than 1:40 (Domenico and Schwartz, 1998).

Geophysical gravity methods enable changes in subsurface water storage to be estimated by measuring changes in the Earth's gravitation field (e.g. Pool et al, 2000) Interferometric Synthetic Aperture Radar (Galloway et al, 2000) and by the GRACE satellites (see Box 1.1).

5.5 GROUNDWATER MOVEMENT

Groundwater taking an active part in the hydrological cycle moves more or less continuously from areas of recharge to areas of discharge. Some groundwater movement is in response to a **chemical** or **electrical gradient**. For example, groundwater moves from dilute to more concentrated pore-fluid solutions, especially in clays, which tend to exhibit osmotic characteristics consistent with those of leaky semi-permeable membranes. Osmotically-driven groundwater movement may play an important role in some arid and semi-arid basins. Mostly, however, groundwater movement is in response to the prevailing **hydraulic gradient**, and it is on this type of movement that the remainder of this chapter concentrates. Hydrological interest is concerned with both the speed and the direction of groundwater movement. Groundwater flow rates are very slow compared with those of surface water and are also very variable. Observed rates of groundwater movement through permeable strata in the UK range from less than 0.1 mm day^{-1} in some fine-grained pervious rocks, to as much as the equivalent of 5,500 m day^{-1} for very short periods through fissured chalk (Buchan, 1965).

The direction of groundwater movement is similarly variable since, like surface water, groundwater tends to follow the line of least resistance. Other things being equal, flow tends to be concentrated in areas where the interstices are larger and better connected, and the hydrologist's problem is to locate such areas, often from rather scanty geological information. Theoretical analyses commonly assume ideal and greatly simplified conditions, and the results from them may be difficult to apply in field conditions. For example, it is often assumed that aquifers are homogeneous and isotropic, and that groundwater flow systems are more or less complete and independent, e.g. bounded by impermeable beds. In many real situations, however, flow systems are bounded by semi-permeable rather than by completely impermeable beds, so that very complex and widespread systems of regional groundwater flow can develop. However, simplifying assumptions are often reasonable and helpful, although it should be emphasized that important 'untypical' groundwater flow systems occur in limestone and volcanic rocks where most groundwater flow occurs through the fracture systems (White, 2002; Price, 1987).

The direction and rate of groundwater movement in a porous medium may be calculated from the prevailing hydraulic gradient and the hydraulic conductivity of the water-bearing material, using the Darcy equation.

5.5.1 DARCY'S LAW

Most groundwater movement takes place in small interstices so that the resistance to flow imposed by the material of the aquifer itself may be considerable. As a consequence the flow is **laminar**, i.e. successive fluid particles follow the same path or streamline and do not mix with particles in adjacent streamlines. As the velocity of flow increases, especially in material having large pores, the occurrence of turbulent eddies dissipates kinetic energy and means that the hydraulic gradient becomes less effective in inducing flow. In very large interstices, such as those found in many limestone and volcanic areas, groundwater flow is almost identical to the turbulent flow of surface water.

While designing a sand filtration system as part of a water treatment works for the city of Dijon, Henry Darcy conducted a series of experiments to study the flow of water through columns of sand. He found that the volume of flow was directly proportional to the head difference, and inversely proportional to the distance that the water passes through (Darcy, 1856). The equation that expresses the dependence of the velocity of capillary or laminar flow to the hydraulic gradient is often referred to as Darcy's law, and is used to describe the movement of groundwater through natural materials. Darcy's law for saturated, laminar flow may be written as:

$$v = -K(h_2 - h_1) / L = -K(\delta_h/\delta_l) \quad (5.2)$$

where v is the so-called 'macroscopic velocity' of the groundwater (m d^{-1}). This is not in fact a velocity but rather a 'volume flux density', i.e. a volume of flow through a cross-sectional area which contains both interstices and solid matrix. K is the saturated hydraulic conductivity and δ_h/δ_l is the hydraulic gradient comprising the change in hydraulic head (h) with distance along the direction of flow (l). K is therefore also a volume flux density per unit hydraulic gradient (when the hydraulic gradient is set at 1). The negative sign indicates that flow is in the direction of decreasing head. Darcy's law assumes **steady** flow, that does not change over time.

Two main components contribute to total hydraulic head (which equates with the elevation of the potentiometric surface) at a given point in the groundwater flow system. These are: (i) the pore water pressure at that point, measured in a piezometer, which determines the **pressure head**, and (ii) height above sea-level, or some selected datum, which determines the **elevation head**. The total head (h) is therefore defined as:

$$h = \psi + z \quad (5.3)$$

where ψ is the pressure head and z is the elevation head above a selected datum

(Figure 5.8). Both pressure and elevation are forms of potential energy, the one possessed by virtue of state and the other by virtue of position. Kinetic energy, the other energy component of fluid flow, is ignored because groundwater flow is so slow. Total head may be converted to potential energy (ϕ) by applying the gravitational constant so that

$$\phi = g\,h \quad \text{or} \quad \phi = \psi_p + \psi_g \tag{5.4}$$

where ψ_p is the pressure potential and ψ_g is the elevation potential. Using this terminology, Darcy's law states that *water will move from a location where the potential energy is higher to one where it is lower*. This is equally applicable to either saturated or unsaturated conditions, as is shown in Chapter 6. Potentials will decease downwards in recharge areas and decrease upwards in discharge areas.

Under hydrostatic conditions, there is no change of groundwater potential with depth below the water table because the increase of pressure potential (ψ_p) is exactly offset by the decrease in gravitational potential (ψ_g) (Figure 5.8). As a result, there is no groundwater movement in the vertical plane. However, a zero change of potential with depth is also characteristic of situations where the direction of unconfined groundwater flow is approximately horizontal. In such cases the lines of equal head or **equipotentials** are approximately vertical and are labelled by the height at which they intersect the water table, and the potential gradient is simply the slope of the water table immediately above the given point. This approximation ignores vertical flows and might be the situation, for example, where an extensive thin permeable bed rests on an underlying horizontal impermeable bed

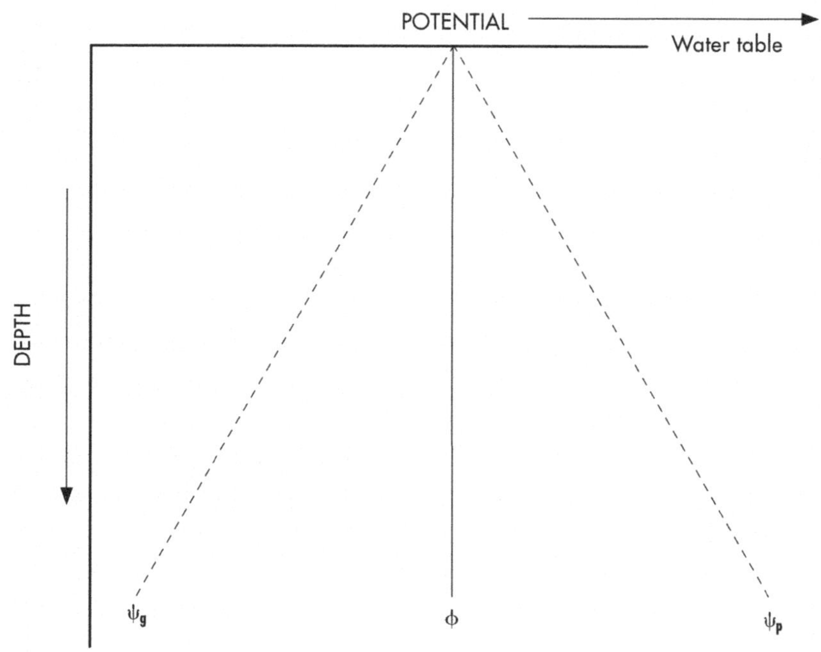

Fig 5.8 Under hydrostatic conditions, the increase in pressure potential (ψ_p) with depth below the water table is offset by the decrease in gravitational potential (ψ_g), resulting in no change in total groundwater potential (ϕ) with depth.

and is known as the Dupuit-Forchheimer approximation.

The **hydraulic conductivity**, K, in the Darcy equation refers to the characteristics of both the porous medium and the fluid. This is virtually synonymous with the earlier term coefficient of permeability. It should not be confused with the **intrinsic** or **specific permeability**, usually denoted as k, which depends only upon the characteristics of the porous medium itself. As indicated in the explanation of Equation 5.2, the Darcy equation yields only an 'apparent' velocity value, the **macroscopic velocity**, through the whole cross-sectional area of solid matrix and interstices. Clearly, the actual flow velocities through the interstices alone will be higher than the macroscopic value. Furthermore, due to the tortuous nature of the flow path of a water particle around and between the grains in an aquifer, the actual velocity is highly variable and the distance travelled exceeds the apparent distance given by the measured length of the porous medium in the average direction of flow. Thus the **effective velocity** of groundwater movement through the interstices is equal to the volume flux density (macroscopic velocity) divided by the effective porosity. Accordingly, the Darcy equation is strictly applicable only to cases in which the cross-section being considered is so much greater than the dimensions of its microstructure that it can reasonably be regarded as uniform.

Another factor complicating the field application of the Darcy equation is that hydraulic conductivity is often markedly anisotropic, particularly in fractured and jointed rock. Extreme flow velocities may also result in deviations from Darcy's law. The effect of turbulence in modifying the relationship between the hydraulic gradient and high rates of flow has already been mentioned. At the other extreme, in certain clay materials the flow rates under very low hydraulic gradients are less than proportional to the hydraulic gradient because much of the water is strongly held by adsorptive electrostatic forces and so may be less mobile than ordinary water.

Despite these restrictions on its strict validity, the Darcy law constitutes an adequate description of most situations of groundwater flow and is equally applicable to both confined and unconfined conditions. An understanding of many groundwater problems demands information not only about the velocity of water movement but also about the quantity of water flowing through a particular aquifer, its **transmissivity** (T), which is important for water supply and is defined as:

$$T = K b \tag{5.5}$$

where K is the hydraulic conductivity and b is the saturated thickness of the aquifer.

By itself, the Darcy law suffices to describe only steady flow conditions, so that for most field applications it must be combined with the mass-conservation (or continuity) law to obtain the general flow equation or, for saturated conditions, the Laplace equation. A direct solution of the latter equation for groundwater flow conditions is generally not possible so that it is necessary to resort to various approximate or indirect methods of analysis.

5.5.2 FACTORS AFFECTING HYDRAULIC CONDUCTIVITY

Fundamental to the application of Darcy's law is a knowledge of the hydraulic conductivity of the saturated medium. The natural variation of K is huge - over 13 orders of magnitude (Freeze and Cherry, 1979). The factors affecting hydraulic conductivity may be conveniently grouped into those pertaining to the water-bearing material itself and those pertaining to the groundwater as a fluid.

An important, though often elusive, **aquifer characteristic** concerns the geometry of the pore spaces through which groundwater movement occurs. Many studies have used an indirect approach whereby the pore space geometry is related to factors such as grain size distribution on the not always very sound assumption that there is a definable relationship between these properties and pore size distribution. Another aquifer characteristic relates to the geometry of the rock particles themselves, particularly in respect to their surface roughness, which may have an important effect on the speed of groundwater flow. Finally, hydraulic conductivity and therefore groundwater flow may be influenced significantly by secondary geological processes such as faulting and folding, which may increase or decrease groundwater movement, secondary deposition, which will tend to reduce the effective size of the interstices and the flow of water, and secondary solution in rocks such as limestone. Indeed, Heath (1982) mapped five types of groundwater flow system in the USA largely on the basis of the way in which porosity has been affected by secondary geological processes.

The chalk in England provides examples at various scales of the effect of aquifer characteristics on groundwater movement. In

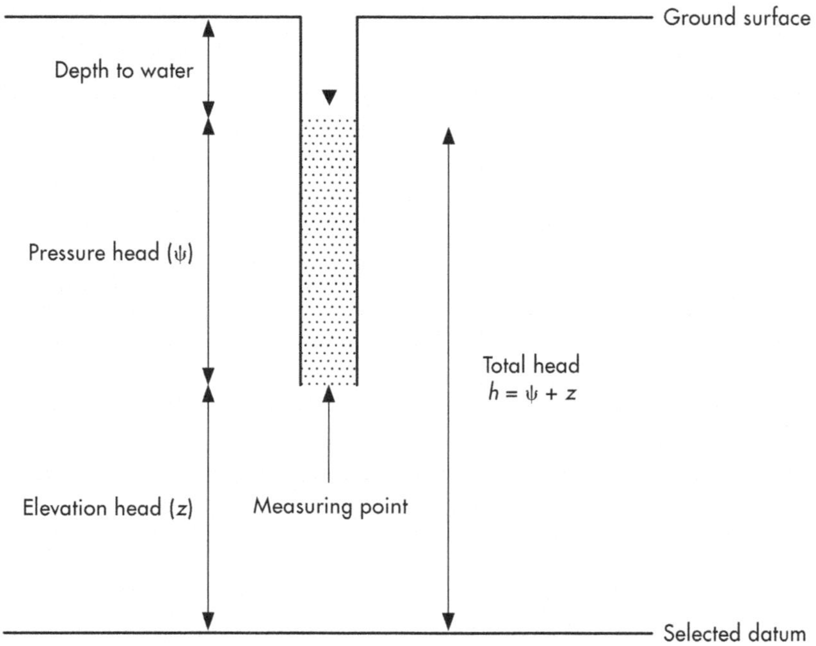

Fig 5.9 Diagram showing the difference between elevation head, pressure head and total head for a particular measuring point.

East Anglia, the areas of high transmissivity tend to be related to topographic valleys, which in turn are associated with fold or fault structures, or with increased fissuring. In the London Basin, compacted synclinal areas in the chalk are associated with low rates of groundwater flow compared with the more open-textured anticlinal areas. Although tranmissivity maps are normally based on well tests or groundwater models, Bracq and Delay (1997) showed that, in Northern France, transmissivity could also be related to surface breaks of slope which reflect underlying vertical fracturing in the chalk aquifer.

The effects of **fluid characteristics**, such as density and viscosity, on hydraulic conductivity tend to be rather less important than the effects of the aquifer characteristics. Certainly, in normal conditions of groundwater flow, the physical properties of the groundwater are likely to be influenced only by changes in temperature or salinity. Temperature, by inversely affecting the viscosity, has a direct influence on the speed of groundwater flow althoigh since most groundwater is characterized by relatively constant temperature this factor is unlikely to be important except in special circumstances. Variations of salinity are also unlikely to be significant in normal groundwater conditions. Where saline infiltration occurs, however, hydraulic conductivities may be affected both by changes in the ionic concentrations of the groundwater and also by the chemical effect of the saline water on the aquifer material itself, particularly where this is of a clayey nature. Increasing salinity will increase water density and so may affect hydraulic heads and gradients.

Finally, groundwater studies traditionally presuppose that hydraulic conductivity is not affected by water content, since the aquifer matrix below the water table is assumed to be saturated. In some materials, such as peat, full saturation may be prevented due to the presence of undissolved gases. Methane is produced as a by-product of the microbial decay of peat in anoxic conditions and by occupying part of the pore space, thereby reduces both water storage and hydraulic conductivity (Baird, 1997). A similar effect has also been observed in mineral soils when air is trapped in soil pores during *rapid* infiltration or after the *sudden* rise of a shallow water table.

5.6 GROUNDWATER FLOW SYSTEMS

Although groundwater flow is three-dimensional, methods of estimating it have often been simplified so as to consider it taking place in just two dimensions. Although quasi three-dimensional groundwater modelling packages are readily available, nevertheless the two-dimensional characterisation of groundwater flow provides a useful conceptual framework for the present discussions. Because of the readier availability of the appropriate data, groundwater flow is normally considered as a two-dimensional problem in the vertical rather than the horizontal plane, and in

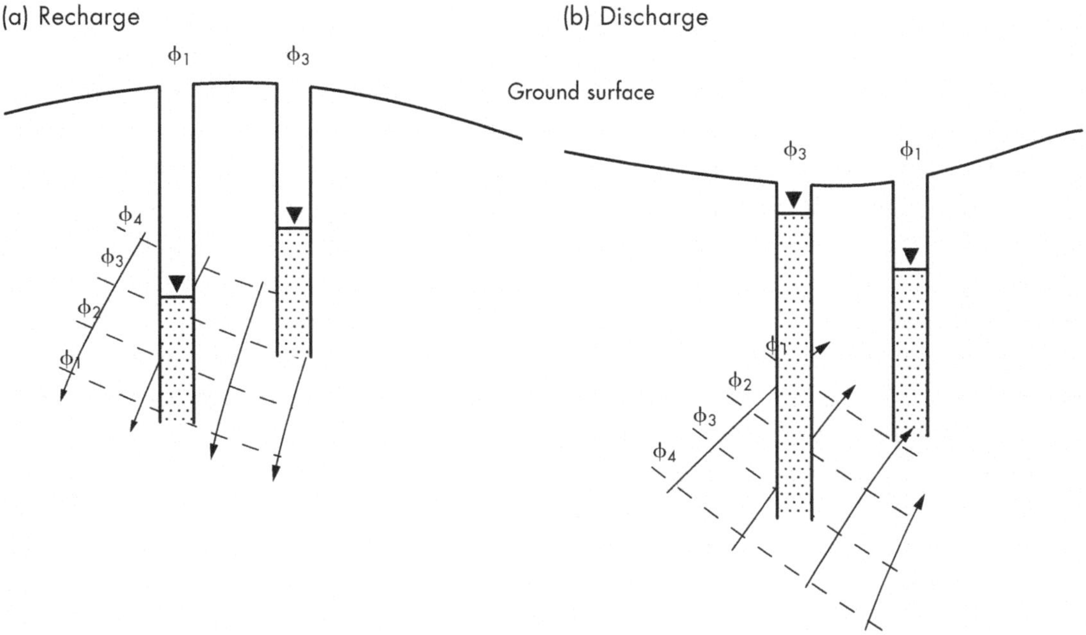

Fig 5.10 Differences in water levels in piezometers of different depths when there is a vertical component of groundwater flow: a) Recharge reflecting a decrease in potential (from ϕ_4 to ϕ_1) with depth causing downward flow, and b) Discharge due to upward flow due to an increase in potential with depth. Note that in each case the higher water level occurs in the piezometer intercepting the higher equipotential line marked ϕ_3 and the lower level in the piezometer ending at ϕ_1.

this case it is information on the rate of change of hydraulic head or potential with depth that facilitates the definition of the groundwater flow pattern.

Water in a piezometer that is open at a particular depth will rise to an elevation equal to the hydraulic head at that point. If groundwater potential increases with depth, flow will be upwards and if it decreases with depth, flow will be downwards (Figure 5.10). Where groundwater potential decreases with depth and water is moving downwards is a **recharge area**, and a **discharge area** occurs where groundwater is moving upward to the water table due to groundwater potential increases with depth.

5.6.1 FLOW NETS

Although groundwater flow cannot be observed directly, it is possible to use the relationship between flow and the hydraulic or potential gradient in order to examine two-dimensional groundwater flow indirectly by reference to the subsurface distribution of groundwater potential. Lines joining points of equal potential (ϕ) are known as **equipotentials**, and the potentiometric surface of an aquifer (confined or unconfined) above a datum plane may be contoured at regular increments of ϕ by a family of such lines. This is shown in Figure 5.11, where the equipotentials decrease in value from ϕ_9 to ϕ_1. In accordance with Darcy's law, groundwater is driven along the maximum gradient of potential, i.e. perpendicular to the equipotential lines. This is depicted by the **streamlines**, which show the direction of the force on the moving water and therefore represent the paths followed by particles of water. Since at any one point the flow can only have one direction, it follows that streamlines never intersect. The network of meshes formed by the equipotentials and streamlines is known as a **flow net**. In this context Figure 5.11 could be viewed as either a potentiometric map in the vertical plane with the streamlines illustrating groundwater recharge and downward movement at the left-hand side, or alternatively as a map in the horizontal plane.

Flow nets show the direction of groundwater movement and can also be used to estimate the rate of flow, either by graphical construction or mathematically. The zone between any pair of neighbouring streamlines is known as a **streamtube**, and at every cross-section of a streamtube the total flow (q) remains the same. If adjacent equipotentials differ by the same increment of potential ($\Delta\phi = \phi_2 - \phi_1 = \phi_3 - \phi_2$, etc.) and the streamlines are chosen to be evenly spaced so as to give the same rates of flow in all streamtubes, Darcy's law can be applied to any of the elements of the flow net having width W and length L so that

$$q = -K\Delta\phi \, (W/L) \tag{5.6}$$

Regional and complex groundwater systems may consist of several aquifers and aquitards, and therefore involve flow through both the aquifers and the confining beds. Hydraulic conductivities in aquifers are usually several orders of magnitude greater than in confining beds and so offer least resistance to flow, the result being that head loss (and hence hydraulic gradient)

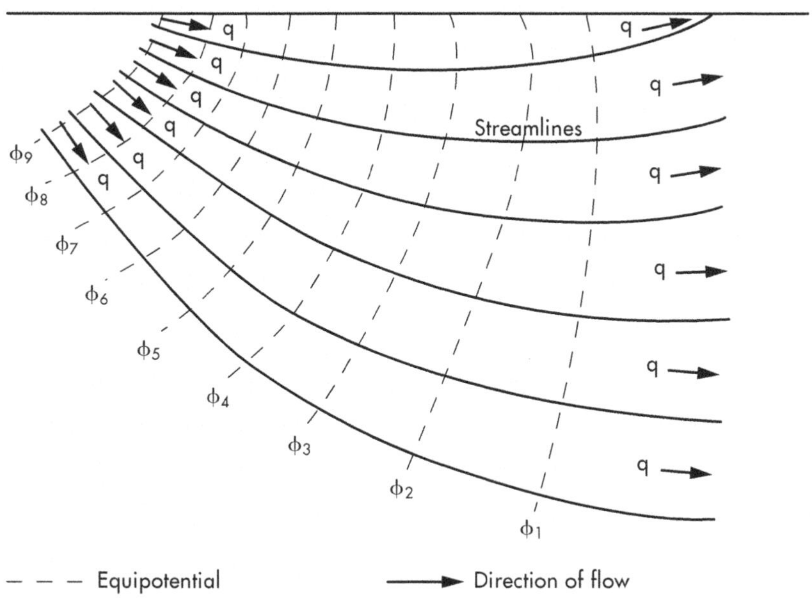

Fig 5.11 Flow net showing equipotentials (decreasing from ϕ_9 to ϕ_1), and streamlines indicating the direction of groundwater flow. See text for further details.

becomes much less in aquifers. In accordance with Darcy's law, other factors being equal, the higher the permeability, the smaller the area required to pass a given volume of water in a given time. Streamlines are thus more widely spaced in low-permeability material and more closely spaced in the high-permeability material.

If the equipotentials are taken to represent water table levels (ϕ_1 having the lowest value and ϕ_9 having the highest value), Figure 5.11 could be regarded as a map, in the horizontal plane, of unconfined groundwater flow in a **non-homogeneous aquifer**. Since the change in potential between adjacent pairs of equipotentials is equal and the hydraulic gradient varies inversely with the distance between equipotentials, then if inflow for any section is just balanced by outflow, the relative steepness of the hydraulic gradient reflects the hydraulic conductivity, as indicated in Darcy's law. Thus the hydraulic conductivity is lowest in the west of the area and increases towards the east where the equipotentials are more widely spaced. Alternatively, this Figure can be viewed as showing the horizontal distribution of potential in a **homogeneous aquifer**. In this case the variable spacing of the equipotentials would reflect a variation in the rate of groundwater flow, with flow decreasing eastwards in response to the decrease in hydraulic gradient in this direction.

5.6.2 CLASSICAL MODELS OF GROUNDWATER FLOW

The usual approach to the solution of groundwater problems, including the application of flow nets, was popularised by Hubbert (1940) and assumes that the water table is a subdued replica of the local topography, that topographical divides may also

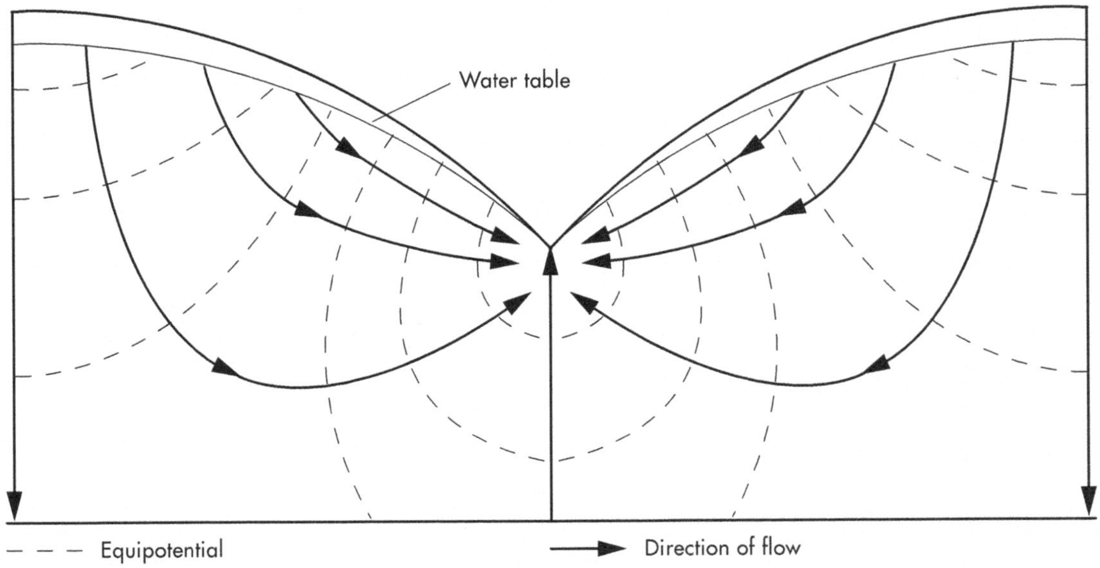

Fig 5.12: Approximate groundwater flow pattern in a uniform, permeable unconfined aquifer.

act as groundwater divides and that groundwater moves from topographical high areas to topographical low areas. Figure 5.12 illustrates the essential features of Hubbert's presentation for a homogeneous, isotropic material. Equipotentials are shown as broken lines and the value of hydraulic head for each line is equal to the elevation of its intersection with the water table. Streamlines, indicating the groundwater flow paths, connect the source areas in which recharge is dominant with the sinks in which discharge is dominant. Under closed conditions, in which rainfall inputs were to cease, groundwater flow would ultimately result in the complete drainage of water from the topographic highs and the production of a flat surface of minimum potential energy (the hydrostatic condition). This tendency, however, is counteracted by continuous replenishment from precipitation. The result of this continuous movement and renewal is the flow pattern shown whose upper surface, the water table, is a subdued replica of the topography. The source areas are the topographic highs and in this diagram the sinks are shown as streams, and each groundwater flow cell is bounded by the lines of vertical flow beneath the groundwater divides and sinks or by widely distributed impermeable beds or by both.

Flow nets assume **steady-state** groundwater flow that does not change over time, which may be acceptable for many purposes as the seasonal change in water table of a couple of metres between the high and low levels of the water table will have little effect on flow patterns, and the relative positions of the high and low points usually remain unchanged (Freeze and Witherspoon, 1966).

Tóth (1962) noted that the Hubbert model predicted all groundwater discharged to the valley bottom, but this did not describe the field conditions that he observed. Hubbert had derived the flow net by graphical construction for one particular case, Tóth adopted mathematical analysis to derive analytical solutions to study this further and extend the approach to consider more general cases. His results confirmed that major stream valleys may indeed be major groundwater sinks (as in the Hubbert model), but indicated that for small valleys and limited topography the groundwater discharge is not concentrated solely at the stream but is broadly distributed on the lower side of a **hinge line** situated between the valley bottom and the groundwater divide (Figure 5.13). The hinge line is a narrow zone of approximately vertical equipotential where groundwater potential does not vary with depth and flow is essentially horizontal. It separates an up-gradient recharge area from a down-gradient discharge area. In this example the hinge line is shown situated in the middle of a symmetrical flow system, which is an artefact of the analytical solution used. For most topographical situations the hinge line actually lies closer to the valley bottom than to the hilltop (Freeze and Cherry, 1979).

The unconfined groundwater models discussed above all assume that the porous medium is hydrologically isotropic and homogeneous, and that the aquifer is unconfined. At the other extreme is the situation where, with alternating beds of markedly different lithology and permeability, groundwater is confined beneath an impermeable layer and the potentiometric surface of the flow field is completely independent of surface topography and of the configuration of the water table in the upper, unconfined groundwater body. What is commonly found in actual conditions is neither a completely confined system nor a completely unconfined system, but rather a system of flow that possesses distinct characteristics of both extremes. Confining beds rarely form an absolute barrier to water movement so that there is normally some degree of hydraulic continuity. This suggests that the potential distribution with depth in a confined groundwater body is partly affected by the potential distribution of the overlying water table. Also, in an apparently unconfined situation, the flow field may possess characteristics of confinement whenever flow is refracted on emerging from a low-permeability bed, so that it proceeds almost tangentially to the lower surface of that bed.

Streamlines are refracted at the boundary between strata of different hydraulic conductivity, in the direction that produces the shortest route through the confining bed. The effect is to make streamlines nearer horizontal in the more permeable layer, and nearer vertical in the lower permeability layer (Freeze and Cherry, 1979). This is illustrated in Figure 5.14, which represents an unconfined system with hydraulic continuity in the vertical direction. Streamlines in the high-permeability bed are almost tangential to the lower surface of the low-permeability bed. There is very little difference in groundwater potential along an imaginary vertical line passing

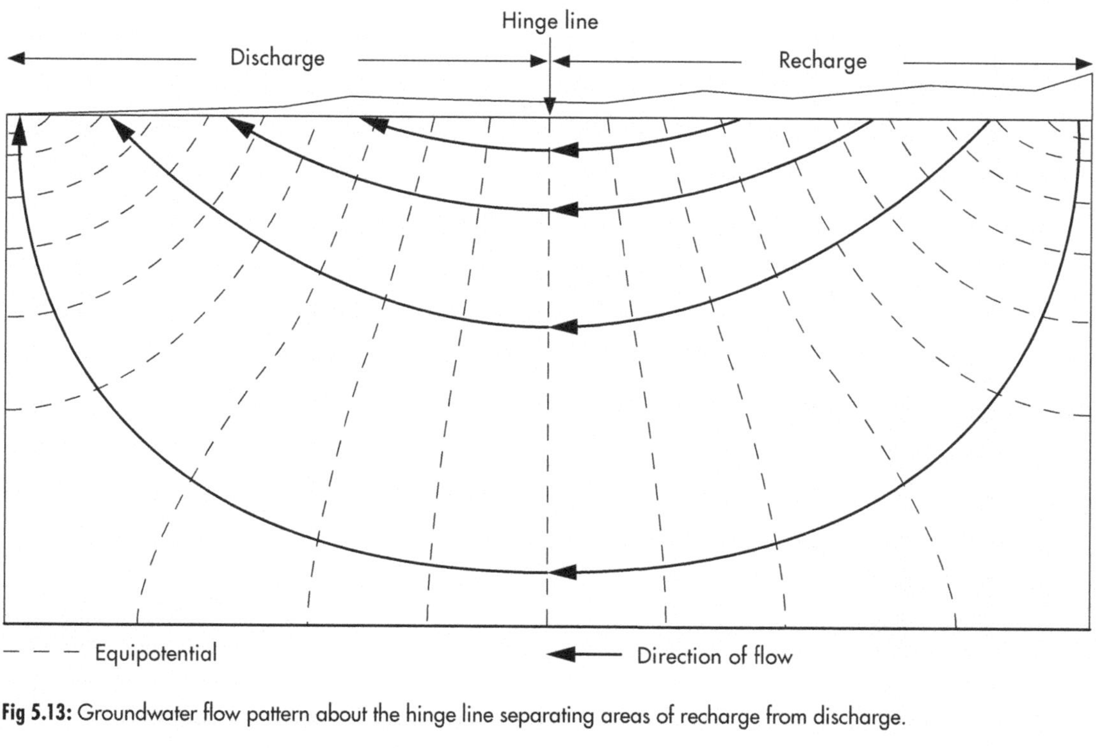

Fig 5.13: Groundwater flow pattern about the hinge line separating areas of recharge from discharge.

through the high-permeability bed, but if this line is extended upwards, it crosses several equipotentials in the low-permeability bed and relatively few in the medium-permeability material. Thus, if a vertical well was drilled from the ground surface, there would be a large increase in potentiometric head when it first entered the high-permeability bed because static levels can establish themselves more rapidly here than elsewhere in the system. This increase in head is often attributed to confinement of water under pressure, although in reality it results from the movement of water through the low-permeability bed. In other words, the conditions implied by the term confinement will arise when a unit of low permeability overlies a unit of high permeability. The so-called artesian condition is an extreme case of this (Price, 1985).

5.6.3 REGIONAL GROUNDWATER FLOW

Recognition that groundwater may vary between confinement and non-confinement as it flows through a sequence of materials having contrasting permeabilities, makes it easier to grasp the concept of 'regional' groundwater flow over substantial distances and possibly involving more than one topographic basin. The pioneering contributions of hydrologists and geologists recognised the possibility of regional groundwater flow systems, but significant progression of these ideas had to await the development of more powerful, computer-based modelling techniques when important contributions were made by a group of Canadian hydrologists including József Tóth, Peter Meyboom and R. Allan Freeze (Tóth, 2005).

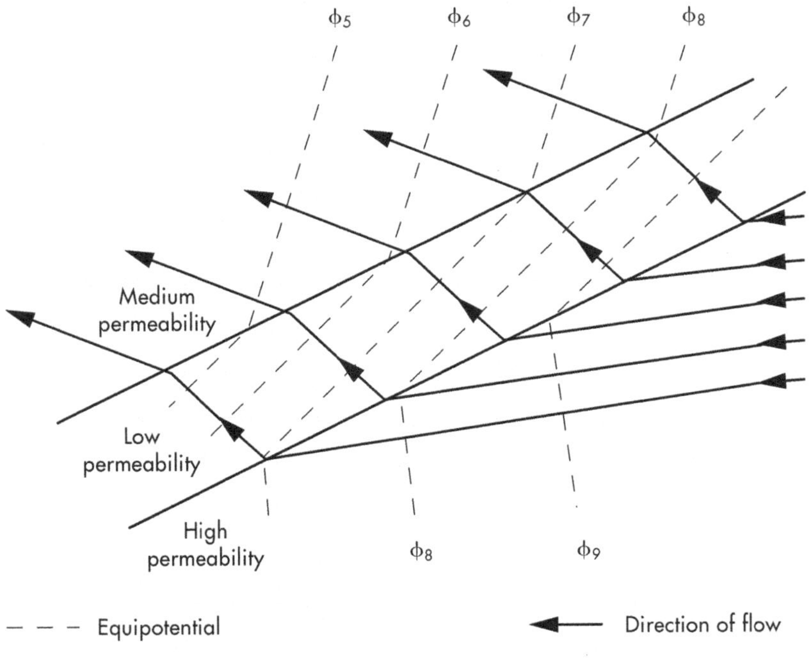

Fig 5.14: Refraction of flowlines at a geologic boundary between two materials of different permeability.

Tóth (1962, 1963) considerably extended Hubbert's work by showing that groundwater flow patterns can be derived as mathematical solutions to formal boundary value problems. He assumed two-dimensional vertical flow in a homogeneous isotropic medium bounded below by a horizontal bed and above by the water table, which is a subdued replica of the topography. The lateral flow boundaries are the major groundwater divides. Tóth (1963) recognized three broad categories of groundwater flow system (defined as one in which any two streamlines which are adjacent at one point remain adjacent throughout the system) illustrated in Figure 5.15. A *local* system, identical to the classic Hubbert model, which extends over distances of a few hundred metres, has its recharge area at a topographic high and its discharge area at an adjacent topographic low. An *intermediate* system, which extends over distances of a few kilometres, has one or more topographic highs and lows located between its recharge and discharge areas. Finally, a *regional* system, e.g. the Hungarian Basin, which may extend over hundreds of kilometres, where the recharge area is at the main topographic high and the discharge area at the lowest part of the basin. A useful review of regional groundwater flow in large sedimentary basins was given by Tóth (1995; 1996).

Even in basins underlain by uniform material, topography can create complex systems of groundwater flow. Under extended flat areas, characterized by local waterlogging and groundwater mineralization from concentration of salts, neither regional nor local systems will develop (Tóth, 1963). However, regional systems will develop if local relief variations are negligible but there is a general topographic slope. With increasingly pronounced local relief, deeper local rather than regional systems will develop so that, in an area of large river valleys and steep divides, extensive unconfined regional systems are unlikely to occur. As Winter et al (2003) note, it would be 'convenient' if groundwater divides lay beneath surface divides, but in practice this commonly is not the case. Furthermore, the superposition of flow systems can result in groundwater discharge to the surface deriving from more than one flow system, with deeper water flow systems that may have travelled long distances far below the surface, then mixing with shallower local waters.

Flow nets assume **steady flow** that does not change over time, which may be acceptable for many situations. The work of Tóth was subsequently generalized by Freeze and Witherspoon (1966, 1967, 1968) whose more versatile numerical modelling technique was able to determine steady-state flow patterns in a three-dimensional, non-homogeneous anisotropic groundwater basin, with any water table configuration, given knowledge of the dimensions of the basin, the water table configuration and the permeability resulting from the subsurface stratigraphy.

Groundwater flow systems can now be modelled rapidly and flexibly using numerical models and software packages, such as the widely available MODFLOW program developed by the US Geological Survey for which the latest version can be downloaded from: http://water.usgs.gov/ogw/modflow/). This model simulates three-dimensional groundwater flow using a finite-difference approach (see review by Loudyi et al, 2014) and is widely-used for simulating and predicting groundwater conditions and groundwater/surface-water interactions. Layers of the groundwater basin can be simulated as confined, unconfined, or a combination of both. The effects on groundwater flow of wells, areal recharge, evaporation, drains and streams can also be simulated.

These models can simulate complex and irregular shapes and time-dependent **transient flow** conditions, but many of the basic concepts and principles of aquifer flow discussed here, including consideration of its geometry, parameters of hydraulic conductivity and porosity, and the boundary conditions are essentially the same. Wang and Anderson (1995) provided a useful and comprehensive introduction to groundwater modelling.

The understanding of regional groundwater systems has also benefitted from an improved understanding of groundwater chemistry and from the increasingly sophisticated use of tracers to reconstruct groundwater flowpaths. It has long been known that there is a broad correlation between the chemistry and residence time and flowpath of groundwater (see also Section 8.5.5). As groundwater moves away from the point of recharge in the outcrop area, its ionic content increases and processes of dissolution are gradually replaced by processes of ion exchange. As well as hydrochemical facies, a wide range of individual tracers, such as chlorofluorocarbons (CFCs),

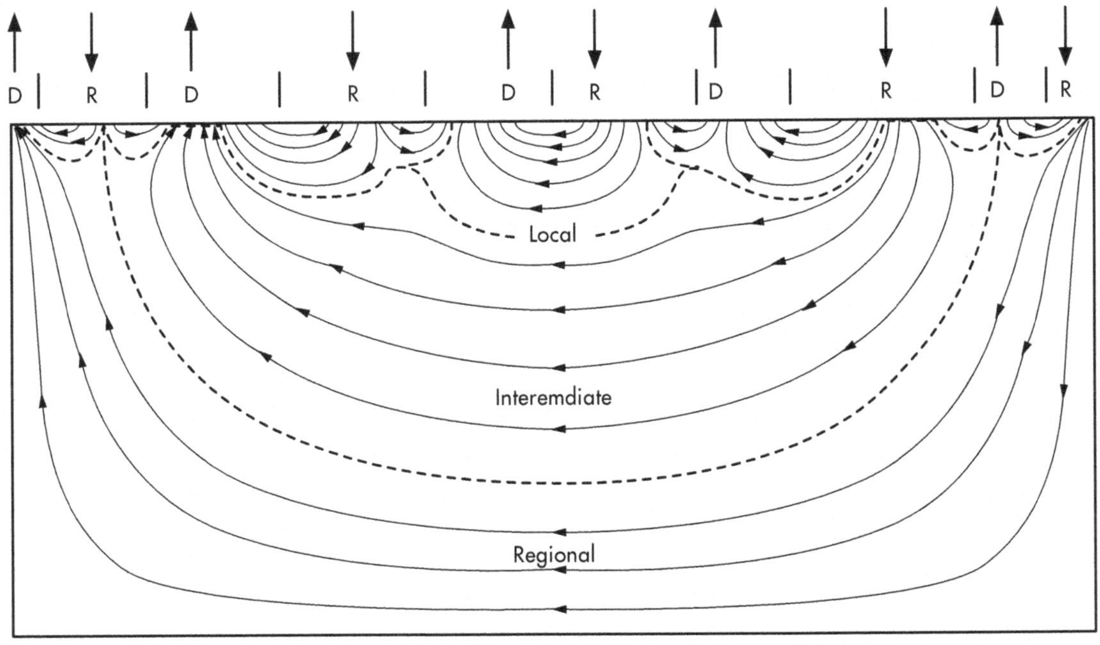

Fig 5.15: Vertical section through a large aquifer system, showing local, intermediate and regional flow systems coexisting. R = Recharge, D = Discharge zones and dashed lines are the boundaries between the three flow systems. Redrawn and simplified from a diagram by Engelen and Kloosterman, 1996 which was redrawn from a diagram by Toth, 1995.

sulphur hexaflouride (SF6), noble gases and radioelements, is now also used to investigate regional groundwater flow systems (Busenberg and Plummer, 2000; Plummer and Busenberg, 2000). For example, CFCs can be used as tracers to date groundwater over periods of up to 40 years and can estimate groundwater flow paths with a better accuracy than traditional hydraulic-based methods.

5.6.5 GROUNDWATER FLOW IN COASTAL AQUIFERS

The discharge of groundwater directly to the sea is not as obvious as river discharge and much more difficult to estimate reliably. Burnett et al (2003) reviewed the literature and concluded that it is small in comparison with the outflow of rivers, amounting to perhaps 6% of the total continental fresh water outflows. However, its importance is almost certainly greater than this low value implies, due to the high density of human population in coastal areas and the susceptibility of coastal aquifers to degradation by salt water intrusion (UNESCO, 1987). Efforts continue, therefore, to determine freshwater discharge into the sea with ever greater accuracy (e.g. Burnett et al, 2006).

In coastal areas there is normally a hydraulic gradient towards the sea and the resulting seaward groundwater flow effectively limits the subsurface landward encroachment of saline water. In unconfined groundwater conditions, with a water table sloping towards sea level at the coast, the groundwater body takes the form of a lens of fresh water 'floating' on more saline water beneath. The position of the interface between the fresh and salt water was investigated independently by Badon-Ghijben (1889) and Herzberg (1901). Rather than saline water being encountered at sea level in wells near the coast, it was found at a depth below sea level of about 40-times the height of the water table above sea level reflecting the hydrostatic equilibrium between lighter fresh groundwater and heavier saline groundwater (Figure 5.16).

Assuming hydrostatic conditions and a negligible mixing zone the **Ghijben-Herzberg relationship** may be written as

$$h_s = (\rho_f/(\rho_s - \rho_f))\, h_f = \alpha h_f \qquad (5.7)$$

where h_s is the depth of the fresh water below sea level, ρ_f, is the density of fresh water, ρ_s is the density of sea water and h_f, is the height of the water table above sea level. Fresh water has a density of 1.00 g cm^{-3} and saline water has a density of 1.02 -1.03 g cm^{-3}, depending on temperature and salinity. For $\rho_s = 1.025$ the ratio $\alpha = 40$.

Hubbert (1940) demonstrated that a state of dynamic rather than static equilibrium must exist at the interface or else there would be no way for fresh water to discharge into the sea. As a result, the interface between saline and fresh water will occur seaward of that estimated from the Ghijben-Herzberg relationship and intersect the sea floor some distance from the shore, leaving a gap or band through which the fresh groundwater can escape into the sea.

In stratigraphically layered aquifers there may be more than one saline wedge. Thus, in the case of a lower, confined aquifer the saline wedge may either intrude even further inland or seaward of the shoreline depending on the freshwater head in the confined layer.

The interface between saline and fresh groundwater is not as sharply defined as in Figure 5.16. Tidal fluctuations as well as variations in recharge and discharge continually disturb the balance between the fresh water and the sea water and cause the interface to fluctuate (Burnett et al, 2003). These fluctuations, together with the diffusion of salt water, destroy the sharp interface and create a transitional diffusion zone of brackish water instead.

Excessive abstraction of groundwater from coastal aquifers will reduce the flow of fresh groundwater and cause a lowering of water table and potentiometric levels. In turn, this may readily lead to the incursion of saline groundwater and to the long-term contamination of the aquifer. This problem has already occurred in many coastal areas of the Netherlands, Spain, Israel, France, the USA, Italy and Britain (UNESCO, 1987). In some cases the existence of saline groundwater in coastal aquifers may represent the residue of an earlier invasion of sea water, which occurred when the land was relatively lower and which now takes the form of a wedge of highly saline water beneath a superficial seaward-flowing body of fresh groundwater. The presence of saline water in the chalk along parts of the East Yorkshire, Norfolk and Suffolk coasts of England may probably be explained in this way, as may some of the saline groundwater along the Atlantic coast of the United States.

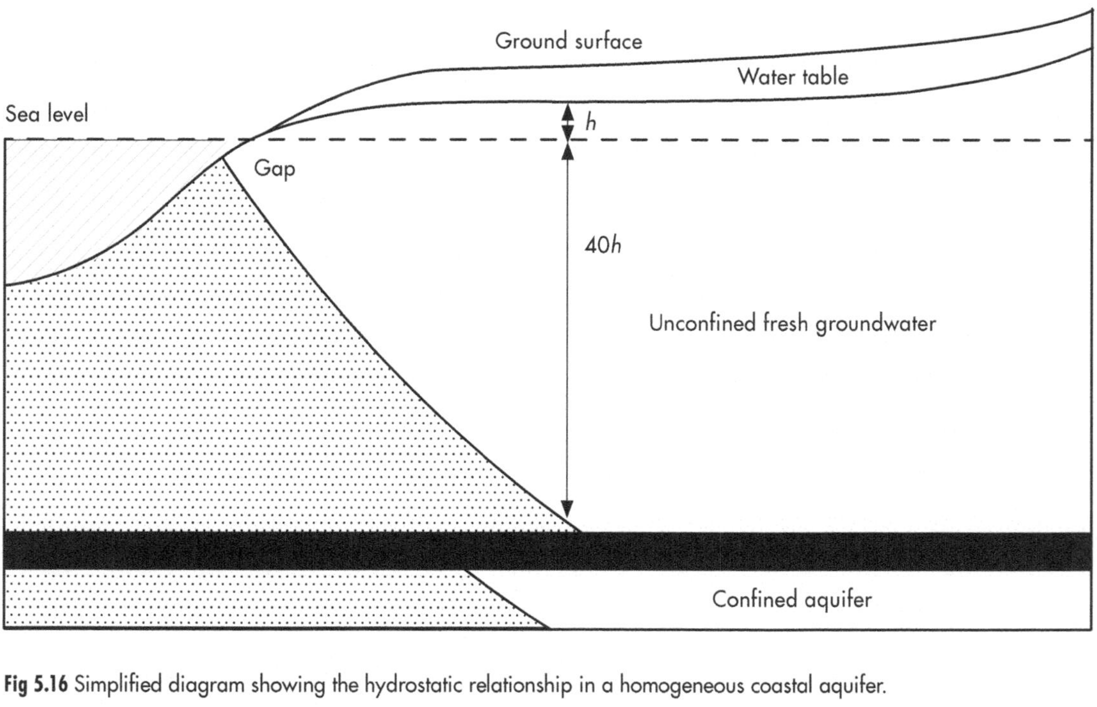

Fig 5.16 Simplified diagram showing the hydrostatic relationship in a homogeneous coastal aquifer.

5.7 GROUNDWATER IN JOINTED AND FRACTURED ROCKS

Most of the foregoing discussion of groundwater storage and movement has implied archetypal conditions of homogeneity and isotropy which are more characteristic of uniform sedimentary strata. It may be helpful, therefore, to conclude this chapter by considering briefly the particular characteristics of groundwater occurrence in jointed and fractured rocks. Together, these underlie about 40% of the Earth's land surface and constitute a major source of water supply in many tropical and sub-tropical areas (e.g. MacDonald et al, 2012; Wright, 1992). Outcrops of carbonate rocks in Europe cover some 3 x 10^6 km^2, i.e. 35% of the area, but in many countries such as Germany, France, Poland, Romania and Russia, major limestone aquifers also exist beneath a thick cover of other rocks (Crampon et al, 1996).

Many carbonate and crystalline rock masses contain systems of faults, solution joints, or fractures. Especially at shallow depths, where these features are most frequent, the rock mass therefore consists of an assemblage of intact rock blocks that are separated by various features which can be generally described as fissures. These fissures will normally be important for groundwater storage and movement, although the significance of their role will depend partly on their contribution to total porosity and partly on the porosity and permeability of the intervening blocks. Matrix permeability may range from $<= 0.1$ m y^{-1} in dense unweathered bedrock to $>= 1.0$ m y^{-1} in soft limestone such as chalk. In addition, some of the fissures may be filled with fine deposits which are more permeable than the intervening blocks, e.g. in crystalline rock, or less permeable than the intervening blocks, e.g. in limestone (see Table 5.1). In most circumstances, however, the movement of water through the fissures will be several orders of magnitude greater than the movement of water through the block matrix. Accordingly, the classical concept of an 'aquifer', in which groundwater storage and movement are related to the formation rather than to the structures within it, is not entirely appropriate in carbonate and fractured rock.

In all **block-fissure systems** there is likely to be a wide range of groundwater flow rates. Where meaningful flow occurs through the blocks it will usually consist of diffuse, laminar flow in accordance with the Darcy law. Through larger, super-capillary conduits and through major fissures, flow will be rapid and turbulent and will not therefore accord with the Darcian flow model. A further complication of such dual-flow systems is that the relative importance of flow through the matrix of the blocks and through the fissures varies depending on whether the system is saturated or unsaturated. Rapid flow through the fissures predominates in saturated conditions, and slow flow through the matrix predominates in unsaturated conditions. Accordingly, there will be significant changes of groundwater flow path and travel time during periods of fluctuating water table level, particularly in more permeable rocks such as limestone and chalk. Depending on water-table depth, it has been estimated that significant pesticide concentrations could reach groundwater through the chalk *matrix* up to 30 years after application (IH, 1997). The groundwater hydrology of Yucca Mountain in southern Nevada, USA, has been intensively studied during assessment of its suitability as a radioactive waste repository. The hypothesis that the fracture systems are barriers to water flow when the matrix rocks are unsaturated and become conduits for water flow only when the matrix is at or near saturation, thereby greatly reducing groundwater travel times, has been challenged (Albrecht et al., 1990). If, alternatively, water moves within fractures even at low values of matrix saturation, this would significantly reduce groundwater travel times and have profound implications for conditions within the unsaturated zone.

Apart from the effects of fluctuating water-tables on groundwater movement, the orientation of fissures may itself vary with depth, particularly in rocks such as granite (Figure 5.17). This means that dominant direction of groundwater flow may change significantly with depth, although of course this depends very much on the continuity of these fissures.

The mix of flow processes in block-fissure systems complicates the investigation of groundwater hydrology and may even make it difficult to define the scale of the investigation. For example, at the microscale, a small sample volume of fractured rock may consist entirely of solid rock or entirely of a single fracture void, resulting in the largest possible potential variation of permeability. As the sample volume is increased larger and more extensive fracture systems may be incorporated, so that there is a tendency for the observed permeability to increase (Figure 5.18). In some cases, a consistent relationship between voids and solid may be reached at the mesoscale, whereby increasing the sample volume further has no further effect on the average permeability. If however the next increase in sample volume includes a major

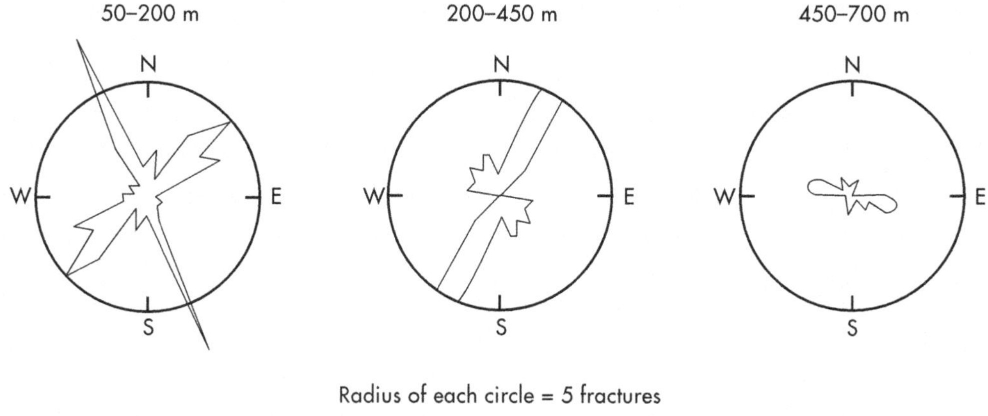

Fig 5.17: Orientation of water-conducting fissures in granite at different depths at Troon in Cornwall, UK. Most of the groundwater flow occurs through these fissures. Based on an original diagram in NERC (1991).

zone of fracturing this may result in a marked increase in average macroscale permeability. Finally, at the megascale, a much larger sample may include many major fracture zones and again exhibit a still larger, but relatively homogeneous, permeability. In other words, the size of the sample used to determine groundwater characteristics in heterogeneous rocks should be of a comparable scale to that of the area under investigation. Since, however, all samples will contain some unrepresentative fissures, some authorities have suggested that there is no such thing as a representative sample volume (Domenico and Schwartz, 1998). And in a comprehensive review of groundwater flow in porous fractured media, Elsworth and Mase (1993) argued that in many cases, despite the costs involved, only initial large-scale testing is likely to overcome the overt scale dependence of the geological medium. Barker (1991) provided a useful general review of flow conditions in fractured rock. Not surprisingly, however, much of the work reported in the literature concerns groundwater flow through specific fissures (e.g. Whitaker and Smart, 1997).

Studies have emphasised the role of fracture systems in determining the rates and directions of maximum groundwater flow in aquifers ranging from the karst of the Yucatan peninsula, Mexico, to the granite of Cornwall in the south-western UK. As the scale of measurement in such rocks is increased, e.g. from laboratory cores, through borehole tests, to basin-wide groundwater flow, there is often an increase in the observed hydraulic conductivity (e.g. Garven, 1985).

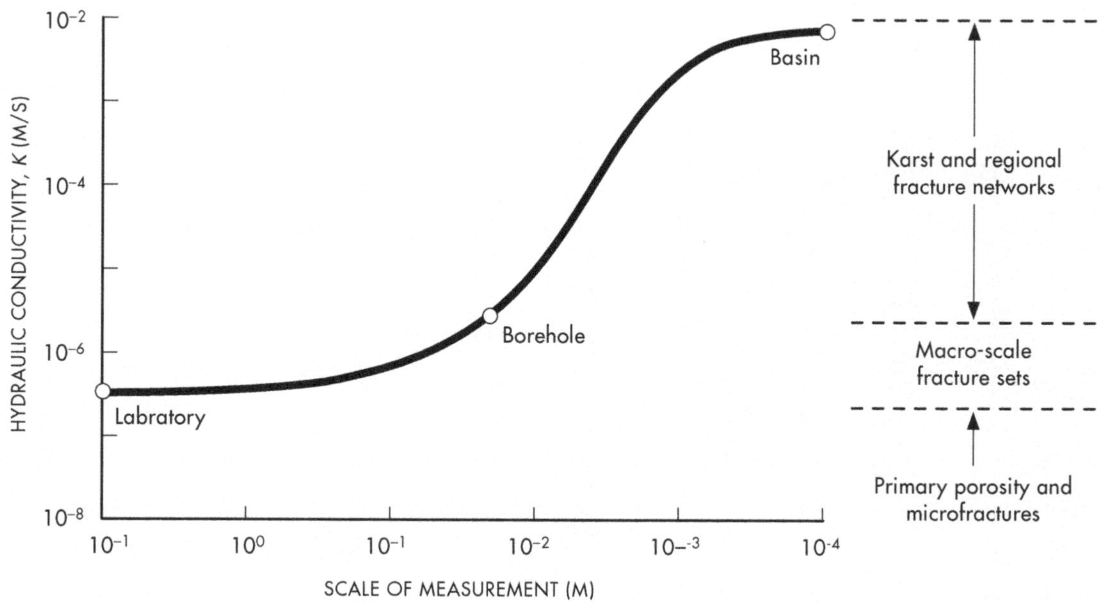

Fig 5.18: Hydraulic conductivity values for carbonate rocks in central Europe, showing values increase with the scale of measurement, largely because of the incorporation into the sample volume of increasingly large and even more extensive fracture systems (Redrawn from Garven, 1985).

High rates of concentrated underground flow from karst geology in Italy would have been known to Leonardo da Vinci (1452 – 1519) who wrote "Very great rivers flow underground".

Because carbonate and fractured rocks may have restricted and shallow water storage, which will tend to deplete rapidly, groundwater in such materials is usually strongly dependent on contemporary recharge. However, in some areas, dating techniques have identified that some of the deeper groundwater must have originated during the later Pleistocene period (e.g. Silar, 1990). In the case of the Lincolnshire Limestone in eastern England, groundwater movement is dominated by new water in the fissures. However, older water stored in the aquifer matrix enters the fissure system by diffusion and as a result of pressure differentials accentuated by groundwater abstraction (Downing et al., 1977). Investigations of higher-than-expected low flows in rivers draining chalk catchments in southern England have indicated that these result from substantial quantities of water stored on and gradually released from the surfaces of the fissure and micro-fissure systems.

Groundwater flow systems in fractured and jointed rocks are still imperfectly understood and whilst much has been achieved, further work remains to be done. However, it is becoming evident that groundwater processes in 'normal' aquifers and in fissured rocks may be more similar than was once thought.

Photo 5.2: Groundwater emerging at a boundary layer from a karst cliff following rainfall – Glasterntal, Switzerland (Photo source: Mark Robinson)

REVIEW PROBLEMS AND EXERCISES

5.1 Discuss the role of groundwater in the hydrological cycle.

5.2 What types of materials make good aquifers and aquitards?

5.3 Explain, with the help of diagrams, what is meant by unconfined, confined, and perched groundwater.

5.4 Define the following terms: artesian, porosity, effective porosity, specific yield, coefficient of storage.

5.5 Explain Darcy's law of groundwater flow.

5.6 Define the following terms: hydraulic gradient, hydraulic conductivity, effective velocity, equipotential, flow net, potentiometric surface.

5.7 Calculate the Darcy macroscopic velocity of groundwater flow through a sand ($K = 10\ m\ d^{-1}$) when two piezometers, installed 10 m apart, give the following data (m):
Piezometer 1: $z = 0$, $\psi = 0.5$
Piezometer 2: $z = -3$, $\psi = 0.15$

5.8 Why is the water table sometimes described as a subdued replica of the surface topography?

5.9 Explain, with the help of diagrams, what is meant by regional, subregional and local groundwater flow.

5.10 Define the following terms: hydrochemical facies, the Chebotarev sequence, the Ghijben-Herzberg relationship.

WEBSITES

ADES Groundwater National Portal (France):
http://www.ades.eaufrance.fr/
British Geological Survey:
http://www.bgs.ac.uk/
Digital Maps of Britain:
http://www.bgs.ac.uk/products/digitalmaps/digmapgb.html
UK Groundwater Forum:
http://www.groundwateruk.org/Default.aspx
www.groundwateruk.org/Groundwater_resources_climate_change.aspx
US Geological Survey: Water Resources of USA:
http://www.usgs.gov/
Natural Resources Canada:
http://www.gw-info.net/
William Smith:
www.williamsmithonline.com
Africa groundwater:
www.bgs.ac.uk/africagroundwateratlas

6

SOIL WATER

"There are holes in Darcy's law – they are called macropores"

ANON

6.1 INTRODUCTION

In large measure, the importance of soil water reflects its vital role as a source of water for plants. Knowledge about the factors controlling water storage and movement in the soil is essential to an understanding of a wide range of processes, including not only the supply of water to terrestrial ecosystems, but also the generation of runoff, recharge to underlying groundwater and the movement and accumulation of pollutants. As well as regulating runoff from land surfaces and transpiration by vegetation, soil wetness also plays an important role in partitioning incoming radiative energy into sensible and latent heat through its control on evaporation. Soil water is thus of interest to investigators in a number of disciplines in addition to the hydrologist, including agronomists, climatologists, ecologists, geomorphologists and civil engineers.

The term **soil water** is normally used to include both the water in the soil profile itself and that in the unsaturated subsoil layers between the soil profile and the water table (see Section 5.1). Thus defined, soil water includes all the water in the unsaturated **zone of aeration** (termed **vadose zone** in the US) which may extend tens or even hundreds of metres below the ground surface. However it is the shallow soil profile proper that is generally of most hydrological importance and will therefore be the main focus of discussion in this chapter. This layer is the most hydrologically active since it is where rainfall enters and where the plant roots are located, so it largely controls the overall seasonal and shorter-term water balance. Deep, permeable soils can absorb and store large quantities of water, providing a reserve through times of drought and helping to produce a more even pattern of river runoff. In contrast, where soils are thin or impermeable, rainfall runs quickly off the surface, and little water is held in the soil to sustain plants and animals until the next rain.

6.2 PHYSICAL PROPERTIES OF SOILS AFFECTING SOIL WATER

The **soil profile**, i.e. a vertical cross-section through the soil, normally comprises a number of layers or **horizons** having different physical characteristics. The nature of the soil profile depends upon a wide range of factors including the original parent material, the length of time of development and prevailing climate, as well as the vegetation and topography.

Three **phases** make up the soil system: the solid phase, or **soil matrix**, comprising the mineral and organic particles of the porous medium, and in the voids between a varying mix of the gaseous phase of the soil air and the liquid phase of the soil water. The latter is sometimes, more correctly, referred to as the soil 'solution' because it always contains some dissolved substances. The soil-water system is complex because soil properties often vary over short distances and may change through time, due to factors such as swelling and shrinking of clays, and compaction and disturbance by plants, animals and humans. For these reasons much of the theoretical basis of soil physics was developed assuming 'idealized' media such as glass beads or conceptual models based on bundles of capillary tubes. The challenge for the hydrologist is to apply these concepts to field situations. This chapter aims to provide an introduction to some of the most important phenomena of soil water theory and their practical consequences for the hydrologist.

The amount of water that can be held in a given volume of soil and the rate of water movement through that soil depend upon both the soil **texture**, i.e. the size distribution of the mineral particles of the soil, and upon the soil **structure**, the aggregation of these particles and larger features including cracks, worm and root cannels that are collectively known as **macropores**. Water may occupy the pores between the particles and the larger voids between aggregates. At high water contents water flow through the latter may be dominant, but as the soil becomes drier they rapidly become less important. In general, the coarser the particles or aggregates, the larger the intervening voids and the easier it will be for water movement. Thus sandy soils tend to be more freely draining and permeable than clay soils, which are both slower to absorb water and slower to drain.

Soil structure results from the aggregation of the primary particles into the structural units, or **peds**. These are separated from one another by planes of weakness which may also act as important flowpaths for water moving through the soil profile. The mechanics of soil structure formation and stability are very complicated and depend on a number of factors. In the surface horizon of the soil the aggregates will alter over time due to weather and soil tillage, but in deeper horizons they will be more constant. Plants are very important for the structure of surface horizons since their roots bind particles to help form stable aggregates. Grasslands for example, are particularly effective due to the high density of roots near the surface (White, 2005). Soil structure is too varied for simple geometric characterization and is usually described qualitatively in terms of form: granular, blocky, prismatic, etc., and by the degree of development, whether structureless or strongly aggregated (e.g. USDA, 2015). However, such descriptions are often only weakly related to soil hydraulic properties, especially in structured clay soils.

Finely divided clay material is the most important size fraction in determining the physical and chemical properties of the soil. In contrast to the sand and silt fractions which comprise mainly quartz and other primary minerals that have undergone little chemical alteration, the clays result from chemical weathering, forming secondary minerals with a great variety of properties (White, 2005; Gregory and Nortcliff, 2013). One key difference is that the clay particles comprise platey sheets and have a much higher **specific surface**, i.e. the surface area per unit volume (Brady and Weil, 2007). Most clays have negatively charged surfaces and are balanced externally by cations which are not part of the clay structure and which can be replaced or exchanged by other cations (see Section 8.5.2). Some types of clay have only weak bonds between adjacent sheets, and the 'internal' surfaces may also be available for taking part in adsorption. Water can enter between these sheets causing them to separate and expand. This is important for the retention and release of nutrients and salts. Many clay soils swell on wetting, and shrink and crack on drying, which can be important for the porosity and hydraulic properties of soils.

This broad description of the main soil properties provides a basis for subsequent Sections. These first describe the processes governing the storage and movement of water in idealised soil conditions (Sections 6.3 and 6.4) and then provide examples of the resulting patterns of soil moisture and flow rates that are found under naturally occurring field conditions (Section 6.5).

6.3 STORAGE OF SOIL WATER

If gravity were the only force acting on soil water then the soil would drain completely after each input of rain so that water would be found only below the water table. If that were the case, plant growth would be restricted to areas with very frequent rainfall or locations where the water table was close to the surface. In fact soils in natural conditions always contain some water, even at the end of long dry periods of many months or even years. This indicates that very powerful forces are retaining water in the soil and playing a vital role in supporting life on Earth.

6.3.1 WATER RETENTION FORCES

The main forces responsible for holding water in the soil are those of **capillarity**, together with **adsorption** and **osmosis**. Capillary forces result from **surface tension** at the interface between the soil air and soil water. Molecules in the liquid are attracted more to each other than to the water vapour molecules in the air, resulting in a tendency for the liquid surface to contract. If the pressures were exactly the same on either side then the air-water interface would be flat, but pressure differences result in a curved interface, the pressure being greater on the inner, concave, side by an amount that is related to the degree of curvature. At the interfaces in the soil pore space, the air will be at atmospheric pressure, but the water may be at a lower pressure. As water is withdrawn from the soil the pressure difference increases across the interfaces which become increasingly curved and can only be maintained in the smaller pores (Figure 6.1). The force with which these small wedge-shaped films of

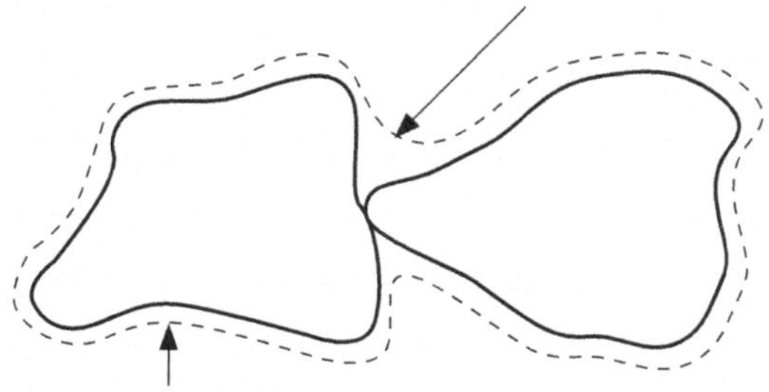

Fig 6.1: Water is held by adsorption as a thin film of water on soil particles, and by capillary forces between them.

water are held will vary, in the same way the capillary rise in a glass capillary tube varies with the radius of the tube, the curvature of the surface meniscus and the surface tension of the water. For a given viscosity and surface tension, water will be held more strongly in smaller pores than larger ones. Hence as the water content of a soil reduces, the larger pores empty at lower suctions than the smaller pores.

In addition to capillary forces, water molecules can be adsorbed upon the surfaces of soil particles mainly due to electrostatic forces in which the polar water molecules are attached to the charged faces of the solids. Since the forces involved are effective only very close to the solid surface, only very thin films of water can be held in this way (Figure 6.1). Nevertheless, if the total surface area of the particles (i.e. the specific surface) is large, and/or the charge per unit area is large, then the total amount of water adsorbed in a volume of soil may be significant.

The magnitude of the specific surface depends upon the size and shape of the particles. Its value increases as the grain size decreases and as the particles become less spherical and more flattened. Clay size particles and organic matter contribute most to the specific surface area of a soil, with values of under 0.1 m^2 g^{-1} for sand, rising to over 800 m^2 g^{-1} for expanding layer clays such as montmorillonites. This helps to explain the very strong retention of water by clays during prolonged periods of drying.

These forces of attraction between water and soil reduce the free energy of the water, so that, in unsaturated soil, the pressure of water in the pores is negative (i.e. less than atmospheric pressure) and so both capillary and adsorption forces may be regarded as exerting a **tension** or **suction** on the soil water. Their comparative importance depends upon soil texture and soil water content and in practice they are in a state of equilibrium with each other and cannot be easily measured separately. It is therefore usual to deal with their combined effect on the way in which water is held in the soil matrix, known as **matric suction** or **matric potential** (See also Section 6.3.5).

A third force which acts to retain water in the soil results from osmotic pressure due to solutes in the soil water. Although often ignored, especially in humid environments, osmotic pressure may be important when there is a difference in solute concentration across a permeable membrane such as at a plant root surface. It acts to allow the movement of water from the more dilute to the more concentrated solution, and so especially in saline soils will make soil water less available to plants (Hillel, 2003).

A combination of matric and osmotic forces are responsible for holding water in the soil, and these retention forces vary with water content. In general, soils contain a wide range of pores of varying shapes and sizes. Those with large entry channels empty at low suctions while those with narrow channels empty at higher suctions. The relationship between soil moisture suction and water content is clearly of fundamental importance to an understanding of soil water behaviour.

6.3.2 SOIL WATER CHARACTERISTICS (RETENTION CURVES)

The relationship between suction and the amount of water remaining in the soil can be determined experimentally in the laboratory using cores of soil. The resulting function is known as the **water characteristic** (or **retention curve** in the case of a drying soil). Examples of water characteristics obtained for different types of soil are shown in Figure 6.2. The shape of the curve is related to the pore size distribution. In general, sandy soils show a more rapid decrease in water content with reducing matric potential than clayey soils since most of the pores are relatively large and once they have emptied there is little water remaining. Clayey soils, in contrast, have a wider distribution of pore sizes and consequently have a more uniform slope. The mechanism of water retention varies with suction. At very low suctions it depends primarily on capillary surface tension effects, and hence on the pore size distribution and soil structure. At higher suctions (lower moisture contents) water retention is increasingly due to adsorption, which is influenced more by the texture and the specific surface of the material. Due to the greater number of fine pores and the larger adsorption, clays tend to

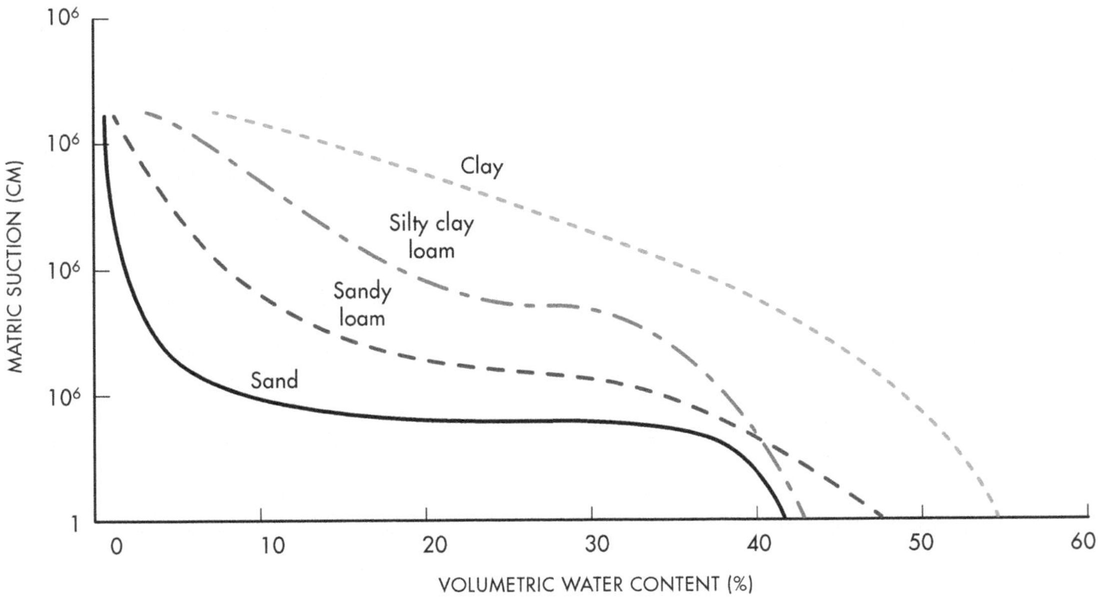

Fig 6.2: Soil water characteristic curves of different soil materials (redrawn from an original diagram by Bouma, 1977).

have a greater water content at a given suction than other soil types.

At matric potentials close to zero, the soil is close to saturation and water is primarily held by capillary forces. As water content reduces, water is held in progressively smaller pores, mainly by adsorptive forces. When an increasing suction is first applied to a saturated soil, little or no water may at first be released. A certain critical suction must be achieved before air can enter the largest pores enabling them to drain. The pressure at which drainage begins during desaturation is called the **bubbling** or **air entry pressure** and is determined by the largest pores. The critical suction will obviously be greater for fine-textured material such as clays since their narrow pores supporting air-water interfaces of much greater curvature will not empty until larger suctions are imposed. As coarse sands often have a more uniform pore size, their water characteristics may show the air entry phenomenon more distinctly than finer-textured soils.

The fact that the suction required to drain a pore varies inversely with its radius means that the slope of the moisture characteristic can be used to indicate the 'effective' pore size distribution of the soil. If a gradually increasing suction is applied to a soil sample, then the volume of water withdrawn from the sample during each increment of suction represents the volume occupied by those pores whose diameter corresponds to that range of suction. In making such an estimate it must be remembered that at high suctions adsorptive rather than surface tension forces predominate and that, due to the often tortuous flow paths through the medium, not all pores of a given size will empty at the same time. A large water-filled pore may be surrounded by smaller pores, and so cannot drain until these smaller pores drain first and air can pass through to the large pore. This phenomenon can lead to jumps in the water characteristic, especially at low suctions.

6.3.3 HYSTERESIS

One of the main limitations to the use of water characteristic curves is that the water content at a given suction depends not only on the value of that suction but also on the previous pattern of moisture changes. It will be greater for a soil that is

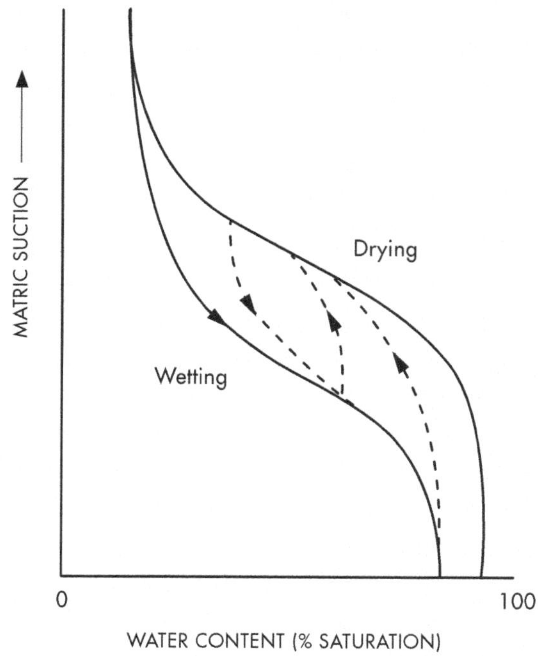

Fig 6.3: Hysteresis in the soil water characteristic, showing the main wetting and drying boundary curves (solid lines), and the intermediate or 'scanning' curves (dashed lines).

being dried than for one that is being re-wetted (Figure 6.3). This dependence on the previous state of the soil water is called **hysteresis**. Pores empty at larger suctions than those at which they fill, and this difference is most pronounced at low suctions and in coarse-textured soils.

Hysteresis arises mainly from differences in the process of emptying and filling individual pores. Two important causes of hysteresis are the 'ink bottle' effect and the 'contact angle' effect, both of which are dependent on pore behaviour. The former results from the fact that during drying a larger suction is necessary to enable air to enter the narrow pore neck, and hence drain the pore, than is necessary during wetting, which is controlled by the lower curvature of the air-water interface across the wider pore itself (Childs, 1969). The 'contact angle' effect results from the fact that the contact angle of the water interface on the soil solids tends to be greater when the interface is advancing (i.e. wetting) than when it is receding (drying), so a given water content tends to be associated with a greater suction in drying than in wetting. Entrapped air will also decrease the water content of newly wetted soil, and the failure to attain true equilibrium in experimental conditions may create or accentuate the hysteresis phenomenon. Investigators have found that the water content at zero suction in field soils may be only 80–90% of the total porosity, with the remaining pore space occupied by air (Klute, 1986b), although normally surface roughness and macropores would be sufficient to allow air to escape. In field experiments people have tended to apply water at very high intensities, often resulting in rapid surface ponding, which would make air entrapment more likely.

In fine-grained clay soils, wetting and drying may be accompanied by swelling and shrinkage. This leads to changes in pore sizes and the overall bulk density, and hence to a different volumetric water content at a given suction than if the matrix had remained stable and rigid. As water is withdrawn from the interstices between the plate-like particles (lamellae), the particles move closer together, thus reducing the overall volume. In some conditions it appears that as the lamellae are drawn closer together, they may reorientate themselves; on subsequent rewetting they may not necessarily return to their original alignments, resulting in a lower volumetric water content. However, the swelling and shrinking behaviour of clay soils reflects not just their water content and suction but also the interaction of attractive and repulsive forces between the lamellae. This, in turn, is affected by the composition and concentration of the soil solution as well as by the type of clay.

Methods for modelling hysteresis intermediate **scanning curves** were reviewed by Pham et al, (2005) and may be applied to soils for which the boundary wetting and drying curves are already known. In practice, however, given the many problems associated with measuring the water characteristic accurately, the hysteresis phenomenon is usually ignored (Hillel, 2003). Also, although the pore size distribution is related more closely to the wetting curve (governed by the size of the pore entry channels), the drying (retention) curve is much easier to measure experimentally and is therefore used more frequently. For this reason the terms *water characteristic* and *water retention* curve are often used interchangeably.

6.3.4 SOIL WATER 'CONSTANTS'

A number of soil water 'constants' have traditionally been used to facilitate comparisons between the hydrological status of different soils. They are assumed to correspond to particular values of matric suction but, since the water characteristic curve is a continuous function, such points are plainly arbitrary and may have little intrinsic significance. Furthermore, under natural conditions rainfall and evaporation will seldom, if ever, permit soil water content to reach a state of equilibrium over the whole soil profile. Nevertheless, despite their arbitrary nature, soil water constants have long been used in solving practical soil water problems such as those concerned with drainage, irrigation and the modelling of watershed hydrology.

The 'constants' most frequently cited are the **wilting point** and the field capacity. The wilting point is defined as the minimum water content of the soil at which plants can extract water. Although this varies between plant species and with their state of growth, the actual difference in the amount of water is quite small at such low water contents. **Field capacity** is normally regarded as the maximum amount of water held in the soil after gravity drainage of water has largely ceased, in the absence of evaporation. Although some soils continue draining over many weeks, field capacity is commonly taken to reflect the soil water content 48 hours after rainfall or irrigation had thoroughly wetted the soil. In practice, permeable soils drain more rapidly than less permeable soils and achieve a state of little further water content change more quickly and at much lower suctions (Smedema et al, 2004). It is also assumed that the water table is sufficiently deep so as to have no influence on the water content of the soil profile. Despite such limitations, soil water 'constants' are widely used (e.g. Rab et al, 2011) and the difference in water content between field capacity and wilting point, for instance, provides a useful and practical approximation to the **available water capacity** for plants growing in different soils (Figure 6.4).

While these concepts are useful for making broad comparisons and generalizations, a fuller understanding of the behaviour of soil water requires consideration of the dynamic nature of the system. This needs to be based on the concepts and laws of soil physics.

6.3.5 SOIL WATER ENERGY (POTENTIAL)

Soil water can be thought of having a **potential energy** resulting from its position and from retention forces. Apart from macropore or pipe flow, soil water moves so slowly that its kinetic energy is insignificant. Differences in potential energy at different points in the soil are very important in determining water movement. Soil water will flow in response to a gradient from points of higher to lower water potential, as will be shown in Section 6.4.

For most practical purposes the total soil water potential or **hydraulic potential**, comprises the sum of the gravitational potential (ψ_g), the matric or pressure potential (ψ_p) and osmotic potential (ψ_0):

$$\phi = \psi_g + \psi_p + \psi_o \qquad (6.1)$$

Apart from the inclusion of osmotic potential, Eq. 6.1 is identical to Eq. 5.4 for groundwater, emphasising that the gradient of potential energy for subsurface water is continuous throughout the full depth of the unsaturated and saturated zones. Gravitational potential increases with elevation and, in the absence of strong retention forces or impeding factors, water will clearly drain downwards from higher to lower elevations. However, because of water retention forces (Section 6.3.1), pressure potential is negative in the unsaturated zone above the water table and it varies depending upon gains and losses of water due to rainfall and evaporation. Osmotic potential resulting from solutes in the soil water is also negative; in dry conditions it acts to retain water in the soil and therefore to lower the total potential.

Water moves from a point where the total potential energy is *higher* to one where it is *lower*. It is thus not the *absolute* value of potential energy that is important, but rather the *relative* values at different points in the soil. The difference in total potential between points depends upon differences in both retention forces and in elevation, which may not necessarily act in the same direction.

In describing the potential energy distribution in the *saturated* zone it is usual to express energy values in relation to an arbitrary datum level set well below the flow field under consideration so that they are always positive numbers (see Section 5.5).

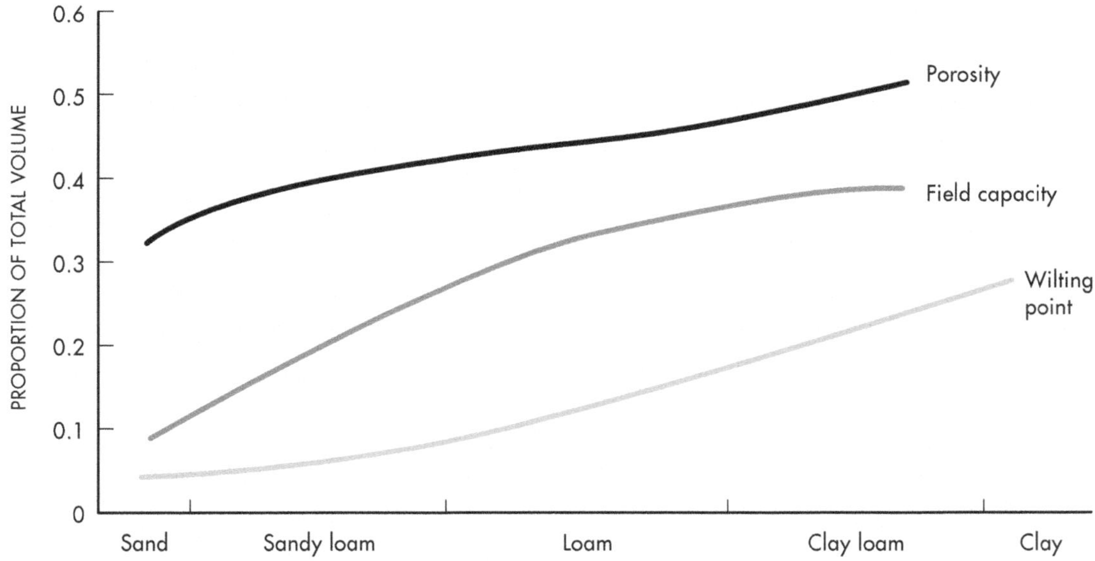

Fig 6.4: Schematic relation between the main oil water constants showing the total porosity, field capacity and permanent wilting points for different soil textures. (From an original diagram in USDA (1955) Yearbook of Agriculture for 1955: Water, USGPO).

In studies of the *unsaturated* zone, however, since they are above a fluctuating water table it is more common to use the ground surface as the datum level for soil water energy values. Exceptionally, for a 'snapshot' view of the flow field, or in conditions of stable water table, the water table level itself can be used, which has the advantage that both gravitational potential and pressure potential are zero at this datum level. However, the dynamic nature of the water table means that this is rarely feasible.

An example of a total energy profile, derived using the ground surface as the datum level, is shown in Figure 6.5. Gravitational potential (ψ_g) declines uniformly with depth below the ground surface in both the saturated and unsaturated zones. All values are therefore negative when referred to a ground surface value of zero. Above the water table the pressure potential (ψ_p) is negative and depends upon water content; below the water table pressure potential increases with depth. Total soil water potential (ϕ) is the sum of gravitational potential and pressure potential, and in this example becomes increasingly negative at the ground surface due to drying by evaporation in the plant root zone. It is less negative in the moister soil layers below the root zone but then, despite increasing soil water content, becomes increasingly negative towards the water table as a result of the declining gravitational potential. Below the water table, conditions are identical to those in Figure 5.8, so that there is no change of total potential with depth as the increase of pressure potential (ψ_p) is exactly offset by the reduction in gravitational potential (ψ_g).

The most widely used technique for measuring matric or pressure potential are **tensiometers**. They comprise a pressure sensor inside a water-filled porous cup embedded in the soil *in situ*, so water can flow between the soil and cup until the pressure potential inside the cup equalizes with

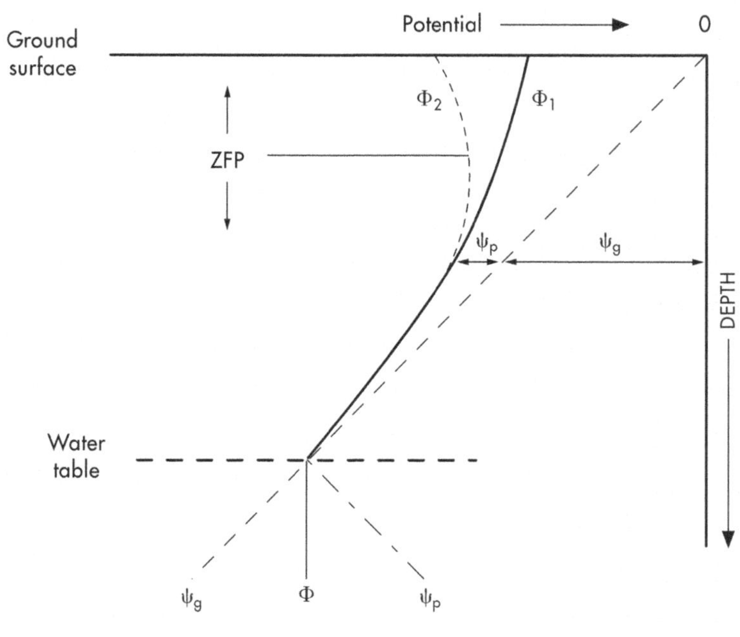

Fig 6.5: Total soil water energy profile (Φ), showing the gravitational (ψ_g) and pressure potential (ψ_p) components below the ground surface.

that of the soil water. Tensiometers can also measure pressure heads below the water table, in which case they operate as piezometers, which can be useful when the water table fluctuates within the depth of interest. The lowest pressure that can be measured by tensiometer is about -800 cm (80 kPa) due to the effervescence of dissolved gases out of the water breaking the film of water covering the porous tip, which allows air to enter the tube. When this occurs, the tensiometer must be refilled with water before it will again function properly.

In the profile of total potential shown in Figure 6.5 the total potential (shown by the solid line, ϕ_1) increases (becoming less negative) upward to the surface. Water will move from higher to lower total potential, so this indicates steady downward infiltration through the unsaturated zone, such as following recent rainfall. In time, however, evaporation from the surface layers will reduce the pressure potential (shown by the short dashed line, ϕ_2) so there will be a depth in the unsaturated zone where there is no gradient of total potential. At this depth, known as the **zero flux plane (ZFP)**, there will be no vertical soil water movement. In this example, the ZFP divides the soil profile into two zones: a zone of upward flux above, and of downward flux below it. The movement of water in the unsaturated zone, and the significance of the ZFP are discussed in the next Section.

6.4 MOVEMENT OF SOIL WATER

It was shown earlier that soil water moves in response to a number of forces. Since gravity is not necessarily the dominant force, unsaturated flow can occur in any direction. There is, however, a tendency for the main controlling forces to operate either from the ground surface (infiltration, evaporation) or from the bottom layers of the zone of aeration (groundwater level fluctuations, capillary rise). This leads to the development of soil water potential gradients in the vertical direction, so that vertical movement of soil water usually predominates in subdued topography or permeable strata. The following Sections deal with the general principles of flow in the unsaturated zone, and with vertical water movement - either upwards or downwards. The factors limiting the rate at which water can infiltrate into a soil are then discussed, as this controls the partitioning of net precipitation into surface and subsurface flow paths. At the end of the Chapter the roles of topography and soil layering are considered, looking at lateral soil water flow down slopes and the resulting spatially variable patterns of soil water content and movement that are observed in the field and are important in many areas for runoff production.

6.4.1 PRINCIPLES OF UNSATURATED FLOW

We know that soil water will move from regions of higher total potential to regions of lower total potential (see Section 6.3.5). For most practical purposes, the total potential (ϕ) is, taken as the sum of the pressure potential (ψ) and the gravitational potential (z). Using the ground surface as the datum level, the way in which these two potentials combine to affect water movement in a soil over a time period is simply illustrated by reference to two tensiometers at depths of, say, 40 and 60 cm (Figure 6.6). The datum level is the ground surface and in this example the soil is initially unsaturated, so both the matric and elevation terms of total potential are negative. Note that at each measurement depth the gravitational potential has a constant value equal to that depth.

At time A, the matric suction is the same at both depths, but due to the difference in elevation, water flow in the soil layer between these points will occur downwards, from higher to lower (i.e. more negative) potential. Over time the upper soil dries more quickly than that at greater depth until, at time B, the difference in matric suction balances the elevation difference. The total potentials are the same, hence there is no flow between these depths, and the ZFP will be at a depth between them. If the upper soil continues to dry, its potential will become more negative than that of the deeper soil and upward flow will occur from 60 to 40 cm depth between times B and C. After an input of rain, the soil becomes wetter, matric suctions reduce and, in this example, the water table rises to within 60 cm of the ground surface. This is indicated by the shaded area at about time D, and demonstrates the usefulness of the pressure potential for sites where both saturated and unsaturated conditions are involved.

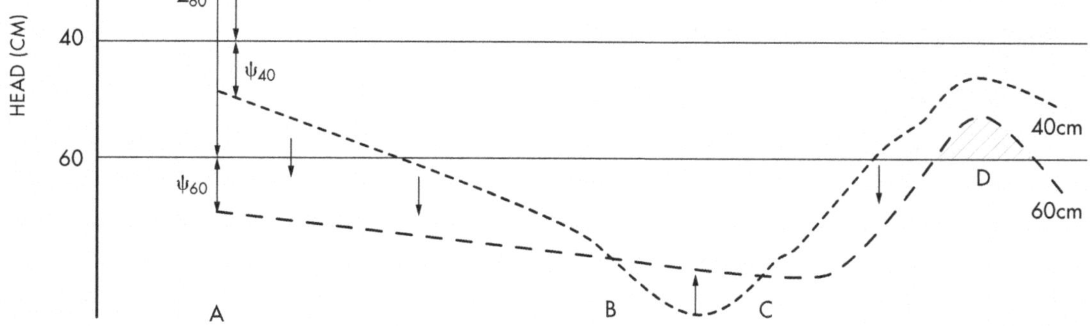

Fig 6.6: Diagram showing the changes in matric and elevation components of total potential over time at two depths: 40 cm (short dashed line) and 60 cm (long dashed lione) and the resulting changes in vertical soil water fluxes between these two depths. (See text for details). (From a diagram in Ward and Robinson, 2000).

The annual changes in soil water movement and ZFP depths in typical temperate climate conditions are shown in Figure 6.7 for an array of tensiometers in the soil profile.

Figure 6.7a, shows the profile at three different times. Time A corresponds to the situation in Figure 6.5, where the total potential is consistently decreasing with depth, indicating downward movement through the whole soil profile to the water table. At time B, as a result of evaporation exceeding rainfall, the upper layers of the soil begin to dry out, leading to an increasingly negative pressure potential. This results in the total potential values reducing near the surface, as well as at greater depths, indicating upward flux to the surface as well as continuing downward drainage in the lower profile. The two zones of the soil are separated by a **divergent ZFP**. Further recharge to the water table will be limited to water already in the lower part of the soil profile below the level of the ZFP.

Such a divergent ZFP develops at the soil surface, and moves downward into the soil during spring and summer as the profile continues to dry out. In typical British conditions, it tends to stabilise at a depth of between 1 and 6 m, depending on climate and soil conditions and the depth of the water table (Wellings and Bell, 1982). As rainfall begins to exceed evaporation in the autumn, the surface soil layers become wetter and their total soil water potential increases. This causes a second, **convergent ZFP** to develop at the surface (Time C) which then move fairly rapidly down the profile until it meets the original, divergent, ZFP. At this point both ZFPs disappear and downward drainage of soil water occurs throughout the profile during the winter months, just the same as at the beginning of the year.

The whole annual cycle of the water flux in a temperate-climate soil profile is illustrated in Figure 6.7 b. Together with measurements of soil water content and soil water balance, such information can be used to quantify both the amount of deep percolation downwards to the groundwater, and also the upward flux due to evaporation (Moser et al., 1986; Ragab et al., 1997). The presence of a ZFP in a soil profile indicates that no direct recharge is taking place from the surface through the soil matrix to the water table, although it does not exclude the possibility of macropore flow (Hodnett and Bell, 1986). An example of flow dynamics for a non-temperate climate with strongly seasonal rainfall, *and* macropore flow, is given later in Figure 6.13.

Darcy's law (Section 5.5.1), stating that the rate of water movement through a saturated porous medium is proportional to the hydraulic gradient, is also applicable to the flow of soil water in unsaturated conditions. The Darcy equation may be simply expressed for unsaturated conditions as

$$v = -K(\theta)(\delta h/\delta l) \qquad (6.2)$$

where v is the macroscopic velocity of water, K is the hydraulic conductivity (which, in unsaturated conditions, varies with the water content, θ, and $\delta h/\delta l$ is the hydraulic gradient comprising the change in hydraulic head (h) with distance along the direction of flow (l). As indicated earlier, the potential gradient and the resulting flow may be in any direction.

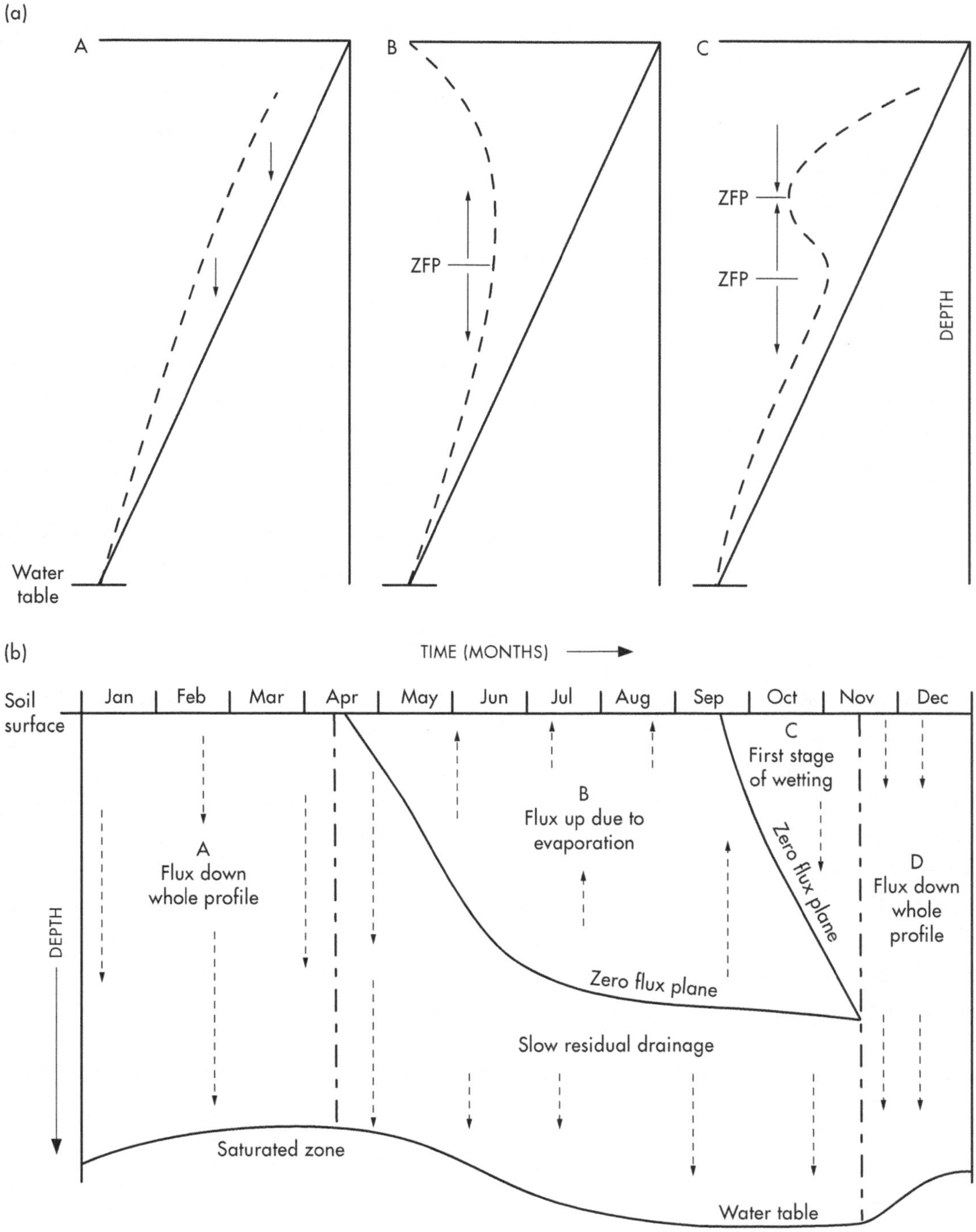

Fig 6.7: Water fluxes in a temperate climate soil profile fluxes. (a) Time sequence following rain – A: drainage throughout the whole unsaturated profile to the water table; B: formation of a divergent ZFP due to evaporation; C: Following rainfall wetting the surface a convergent ZFP is formed. (b) Annual cycle of water movement and zero flux plane development redrawn (after Wellings and Bell, 1982).

The flux calculation may be illustrated in the two simplest cases for:

a. purely *horizontal* flow in the *x* direction, i.e. only a matric potential gradient ψ, and no gravity gradient:

$$v = -K(\theta)\ \delta\psi / \delta x \qquad (6.3)$$

and

b. purely *vertical* flow downwards in the *z* direction, and with a matric potential gradient:

$$v = -K(\theta)\ \delta(\psi - z)/\delta z$$

$$= -K(\theta)\ [(\delta\psi / \delta z) - 1] \qquad (6.4)$$

The application of the Darcy equation, or equations derived from it, to water movement in the unsaturated zone is subject to most of the stipulations noted in Section 5.5.1, not least in respect of its description of 'macroscopic' flow, and its restriction to steady-state situations, where the gradient and flux do not change, or change only slowly over time. Water movement may alter the gradient and the value of hydraulic conductivity and so, for field situations in which flow varies with space and time, Richards (1931) combined Darcy's equation with the continuity equation ($\delta\theta/\delta t = -\ \delta v/\delta z$) to yield the important non-linear equation that bears his name:

$$\delta\theta/\delta t = \delta/\delta z\ [K(\theta)\ (\delta\psi/\delta z) - 1] \qquad (6.5)$$

where *t* is time, and flow is vertically downwards.

Water vapour also moves through the soil matrix, from warm soil to cold soil, as a result of vapour pressure differences. This is much smaller than the mass flow of liquid water and so its contribution to total soil water movement can usually be ignored.

Hydraulic conductivity

The hydraulic conductivity of saturated soil, like that of other saturated porous materials, depends mainly upon the geometry and distribution of the pore spaces (see Section 5.5.2). These include, not only textural voids but also macropores, such as interstructural cracks and root channels, which may greatly influence the hydraulic conductivity. This can be illustrated for a hypothetical clay soil with a **textural porosity** of 1% and a matric conductivity of 0.25 cm day^{-1}. Macropores, such as 1 mm wide cracks at 10 cm intervals, could develop seasonally and contribute an additional **structural porosity**. Although this might also be only 1%, its effect could be to increase the saturated hydraulic conductivity of the soil by a factor of 10,000, to 2,500 cm day^{-1} (Childs, 1969). Clearly then, even though saturated hydraulic conductivity has been correlated in some studies with soil texture and with descriptions of structure (McKeague et al., 1982; Rawls et al., 1982), such **pedotransfer function** relations can be risky, and the resulting estimates of K may be seriously in error.

By definition, the zone of aeration is not normally saturated, so that soil water movement is usually controlled by the *unsaturated* hydraulic conductivity ($K(\theta)$), which varies with soil water content (equation 6.2). Whereas saturated hydraulic conductivity

(K) may be regarded as more or less constant for any given material, K(θ) will vary with soil water content and therefore with matric suction. Figure 6.8 shows that, for several soil types, hydraulic conductivity is largest at or near saturation and decreases rapidly with reducing water potential. This reduction results from the fact that soil water movement can take place only through existing films of water on and between the soil grains. In saturated soil, all the pore spaces form an effective part of the water conducting system. In unsaturated soil, air-filled pores act as a non-conducting part of the system, reducing the effective cross-sectional area available for flow. The greater the decrease in soil water content, the more that flow is confined to the smallest pores and the greater the reduction in effectiveness of the conducting system and the smaller the value of hydraulic conductivity.

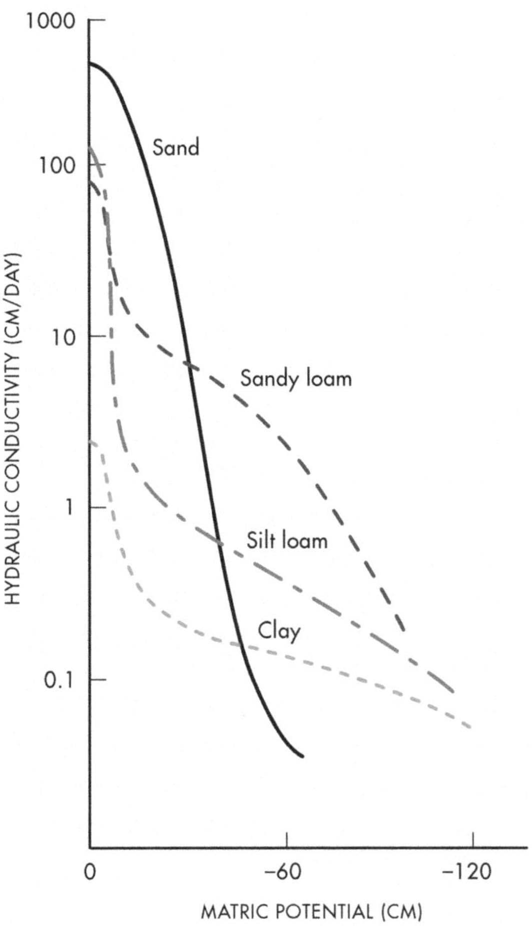

Fig 6.8: Unsaturated hydraulic conductivity as a function of pressure potential (from Bouma, 1977).

The way in which K(θ) varies with soil water content is greatly influenced by the pore size distribution of the soil. In wet soils, conductivity is broadly related to soil texture, and increases as the latter becomes coarser because water is transmitted more easily through large water-filled pores. Near-saturated sand has a high proportion of large water-filled pores, so that K(θ) is larger than in clay soils. As the soil dries, the larger pores are the first to empty, and K(θ) falls rapidly. As the suction increases and the water content decreases, the relationship between conductivity and texture is reversed so that, in dry conditions, clay soils are likely to have a higher conductivity than loamy or sandy soils. This reflects the fact that, at low potentials, finer-textured soils have more water-filled pores and, therefore, a larger cross-sectional area through which flow can take place, than coarser soils in which only a small proportion of the pores contain water. In shrinking soils, the increased suction that accompanies drying, reduces the size of the pores that remain full of water (Section 6.3.3), and this further helps to reduce the hydraulic conductivity.

The measurement of soil hydraulic conductivity is discussed in various texts (e.g. Cooper, 2016). Some studies have shown that values obtained from small soil cores in the laboratory may be unreliable as they depend on the chance inclusion of macropores in the sample volume. The field measurement of K(θ) in unsaturated soils is difficult and expensive, so estimates, based on more readily available soil properties are often used. A simple approach is that based on soil texture, but it is clear, that while texture is a major determinant, many other factors may be important for a particular case. Better estimates of K (θ) may be expected where the water characteristic is available, and many studies have attempted to relate the two (e.g. Van Genuchten, 1980; Mualem, 1976). Ideally, predicted conductivities should be matched with measured values near saturation, although of course, none of these methods is applicable to soils in which the hydraulic conductivity under saturated conditions is dominated by water flow through macropores rather than through the soil matrix.

6.4.2 INFILTRATION OF WATER INTO SOILS

The term **infiltration** is used to describe the process of water entry into the soil through the soil surface, as distinct from percolation or groundwater recharge that refer to water movement down at deeper depths. The maximum rate at which water soaks into, or is absorbed by the soil, its **infiltration capacity**, occurs when the soil at the surface is saturated. This may in certain circumstances be very important in determining the partitioning of precipitation falling on a catchment area. If the rainfall intensity exceeds the infiltration capacity then the excess rain will flow over the ground surface, possibly directly into streams and rivers. Once in the soil the water may move laterally as throughflow, or be retained temporarily before moving downwards as percolation or upwards to become evaporation to the atmosphere.

Although sometimes used interchangeably, the terms **infiltration capacity** and **infiltration rate** should be differentiated. The actual infiltration rate will equal the lower of either the infiltration capacity of the soil surface to absorb water, or by the rate of supply of rainfall and irrigation. It is important to remember that neither term indicates the effective velocity of vertical water movement in the soil (see also the discussion of groundwater movement in Section 5.5.1).

Many early field investigations of infiltration (e.g. Horton, 1933, 1939) were carried out in semi-arid areas, where it is quite common for rainfall intensity to exceed infiltration capacity, resulting in widespread surface ponding and overland flow. As a result, for many years, undue importance was attached to the hydrological role of infiltration capacity. Today, it is recognized that in well-vegetated areas, both temperate and tropical, most soils can absorb the rainfall of all but the most intense storms and that widespread overland flow is uncommon (see Section 7.4). Nevertheless, an understanding of the reasons for temporal and spatial variations of infiltration capacity is still important in situations with very high rates of water inputs, such as during artificial irrigation or in extreme rainstorms.

Infiltration capacity

The infiltration capacity of a soil generally decreases during rainfall, rapidly at first and then more slowly, until a more or less stable value has been attained (Figure 6.9). This decline of infiltration capacity is determined by a number of factors, including limitations imposed by the soil surface, surface cover conditions and the rate of downward movement of water through the soil profile.

Soil surface conditions may impose an upper limit to the rate at which water can be absorbed, despite the fact that lower soil layers are able to receive and to store more water. In general the infiltration capacity is reduced by surface sealing resulting from compaction, the washing of fine particles into surface pores and by frost (e.g. Romkens et al., 1990). Infiltration capacity increases with the depth of standing water on the surface, and the number of cracks and fissures at the surface. Cultivation techniques may either increase or decrease infiltration capacity, and vegetation tends to increase the infiltration capacity of a soil by slowing surface water movement, stabilising loose particles, reducing raindrop compaction and improving soil structure (Section 6.2).

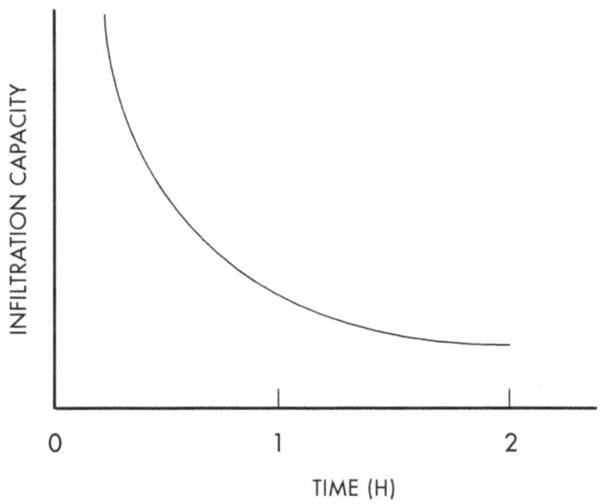

Fig 6.9: Schematic diagram showing the typical decline in infiltration capacity.

These effects are clearly demonstrated by the way in which a dense vegetation cover encouraging good soil structure normally results in higher infiltration. In contrast, frozen soils and soil compaction generally reduce infiltration capacity.

Water cannot continue to be absorbed by the soil surface at a given rate unless the underlying soil profile can conduct the infiltrated water away at a corresponding rate. The ability of a given soil to conduct water depends upon its properties, including texture and structure, and so is normally greater for permeable, coarse-textured, soils than for slowly permeable clays. Other important factors include soil stratification, the initial gradient of the soil water potential and the lengthening flowpaths of water as infiltration proceeds during a rainfall or irrigation event.

Soil water movement during infiltration

Soil conditions in the field are notoriously heterogeneous. In addition to the spatial variability of soil properties, there is frequently both spatial and temporal variability of soil water content, hysteresis, changes of various soil and boundary conditions over time and the existence of two- or three-dimensional flows. As a result, infiltration capacities and infiltration rates typically vary enormously in both space and time. For this reason, the development of infiltration theory was largely based on the investigation of water entry into homogeneous and sometimes artificial soils, assuming a uniform initial water content and the presence of ponded water at the soil surface.

Soil water profiles of the entry of water into such soils were derived from theoretical considerations by Philip (1964). Figure

6.10 shows computed profiles, during sustained surface flooding of a clay loam soil, as a function of the time since the start of infiltration. The increase in profile water over time consists mainly of an extension of a nearly saturated transmission zone. The sharp change in water content in the wetting front is a consequence of the dependence of hydraulic conductivity on water content. From Darcy's law it is clear that a steep hydraulic gradient is necessary in this zone to achieve a flux equal to that in the (near-saturation) transmission zone.

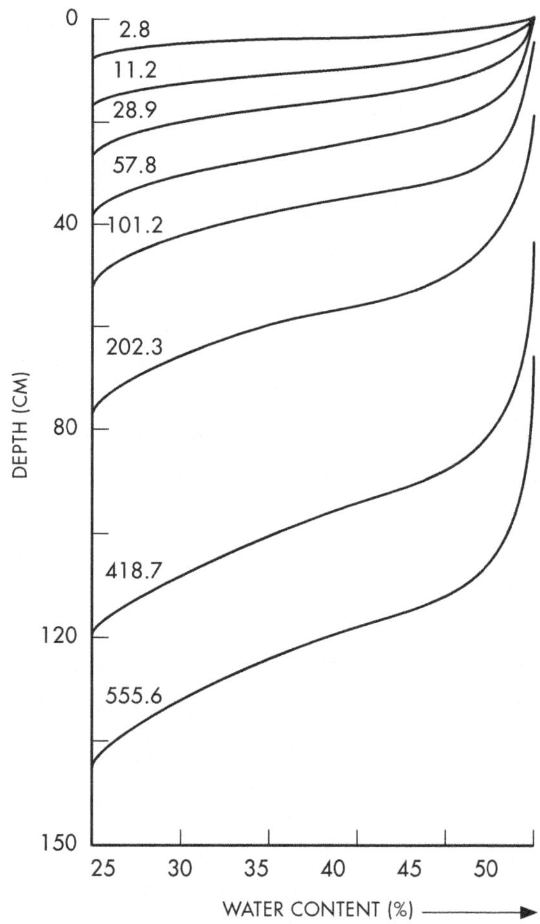

Fig 6.10: Computed soil water profiles (at times in hours) during ponded water infiltration into a clay loam soil (from original diagram by Philip, 1964)

The preceding discussion suggests that, in the early stages of infiltration into a uniformly dry soil, the matric suction gradient in the surface layer will be very steep, and is likely to be the most important factor determining the amount of infiltration and downward movement of water. Initial rates of movement are likely to be high, and the resulting rapid penetration of the water profile is clearly illustrated in Figure 6.10. As the wetting proceeds the transmission zone lengthens and the gravitational gradient becomes relatively more important than the matric suction gradient. The rates of infiltration and downward movement decrease until the infiltration tends to settle down to a steady, gravity-controlled rate which approximates the saturated hydraulic conductivity (Hillel, 2003).

In well-vegetated areas, the soil infiltration capacity usually exceeds the rainfall intensity so that no surface ponding occurs. In these circumstances the surface soil does not become saturated and the water content increases until it reaches a value at which the unsaturated hydraulic conductivity becomes equal to the rainfall rate. Increasing the rainfall intensity results in a higher water content throughout the profile, and hence a larger conductivity. As in the ponded infiltration case, the water content gradient at the wetting front remains steep. Thus, although reached in different ways, the soil water profiles developed by both ponded infiltration and rainfall-limited infiltration are similar in shape.

Some of the different techniques for measuring infiltration capacity including ponded water ring infiltrometers and rainfall simulators were reviewed by Youngs (1991).

Time variations in infiltration capacity

The typical curve of declining infiltration capacity over time (Figure 6.9) results directly from the reduction in the gradient of water potential in the soil profile. In addition, factors operating at the soil surface, including swelling of clay particles, raindrop impact and inwashing of fine material, may also be important in the case of some clay soils or soils with a sparse vegetation cover. The high initial infiltration capacity reflects the initial steep gradient of soil water potential and the subsequent rapid reduction of that gradient, generally within the first hour or so of rainfall, leads to a near-constant value of infiltration capacity, closely approximating the saturated hydraulic conductivity. Where the soil profile contains a relatively impermeable layer at some depth below the surface, the curve of infiltration capacity may show a further sudden reduction once the available storage in the surface soil horizons has been filled, and infiltration is governed by the rate at which water can pass through the underlying layer of lower saturated hydraulic conductivity.

Hydrologists have developed several semi-empirical equations to describe the variation in infiltration capacity for a uniform soil, and to account for the initial rapid decrease leading to an asymptotic approach to a constant value. Most assume that, if sufficient water is applied to maintain surface ponding conditions, infiltration will be function of either the total amount of water infiltrated (i.e. storage-based) or of time (i.e. time-based). One of the earliest storage-based infiltration equations is that by Green and Ampt (1911). This identifies two flow components: Conductivity flow A is the steady rate of flow driven by the gradient of soil water potential between the wet or flooded surface and the drier soil below. Diffusivity flow B occurs as water spreads into the drier soil ahead of the advancing wetting front. The rate of infiltration f is given as

$$f_t = A + B / S_t \qquad (6.6)$$

where S_t is the volume of water stored in the depth of soil saturated by infiltration at time t, and A and B are empirical constants for given soil texture and moisture conditions.

Although it applies best to simple sands with a narrow range of pore sizes, and a sharply defined stepped wetting profile the equation has been widely used, not least by modellers, and because it is physically based, the parameters can either be evaluated experimentally from infiltration data (e.g. Brakensiek and Onstad, 1977) or estimated, less reliably, from basic soil property measurements (Rawls et al., 1983).

The time-based equation developed by Philip (1957) is an extension of the Green-Ampt approach which allows for a deceleration of diffusivity flow over time as the wetting front nears the water table or the base of the soil profile. The rate of infiltration is given as

$$f_t = A + (S)\, t^{-0.5} \qquad (6.7)$$

where parameter S is called the sorptivity and is the ability of the soil to absorb water by matric forces in the early stages of infiltration. This approach has proved useful in predicting infiltration rates for short periods of time, including the initial decline in infiltration capacity. For longer periods it is necessary to use a more complex approach based on the Richards equation (Kirkby, 1985).

The preceding discussion of the infiltration process has, for simplicity, assumed a distinct and continuous wetting front, and uniform initial soil water content. In practice, although widely used in models, these conditions are rarely found in the field. The macropores present in most soils allow variable penetration of infiltrating water and their role in soil water hydrology is discussed further in Section 6.5.2. In addition, the continued movement of soil water following a period of infiltration means that a uniform soil water profile is unlikely to be present at the onset of the next infiltration event.

6.4.3 SOIL WATER REDISTRIBUTION FOLLOWING INFILTRATION

Continued downward movement of water, caused by the gradients of gravitational and matric potential, may continue long after infiltration at the surface ceases. During this period of soil water redistribution, the transmission zone which existed during infiltration becomes a draining zone as water moves from the infiltration-wetted upper layers of the soil to deeper, drier, layers. This process is important since it controls the quantity of water retained in the plant root zone, the available air-filled porosity for subsequent storage of water in the next rainfall or irrigation event, and recharge to groundwater in the saturated zone.

The rate of soil water redistribution slows down, partly because hydraulic conductivity in the former wetted zone diminishes as a result of decreasing water content, and partly because the gradient of matric potential weakens as the soil water content becomes more uniform. The wetting front continues to move down the profile, but its advance becomes progressively slower and less distinct. After a few days the water content changes only very slowly and the soil is said to be at 'field capacity' (Section 6.3.4). This process is illustrated in Figure 6.11 by the successive soil water profiles found in a fine sandy loam during redistribution in the absence of both evaporation at the soil surface and a water table at the base of the soil column. Since redistribution is a continuing process, there is no unique moment when field capacity is reached. Instead, field capacity is normally inferred when successive soil water profiles show little change or when the rate of change becomes less than a predetermined value.

If there were no interruptions, redistribution could theoretically continue until gravity and retention forces were in balance. Then total soil water potential would be equal at all depths, and there would be no further soil water movement. This is rarely, if ever, observed in field situations either because more rainfall occurs or in the absence of rainfall, evaporation dries the surface soil layers. Successive inputs of new infiltration during the early stages of redistribution complicate the idealised model of soil water distribution just described. A conceptual model for infiltration in such conditions was applied to a variety of soil types by Corradini et al. (1997).

In the case of evaporation from the root zone and soil surface, the resulting gradient of total potential encourages the upward movement of water from the draining zone. With simultaneous evaporation and drainage, hysteresis severely complicates attempts to measure or estimate the process of water redistribution since layers first wet up and then start to dry out. Hysteresis presents a more general problem, however, in the sense that while the upper part of the profile is drying through drainage and evaporation, the lower part is still becoming wetter. The relation between water content and suction will therefore be different at different depths, depending on the history of wetting and drying that takes place at each level in the soil (see also Section 6.3.3).

Studies of the post-infiltration redistribution of water have often assumed deep profiles unaffected by the water table. However, in many soils the depth to the saturated zone may be fairly small and

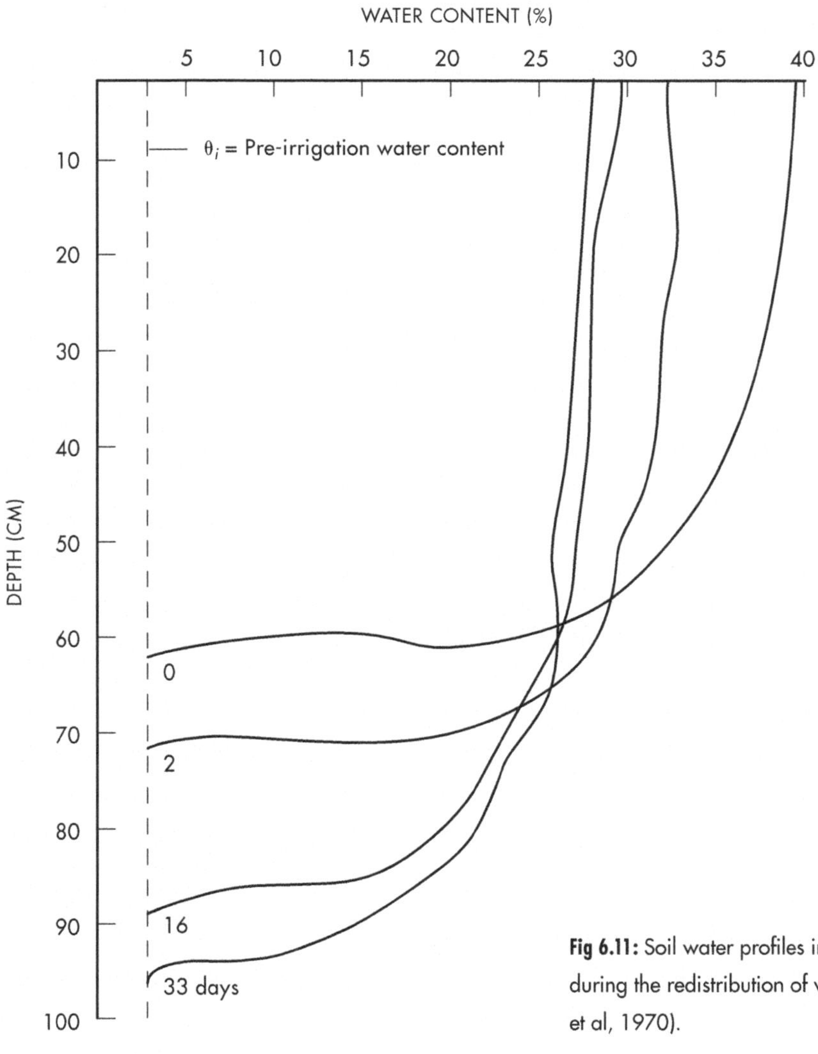

Fig 6.11: Soil water profiles in the absence of evaporation during the redistribution of water after irrigation (Gardner et al, 1970).

shallow water tables may exert a considerable influence on the distribution of water in a soil profile. The upward flux of water through the unsaturated zone to the surface or to plant roots is particularly important where the water table is relatively shallow. In addition, capillary tension keeps the soil pores full of water up to a certain height above the water table. The capillary fringe (see Section 5.3) is of interest to the hydrologist both for its vertical extent, which affects the overlying soil water profile, and for the rate of capillary flow, which determines the ability of the soil to supply groundwater to the unsaturated zone for evaporation. The thickness of the capillary fringe corresponds to the air entry value, which is the suction necessary for air to enter the largest pores (Section 6.3.2). If pore size is small and relatively uniform, it is possible that soils can be completely saturated with water for up to a metre above the water table. In areas where the top of the capillary fringe is near the ground surface it may play an important role in generating the quickflow component of runoff (Section 7.4.2), since only a small addition of water will reduce the suction to zero causing a large and sudden rise in the water table (Sklash and Farvolden, 1979; Gillham, 1984; Jayatilaka and Gillham, 1996).

6.5 SOIL WATER BEHAVIOUR UNDER FIELD CONDITIONS

Much of the preceding discussion of soil water retention, storage and movement has been based on studies of idealized soil systems. These often involve homogeneous and sometimes artificial soils, uniform initial water content, ponded water at the soil surface and a controlled environment in which plants, evaporation losses and water table influences are usually absent and where 'rain' falls at a constant rate. In contrast, the soil profile in the field is a very complex and heterogeneous system to which these restrictions and simplifications are applied. The remaining Sections of this chapter are based on more realistic laboratory experiments and on actual field situations. It is hoped that they will provide at least a partial insight into some of the phenomena that are relevant to the treatment of soil water physics of a particular field site. Many of these phenomena are interrelated and have important implications for the transport of solutes (see Section 8.5.3).

Depending upon the hydrological problem under study, variation in soil properties may be viewed at a number of different scales. These include differences down a soil profile and differences within a small volume of soil in a single soil horizon, up to areal variations across a catchment or between catchments.

6.5.1 SOIL LAYERING

Many soil profiles consist of a sequence of distinctively different layers, or horizons, which are caused by natural soil forming processes. Because these horizons may differ markedly in terms of, say, hydraulic conductivity and porosity, such stratification may greatly affect the movement of water through the soil profile. In many humid areas, for example, minerals and fine particles leached from the surface soil layers are deposited at greater depths, leading to a marked decrease in the number of large pores in the zone of accumulation. In extreme cases, this deposition results in the formation of an iron pan or hard pan of near-zero hydraulic conductivity and consequential waterlogging of the surface layers. Even without hard-pan formation, the subsoil tends to have a lower saturated hydraulic conductivity than the surface horizons. However, this is not always the case so that any soil horizon may limit the overall capacity of the complete profile to transmit water.

In general, layering of the soil profile reduces the infiltration capacity at the soil surface. In the case where a coarse layer of higher saturated hydraulic conductivity overlies a finer-textured layer, the infiltration capacity is initially controlled by the coarse layer. Once the wetting front reaches the finer layer, however, it is the latter that controls the rate of water movement and so infiltration capacity falls sharply. Then, if infiltration is prolonged, a perched water table may develop in the coarse soil immediately above the impeding layer or, in sloping terrain, throughflow may result (see Section 7.4.2 and Figure 7.6b). In the opposite case, where fine material overlies coarse, the infiltration capacity, although lower, is again controlled initially by the upper layer but may fall even further when the wetting front reaches the underlying coarser material which forms a **capillary barrier**. This apparently surprising effect occurs because the soil moisture suction at the wetting front is too high to allow water to enter the larger pores of the coarse material. Continued infiltration raises the moisture content in the upper layer until the matric potential has reduced sufficiently for water to penetrate the coarser layer. A layer of sand in fine-textured soil may, therefore, actually impede, rather than assist, water movement through the soil profile (Brady and Weil, 2007).

Patterns of water content and potential during constant rate irrigation of a layered soil (fine sand overlying coarse sand) in which the saturated conductivity of both layers was considerably greater than the rate of application, are shown in Figure 6.12. Since water transfers between the two layers, the soil water potential must be equal at the boundary as any pressure discontinuity would imply an infinite hydraulic gradient. Similarly, the flux must be equal across the boundary. The sudden jump in water content, in contrast, results from the difference in the water characteristic of the two materials. As infiltration continues, the wetting front moves down and the water content of each layer increases until its hydraulic conductivity becomes sufficient to carry the flow (see Section 6.4.2).

Outside the humid regions, climate also broadly determines the sequence of horizons in the soil profile. Where evaporation exceeds precipitation the net upward movement of water and solutes may result in surface deposition and the formation of a crusted or indurated surface layer. Indurated layers, which are widespread in Africa and Australia, may also result from compaction by raindrops or from the breakdown of soil aggregates during wetting. Even a thin surface crust can considerably impede infiltration. A similar effect is found in high latitudes, and in some high altitude areas, where frozen soils may impede infiltration and water flow through the soil profile. The effect of frozen soils on the rate of infiltration depends largely on the soil water content at the time of freezing. The wetter the soil is when freezing takes place the greater the number of ice-blocked pores and the lower the rate of infiltration, so that, when a saturated soil freezes, its intake rate will be virtually zero.

6.5.2 MACROPORES

One of the main reasons why water in natural soils does not always move in the manner predicted from Darcy's equation is that, in certain circumstances, water movement may be dominated by flow through a few large openings or voids rather than by the bulk flow through the microstructural interstices of the soil matrix. These large openings include structural cracks and fissures and quasi-cylindrical voids caused by earthworms and other burrowing creatures, and by the decay of plant roots. These are collectively known as **macropores**. Interconnected voids with diameters greater than about 30 μm can significantly increase the rate of soil drainage and may even allow sufficient flow velocity to result in some erosional smoothing (Jones, 1997).

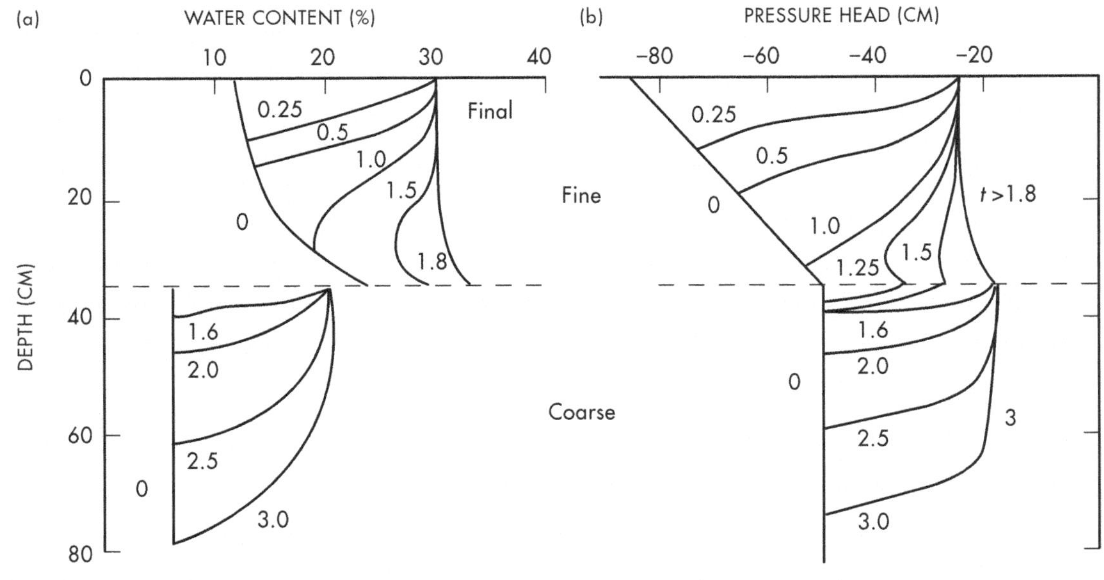

Fig 6.12 Changes over time (hours) in a layered soil in the profiles of (a) water content, and b) matric potential during a constant rate of water application to a layered soil (redrawn after Vachaud et al, 1973).

Larger interstices are even more effective in these respects and some workers have proposed a lower limit for macropores of 60 μm. Others have related the threshold for macropore flow to capillary tension rather than to pore diameter. Thus Beven and Germann (1982), in a seminal review of the influence of macropores on the flow of water through soils, proposed a soil water potential of 0.1 kPa, which is equivalent to pores larger than about 3 mm in diameter. Jones (1997) included naturally occurring soil pipes in peaty soils in a consideration of macropores, thereby considerably extending their size range.

The inclusion of soil pipes usefully underlines the fact that, whether defined by pore diameter or by capillary tension, the hydrological influence of macropores depends much more upon their interconnectivity and continuity than upon their abundance since, in most cases, macropores are likely to comprise only 1 or 2% of the bulk soil volume. Their hydrological role will also vary with soil water content and potential. For example, when the soil is drier than Beven and Germann's proposed threshold potential of 0.1 kPa, which will be the majority of the time for most soils, rapid water flow through macropores will

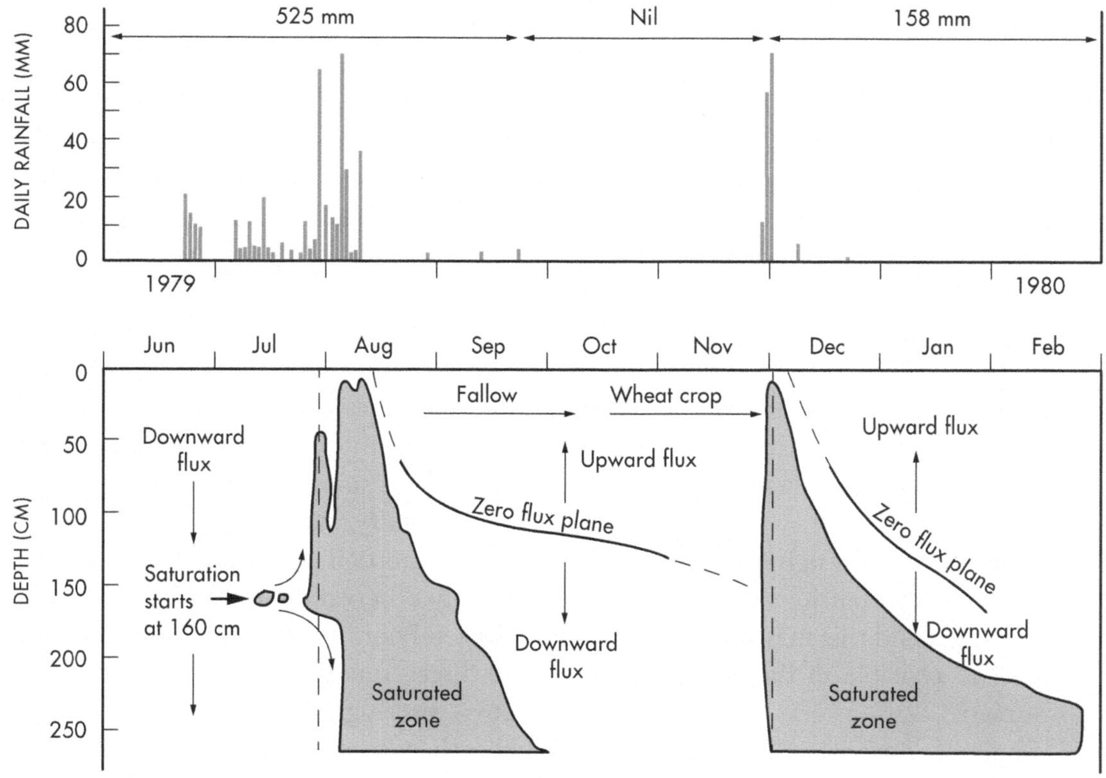

Fig 6.13: Annual cycle of water movement in a swelling clay soil in Monsoon India showing the development of saturated conditions beginning at 1.6 m depth due to the rapid infiltration of 'bypass' flow down shrinkage cracks (after Hodnett and Bell, 1986.)

not take place. In other words, their role and importance in infiltration and redistribution of water in the soil profile is largely confined to brief periods of heavy rainfall, or artificial irrigation. In these circumstances, when the rate of water supply exceeds the infiltration capacity of the adjacent soil peds, water will flow through the macropore system to reach the subsoil rather than moving down the profile as a well-defined wetting front.

A further example of the way in which the pattern of wetting of the soil profile is altered when macropores allow infiltration directly from the surface to the subsoil, was provided by Hodnett and Bell (1986). Working on a swelling clay soil in central India, they found that such flows during the monsoon rains led to initial saturation at about 1.6 m depth rather than the 'classic' pattern of saturation moving steadily downwards from the surface (Figure 6.13).

Rapid, preferential movement of water through the soil profile, which typifies macropore flow, may have a number of hydrological consequences:

- Macropores may considerably increase the infiltration capacity of soils and help to reduce the incidence of overland flow.
- Water flowing through macropores will bypass, and so fail to wet, most of the soil matrix and this may mean that less water is available in the surface layers for plants.
- Macropore flow may allow deep percolation and recharge to groundwater, even when the overlying soil is dry. This is especially likely where macropore continuity extends through the full depth of the soil profile, as for example when shallow clay soils are fully penetrated by shrinkage cracks into more permeable material beneath (Hodnett and Bell, 1986).
- Since macropore flow bypasses the natural filtration of the soil matrix, it may lead to contamination of groundwater supplies.
- In some soil and slope conditions, especially during heavy rainfall, macropore flow may play an important role in the development of throughflow (see Section 7.4.2).

The continuity of macropores is very important for flow processes, and will be less effective where most macropores have closed bottom ends and so fill up with water during rainfall, than where most are connected to large air-filled cavities such as soil pipes, animal burrows or artificial drains. The hydrological importance of macropores is likely to vary over time; seasonal variations may be significant, especially in shrinking soils. Robinson and Beven (1983) found that, in a clay soil exhibiting seasonal shrink-swell, flows from artificial drains were much more responsive to storms in summer, when the ground was dry and cracked, than in winter when the topsoil was close to saturation. Similarly, observations on the Keuper marls in Luxembourg (Hendriks, 1993) showed that direct runoff during summer storm events was higher from forested catchments than from grassland or arable catchments. This was due to the greater development of macropores including shrinkage cracks and animal burrows. Macropore flow dominated soil water flow processes during, and shortly after, rainfall but were less important than matrix flow during the subsequent period of redistribution.

It is now well-established that the presence of interconnected macropores means that soil water behaviour cannot be described adequately by equations based on Darcy's law, since the necessary assumptions of homogeneous soil hydraulic properties and a well-defined hydraulic gradient will no longer apply (Beven and Germann, 1982, 2013). However, the very irregular hydraulic gradients observed during infiltration by macropore flow, and the generally variable nature of macropore flow systems means that the search for an appropriate model of soil water behaviour in such conditions is still an active research topic. Further work is necessary before a full understanding is developed of macropore flow systems and their role, not only in surface and groundwater hydrology but also in respect of water quality issues relating to, say, the leaching of nitrates and other constituents from agricultural land.

Spatial variability

Soil properties usually vary continuously over the Earth's surface, with very few sharp changes. Consequently, the soil units identified by pedologists are often based on arbitrary boundaries within a gradual continuum of soil characteristics. Such units are often intended primarily for agricultural use and tend to reflect features which can be readily and unambiguously noted by surveyors in the field. This means that those physical properties of most interest to the hydrologist may play only a minor role in their classification (Warrick and Nielsen, 1980). Even so, soil units may be useful 'carriers' of certain types of basic data and broad groupings of soil types have been widely used in hydrology (e.g. Bouma, 1986). Care must be taken to scale up from these values: soil hydrological property effects are non-linear so it is not valid to simply take the areally-weighted mean value of, say, hydraulic conductivity measurements to obtain a grid square or catchment average value.

6.5.3 TOPOGRAPHY

Up to now discussion of soil water behaviour has considered predominantly vertical movement in an implicitly horizontal soil profile. In such conditions, soil water behaviour is likely to be dominated by soil properties but this is less likely to be true for sloping ground, where it may be more closely correlated to topography. In particular, ground slope influences the magnitude, speed and direction of lateral flow (e.g. throughflow) in the soil, and this tends to promote large-scale spatial variation in soil water content within drainage basins and especially a downslope increase of soil water content.

Downslope water movement may be particularly marked in layered soils, where water tends to be deflected 'downdip' at horizon boundaries, especially where there are large permeability contrasts between the layers (Warrick et al., 1997). However, tracer studies have confirmed the importance of flow parallel to the soil surface, even in permeable stony soils, providing the angle of slope is sufficiently great (e.g. Buchter et al., 1997). Soil scientists have, in fact, traditionally recognized a 'hydrologic sequence' of progressively deteriorating drainage condition downslope (White, 2005). Both topographic and soil maps may therefore provide hydrologists with a useful starting point in the study of soil water distribution within a catchment area).

The changing patterns of soil water suction and flow in a uniform hillslope soil profile, during the course of a storm, are shown, in a generalized form, in Figure 6.14. Initially the soil water state is close to complete gravity drainage (a) and suction increases with elevation, approximately balancing the gravity potential, except near the base of the slope, where there is some saturated lateral flow. Once rain begins (b), the surface layers of the soil become wetted

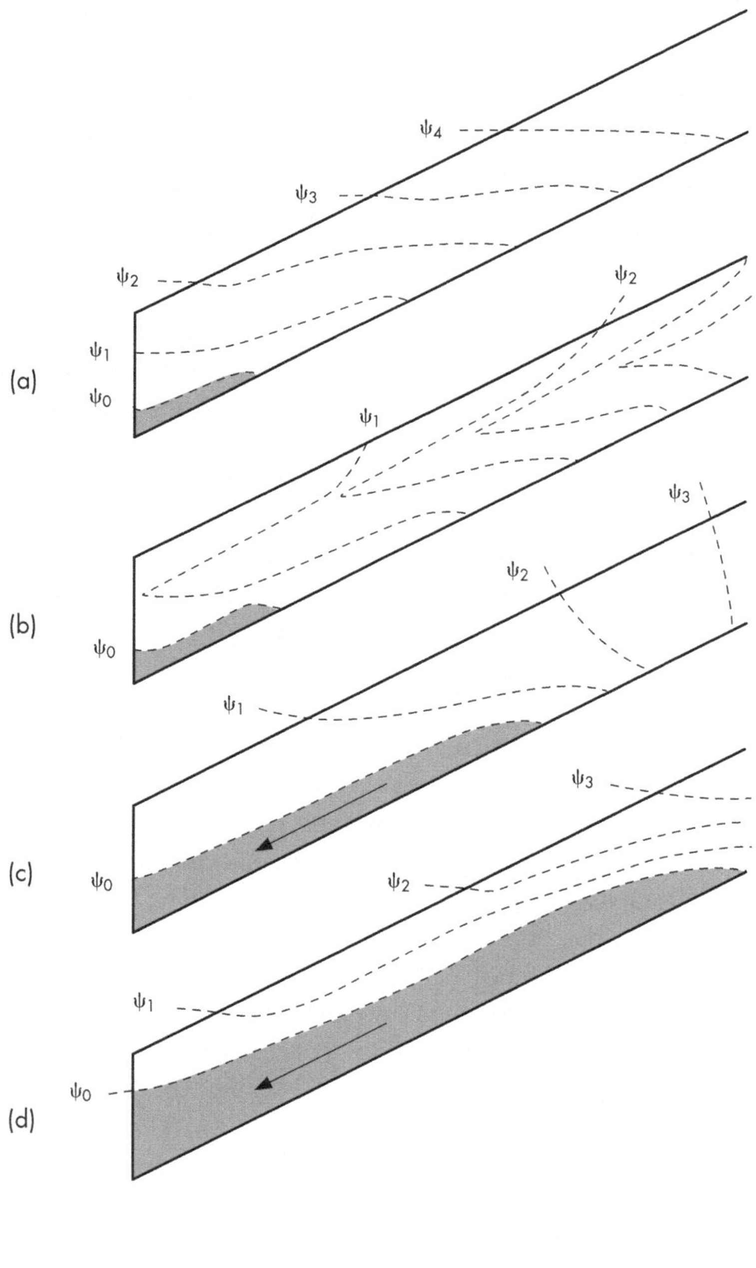

Fig 6.14: Generalized pattern of pressure potential in a straight hillslope. See text for details (redrawn after Weyman, 1973)

and suction is reduced. Then percolation wets deeper layers (c) and the saturated layer begins to grow as it is fed by unsaturated vertical and lateral flows. After the end of rainfall (d), drainage of water from upper to lower layers continues, with drying of the upper layer and further expansion of the saturated conditions. The extent of this saturated zone at the base of the slope is an important factor in the generation of storm runoff (Section 7.4). Finally, with continued drainage of soil water after the storm, the soil water pattern reverts to that of the initial state (a).

Simple two-dimensional studies of downslope soil water movement provide a useful starting point for the explanation of soil water distribution in catchment areas. However, they normally fail to incorporate the effects on soil water distribution of flow convergence caused by slope curvature. The links between flow convergence and surface saturation are discussed in the context of runoff generation in Section 7.4.2. At this point it is sufficient to note that slope concavities, in both plan and cross-section, have long been recognised (e.g. Kirkby and Chorley, 1967) as a cause of flow convergence which may lead to soil water flows that exceed the transmission capacity of the soil profile. Slope concavities tend, therefore, to be associated with high soil water contents (e.g. Zaslavsky and Sinai, 1981), especially in steeply sloping terrain, and in extreme cases the 'surplus' water tends to move towards the soil surface, causing surface saturation and overland flow. Similar effects occur in areas of thinner soil (see also Figure 7.6).

The changing spatial pattern of soil water distribution, including areas of surface saturation, on a hillside was studied by Dunne and Black (1970), and subsequently Anderson and Burt (1978) obtained detailed measurements of pressure potential at a grid of points (Figure 6.15). Similar distributions of soil water on hillslopes were observed by Dunne (1978) and were generated from a computer model by O'Loughlin (1981, 1990) and by the more widely used TOPMODEL (e.g. Beven, 1997) (see Section 7.4.2).

High soil water contents, including areas of surface saturation, are also often observed close to stream channels. These moister conditions are typically associated with a slope-foot concavity but may sometimes reflect the presence of a water table maintained at shallow depth by the level of water in the adjacent channel. In addition, water can only flow out from a soil through seepage faces, such as channel banks, field drains or even large macropores, if the soil water pressure at the boundary exceeds atmospheric pressure (Freeze and Cherry, 1979). This means that water must accumulate until a zone of saturation is reached before outflow can occurs. This 'boundary effect' is especially marked at times of baseflow between storms (Troendle, 1985).

In hilly areas the topography of a catchment area may therefore impose some element of order and predictability on an otherwise heterogeneous spatial distribution of soil water content, especially through the effects of slope gradient and concavity on the discharge and seepage of subsurface water, In general, however, and especially in flat areas or areas of gentle relief, soil water patterns are usually difficult to predict and even to observe (e.g. Baird, 1997; Ragab and Cooper, 1993).

Natural vegetation and soil properties may provide useful indicators of the likely location and extent of saturated conditions in a catchment (Dunne et al., 1975), and remote sensing techniques (e.g. Neale and Cosh, 2012; Albergel et al, 2012) provide a means of studying spatial variations over large areas.

6.5.4 HUMAN INFLUENCES

Human influences can alter soil water conditions in a large number of ways, ranging from irrigation schemes, which considerably increase the amount of water entering the soil, to the construction of large impermeable surfaces in urban areas, which

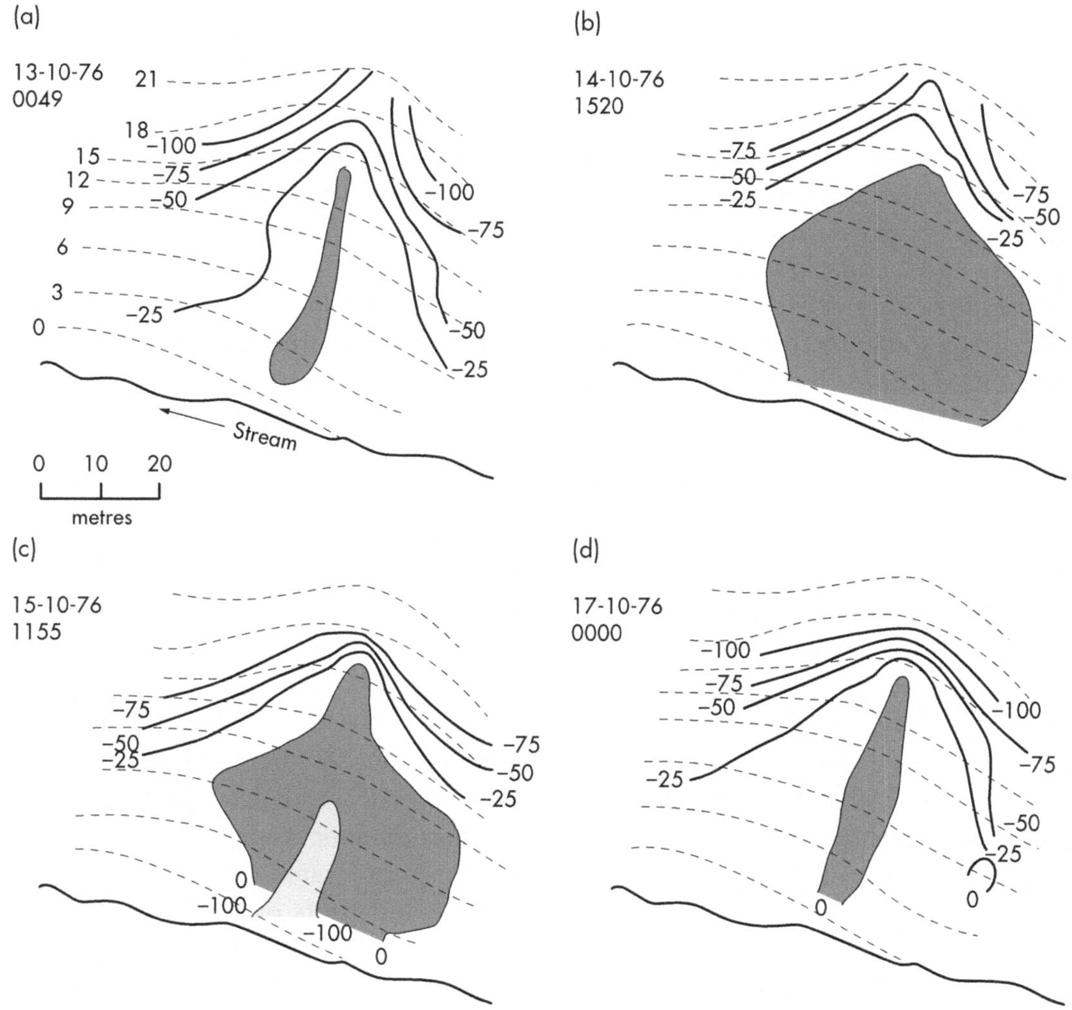

Fig 6.15: Soil water potential (cm) at 60 cm depth (solid lines) over a hillside (m contours shown as dashed lines) during one storm. (a) prior to rainfall, (b) near the end of rain, (c) at stream hydrograph peak,, and (d) two days after the storm. (Anderson and Burt, 1978).

exclude water from the soil beneath. Irrigation and artificial drainage are used over large areas throughout the world as essential tools to increase crop production (FAO, 2016; WRI, 1996). Some of the hydrological effects are discussed at global and regional scales in Section 9.2, and accordingly, only a brief account of a few examples is given below.

Feedbacks may occur between the land surface and the atmosphere. It has, for example been shown that large-scale irrigation in India has lowered the air temperature sufficiently to reduce the land-sea temperature gradient and alter the circulation pattern of the S Asian summer monsoon (Saeed et al, 2009).

In many areas soil water content is modified by human activity, including the drainage of agricultural land. Agricultural drainage schemes comprise open ditches or subsurface pipes (Smedema et al, 2004). These are deeper and closer together than the natural stream channels, so increasing the hydraulic gradient in the soil and lowering the water table more rapidly between storms than would otherwise occur. A detailed account of the distribution and purpose of field drainage in England and Wales, one of the most intensively drained parts of Europe, was given in Robinson and Armstrong (1988), and a discussion of the hydrological impacts of drainage on streamflow downstream is given by Robinson and Rycroft (1999).

Tillage and cultivation operations may also alter the movement and distribution of soil water. Ploughing increases the pore spaces in the upper soil, and may encourage lateral flow in the topsoil, with less downward flow into the subsoil (Goss et al., 1978). It has been shown by tracer studies that ploughing disrupts the vertical continuity with pores in the soil below (Douglas et al., 1980). Infiltrating water was found to penetrate to greater depths on land that had not been ploughed.

A change in agricultural land use from grassland to arable cropping may also affect interception and evaporation losses, especially if the arable farming leaves the soil bare at times of the year. Heavy rainfall on land with little vegetation cover may lead to crusting and sealing of the soil surface, reducing infiltration and increasing runoff. Forestry may have a large effect on interception and evaporation losses, causing soils under trees to be much drier than under other types of vegetation (see Sections 3.5, 4.5 and 4.6). In areas where the natural water table is close to the ground surface, groundwater abstraction lowers the water table, and may cause significant drying of the soil and a reduction in plant growth (Van der Kloet and Lumadjeng, 1987). The most extreme case of human influence on soil water conditions, however, is perhaps found in areas of steep topography, where deforestation and bad farming practices lead to accelerated erosion, and in severe cases may ultimately result in the complete destruction of the soil.

In Western Australia the removal of deep-rooted perennial vegetation and their replacement by shallow-rooted annual crops has led to a rising water table and a severe salinity problem. This will require the restoration of deep rooted plants to lower the water table and take the salts back to deeper depths.

REVIEW PROBLEMS AND EXERCISES

6.1 Why is the importance of soil water described as being 'out of all proportion to its small total amount'?

6.2 Describe the hydrological role of the following: soil matrix, soil texture, soil structure.

6.3 Define the following terms: surface tension, adsorption, matric potential, soil water characteristic.

6.4 Explain the role of total potential in soil water movement.

6.5 Describe the process of infiltration and the main attempts to model it mathematically.

6.6 Explain the difference between saturated and unsaturated hydraulic conductivity.

6.7 What are the main factors affecting the redistribution of soil water after infiltration has ceased?

6.8 To what extent has the development of soil water theory been limited by its failure to account for the heterogeneity of soil conditions in the field?

6.9 Discuss the role of macropores in soil water movement.

6.10 Discuss ways in which an improved understanding of soil water movement has helped attempts to estimate groundwater recharge.

PRINCIPLES OF UNSATURATED VERTICAL FLOWS:

1. Flow of water in a soil or an aquifer occurs from zones of higher to lower total potential. In the unsaturated zone, the total potential represents the combined effects of gravity and of the matric potential (comprising adsorption and capillary suction).
2. In the unsaturated zone the matric potential increases (i.e. it becomes less negative and approaches zero) with increasing water content. Therefore, zones of higher water content have a less negative (i.e. higher) matric potential than adjacent zones of lower water content.
3. Within a soil profile, a divergent zero flux plane (ZFP) divides the portion of the soil profile where the soil moisture is moving upwards from the portion of the profile below where it is moving downwards. A ZFP may not be present throughout the year, but when one is present infiltrated rainwater will not reach the aquifer, except via macropore flow.

TERMS AND UNITS FOR POTENTIAL ENERGY

Potential energy is defined as the work per unit quantity necessary to transport reversibly and isothermally an infinitesimal quantity of water from a pool of pure water at a specified elevation (normally the soil surface) and at atmospheric pressure to a given soil water location (ISSS, 1976).

Potential energy (see also Section 5.5) is expressed using different systems of units:

- Energy per unit mass (ϕ)
- Energy per unit volume (i.e. pressure) (P)
- Energy per unit weight (i.e. hydraulic head) (H)

These expressions are equivalent and translate directly into one another: i.e.

$$\phi = P/\rho = Hg$$
$$P = \rho\,\phi = H\,\rho g$$
$$H = P/\rho g = \phi/g$$

where ρ is the density of water, and g is the acceleration due to gravity.

Commonly used measurement units for groundwater and soil water potential are:

- bars (1 bar = 0.99 standard atmosphere)
- pascals (Pa), (1 newton/m^2);
- centimetres of water (cm); pF ($\log_{10}$ (cm) avoids large numbers at high suctions, and is commonly used to express values of soil water potential.

1 bar = 10^5 pascals =
100 kPa = 1020 cm water

The terms tension, suction and pressure are used interchangeably in soil water studies. Tension or suction are negative pressures (i.e. pressures that are less than atmospheric). Tension/suction is expressed as a positive quantity for unsaturated conditions (e.g. 100 cm). The same value expressed as a pressure would be a negative quantity (-100 cm) or could be expressed as pF = 2.0. Groundwater potential is positive and is, therefore, expressed as a pressure. Many different symbols are used in the soil water literature and it is necessary to understand the definition and context of symbols given in a particular text.

7

STREAMFLOW

"And a river went out of Eden to water the garden"

GENESIS 2:10

7.1 INTRODUCTION

Streamflow or **runoff** comprises the gravity movement of water in channels varying in size from those containing the smallest ill-defined trickles to the largest rivers such as the Amazon, the Congo, and the Yangtze. Streamflow may be variously referred to as river **discharge**, or **catchment yield**. It is estimated that globally an average of 36% of the total precipitation falling on the land areas reaches the oceans as runoff.

At a general level the relationship between streamflow and precipitation can be expressed in terms of the continuous circulation of water through the hydrological cycle. More specifically, we can recognize that in simple situations, where topographic and groundwater basins coincide, each river receives water only from its own drainage basin or catchment area. Each catchment can, therefore, be regarded as a system receiving inputs of precipitation and transforming these into outputs of evaporation and streamflow. Allowing for changes of storage within the system, input must be equalled by output. In all but the driest areas, output from the catchment system is continuous but the inputs of precipitation are discrete, and often widely separated in time. As a result, the annual hydrograph typically comprises short periods of suddenly increased discharge, associated with rainfall or snowmelt, and much longer intervening periods when streamflow represents the outflow from water stored on and below the surface of the catchment, and when the hydrograph takes the exponential form of the typical recession curve (see Figure 7.1).

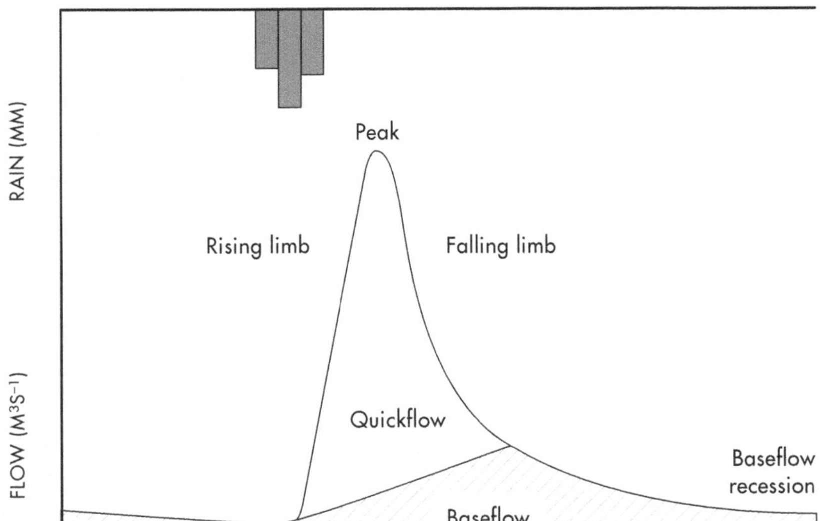

Fig 7.1: Typical stream hydrograph showing a steep rising limb to a peak, and a longer falling limb, with an arbitrary division into quickflow and delayed flow.

7.2 QUICKFLOW AND DELAYED FLOW

The immediacy of streamflow response to a rainfall event indicates that part of the rainfall takes a rapid route to the stream channels (termed **quickflow**); equally, the subsequent continuity of flow through often prolonged dry-weather periods indicates that another part takes a much slower route as **delayed flow**, which is more usually referred to as **baseflow**. These two components of flow are apparent in rivers of all sizes. However, in large river systems lag effects, both within and outside channels, and the multiplicity of flow contributions to the main channel from numerous tributary streams complicate interpretation of the hydrograph response of major rivers to precipitation. Accordingly much of the initial discussion of runoff processes in this chapter attempts to explain the response to precipitation of **headwater streams** draining catchment systems that are comparatively small and simple. In such situations the response of catchments to precipitation is often very rapid, but the proportion of precipitation that appears quickly as streamflow in the storm hydrograph differs widely from storm to storm, depending upon the precipitation characteristics and catchment conditions,

There have been many attempts to classify the hydrological response of catchments to distinguish between the short-term responses to a rainfall and the longer-term delayed flows from storage. This can be of value for a better understanding of the reliability of water supply, the design of water storages, hydroelectric power generation, ecosystem water requirements and contaminant dilution. Accordingly a distinction is often made between 'quickflow' under the storm hydrograph, and delayed or baseflow – usually on the basis of timing and some form of baseflow separation. While this division into quick and slow flow components, can provide valuable, albeit crude, numerical catchment flow descriptors (e.g. BFI, see Section 7.8.2), it should *not* be interpreted as having a real physical meaning and does not separate the contributions from different sources and processes. Runoff generation processes are far more complex, with no distinct separation in time of the different flow contributions to the hydrograph, as the following discussions will demonstrate. It should be noted that the term 'runoff' is widely used interchangeably with 'flow', and does not necessarily imply a restriction to fluxes over the ground surface.

7.3 SOURCES AND COMPONENTS OF RUNOFF

The variable response of streamflow to precipitation, both spatially and with time, reflects the contrasting flowpaths of precipitation towards the stream channels. The rate or quantity of flow has traditionally been represented by hydrologists by the letter Q. Figure 7.2 shows that precipitation may arrive in the stream channel by a number of flowpaths: direct precipitation onto the water surface, Q_p; overland flow, Q_o; shallow subsurface flow (throughflow), Q_t; and deep subsurface flow (groundwater flow), Q_g. Accumulations of snow will, upon melting, follow one of these four flowpaths.

These terms are used widely and relatively unambiguously in the literature, although misuse of other terms such as direct runoff has resulted in unnecessary confusion and ambiguity. Accordingly Figure 7.3 attempts to provide a consistent and logical terminology. This shows that surface runoff is the part of total runoff that reaches the drainage basin outlet via overland flow and the stream channels, although it may in some circumstances also include throughflow that has discharged at the ground surface at some distance from the stream channel. Subsurface runoff is the sum of throughflow and groundwater flow and is normally equal to the total flow of water arriving at the stream as saturated flow through the channel bed and banks. Quickflow, or direct runoff, is the sum of channel precipitation, surface runoff and quick throughflow, and will represent the major runoff contribution during storm

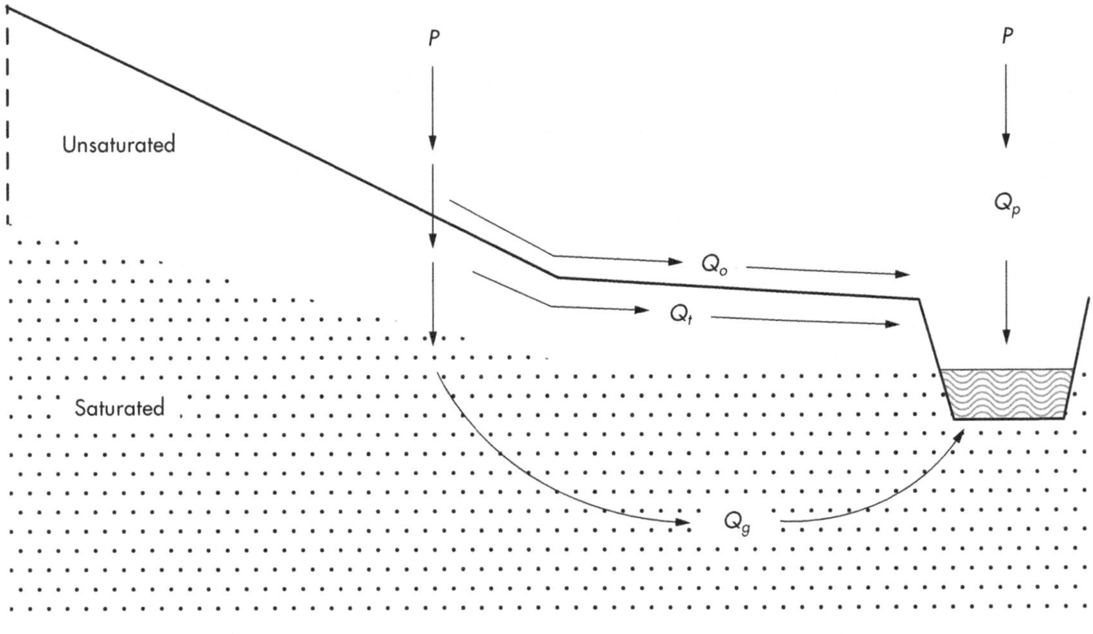

Fig 7.2: Flowpaths of the sources of streamflow: Q_p is the direct precipitation onto the water surface, Q_o is overland flow, Q_t is throughflow and Q_g is groundwater flow (from a diagram in Ward and Robinson, 2000).

periods and most floods. It will be observed that quickflow and surface runoff as defined above are not synonymous.

Baseflow or delayed runoff is the sustained component of runoff which continues even through dry-weather periods. It is normally regarded as the sum of groundwater runoff and delayed throughflow, although some hydrologists prefer to include the total throughflow as illustrated by the broken line in Figure 7.3. Again baseflow and groundwater flow, as defined above, are not synonymous; indeed, in some steep mountain drainage basins, baseflow may consist almost entirely of unsaturated lateral flow from the soil profile.

The relative importance of these runoff sources may vary spatially, depending upon drainage basin characteristics, such as soil type, the nature and density of the vegetation cover, and the precipitation conditions. In addition, the importance of individual runoff sources may vary seasonally, and may also change quite dramatically during individual storms or sequence of rainfall events in response to changes of infiltration capacity, water table levels, and surface water area.

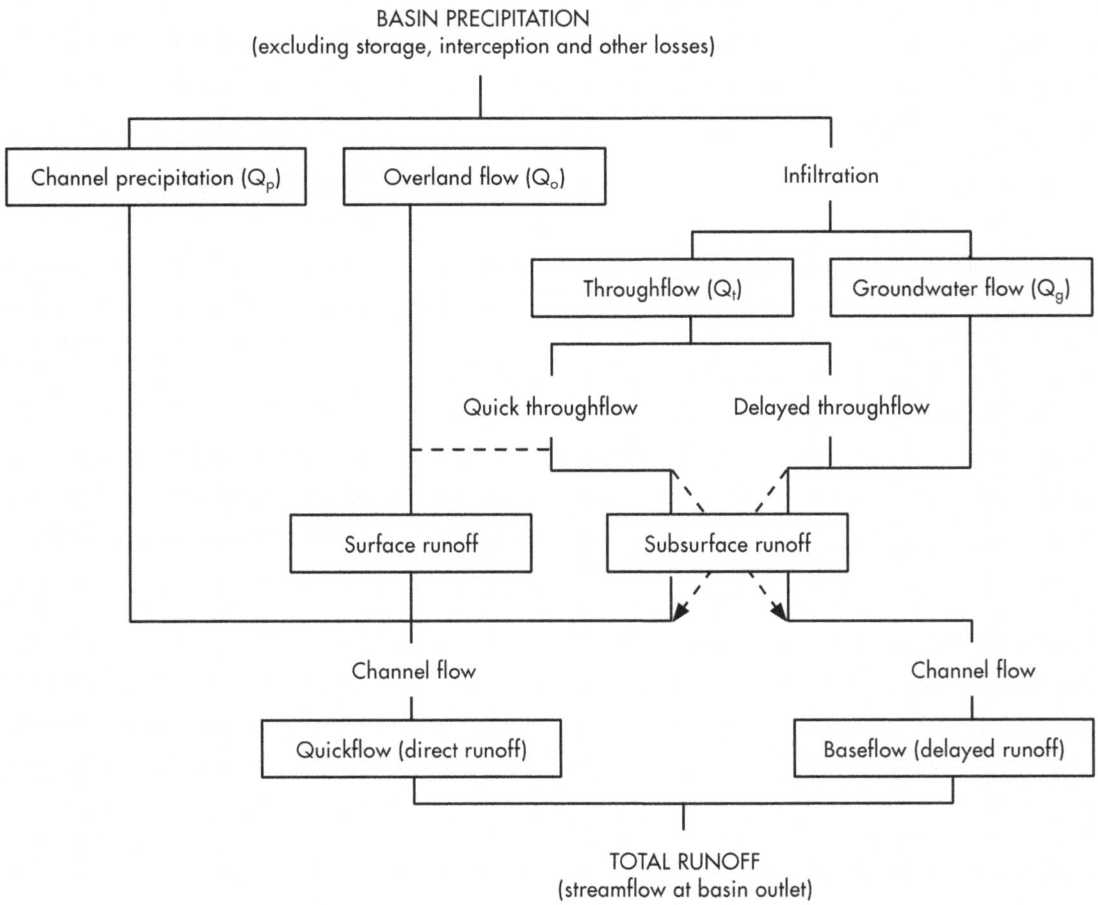

Fig 7.3: Diagrammatic representation of the runoff process (from a diagram in Ward and Robinson, 2000).

7.3.1 CHANNEL PRECIPITATION (Q_p)

The contribution of precipitation falling directly on to water surfaces is normally small because the perennial channel system occupies only a small proportion (typically 1–2%) of the area of most catchments. Even so, for some small precipitation events, channel precipitation may be the *only* component of the hydrograph. Where catchments contain a large area of lakes or swamps the channel system will be more extensive, and in these circumstances Q_p will tend to be a more dominant component of runoff. In addition, Q_p may increase significantly during a prolonged storm or sequence of precipitation events, as the channel network expands (see Section 7.4.2), and in some small catchments may temporarily account for most of total runoff.

7.3.2 OVERLAND FLOW (Q_o)

Overland flow comprises the water that flows over the ground surface to stream channels either as quasi-laminar sheet flow or, more usually, as flow anastomosing in small trickles and minor rivulets. One cause is the inability of water to infiltrate the surface as a result of a high intensity of rainfall and/or a low value of infiltration capacity. Ideal conditions for **infiltration-excess overland flow** are found on moderate to steep slopes in arid and semi-arid areas where vegetation cover may be sparse or non-existent, exposing the surface to raindrop impact and crusting processes. In such areas virtually all runoff may occur as overland flow. Other conditions in which Q_o assumes considerable importance involve the hydrophobic nature of some very dry and sodic soils, the freezing of the ground surface and compaction of soils by agricultural practices.

In humid areas, vegetation cover is thicker and more substantial, and because of the high value of infiltration which characterises most vegetation-covered surfaces, overland flow is a rarely observed. However there are many areas, both humid and sub-humid, where the effects of topography, or the nature of the soil profile, facilitate the rise of shallow water tables to the ground surface during rainfall or throughflow events. In such conditions the infiltration capacity at the ground surface falls to zero and **saturation overland flow** (Q_o(s)) results (see Section 7.4.2).

7.3.3 THROUGHFLOW (Q_t)

Water that infiltrates the soil surface and then moves laterally through the upper soil horizons towards the stream channels, either as unsaturated flow or, more usually, as shallow perched saturated flow above the main groundwater body, is known as throughflow. Alternative terms in the literature include **interflow** and **subsurface stormflow**. Throughflow is liable to occur when the lateral hydraulic conductivity of the surface soil horizons greatly exceeds the overall vertical hydraulic conductivity through the soil profile. Then, during prolonged or heavy rainfall on a hillslope, water entering the upper part of the profile more rapidly than it can drain vertically through the lower part, will accumulate and form a perched saturated layer from which water will 'escape' laterally in the direction of greater hydraulic conductivity.

This process is the one most commonly found. Even in a deep relatively homogeneous soil, the hydraulic conductivity tends to be greater in the surface layers than deeper down in the profile, thereby encouraging the generation of throughflow. Still more favourable conditions exist when: i) thin permeable soil overlies impermeable bedrock, ii) the soil profile is markedly stratified, or iii) an ironpan or ploughpan occurs at a short distance below the surface. There may be several levels of Q_t below the surface, corresponding to textural changes between horizons and to the junction between weathered mantle and bedrock. In addition, water may travel vertically into the soil profile and then move downslope through macropores and macrofissures (see Section 6.5.2), and in some circumstances, soil biological activity may play an important role in runoff generation (Bonell et al., 1984).

The various mechanisms of throughflow formation result in different rates of water movement to stream channels. Accordingly, it is sometimes helpful to distinguish broadly between 'quick' and 'delayed' throughflow (see Figure 7.3). However, apart from flow through interconnected macropores, the rapid arrival of throughflow at stream channels, which has been observed by some investigators, is likely to result from 'piston displacement' (see Section 7.4.2). Some throughflow does not discharge directly into the stream channel but comes to the surface at some point, before flowing over the surface to the stream. This component probably should be considered as subsurface runoff, although it is sometimes regarded as an addition to overland flow and surface runoff, as indicated by the dashed horizontal line in Figure 7.3. The role of throughflow in total runoff is discussed in more detail later in this chapter.

7.3.4 GROUNDWATER FLOW (Q_g)

Away from the relatively steeply sloping terrain of the headwaters, where subsurface runoff is dominated by throughflow, most of the rainfall that infiltrates the catchment surface will percolate through the soil to the underlying groundwater, and so eventually reach the main stream as groundwater flow. Since water movement at depth is usually very slow, the groundwater input to the stream channels may lag behind the occurrence of precipitation by several weeks, months or even years. Groundwater flow tends to be very regular, representing the outflow from a slowly changing reservoir, although in certain circumstances it may show a rapid response to precipitation. Indeed, the 'piston displacement' mechanism discussed in section 7.4.2, results in a rapid response of groundwater flow to precipitation during individual storms, and this is represented by the thin dashed line in Figure 7.3. Since this can operate only in moist soil and subsoil conditions, however, the replenishment of large moisture deficits created during summer conditions may result in a considerable lag of groundwater outflow after precipitation immediately following prolonged dry periods. In general, Q_g represents the main long-term component of total runoff and is particularly important during dry spells when surface runoff is absent.

7.4 EVENT-BASED VARIATIONS

The nature of streamflow response to precipitation at differing times scales (e.g. single precipitation event, seasonal and annual) is determined by the ways in which the various sources and components of runoff combine in specific circumstances. It has long been recognised that the balance between the quickflow and baseflow components is the essential determinant of the hydrograph shape. Early attempts to explain the variation of streamflow with time, especially through a precipitation event, concentrated almost exclusively on the overland flowpath. Later work clarified the varied nature of overland flow, and showed that throughflow and even groundwater flow may also play an important role. Many hydrologists contributed to our present understanding of the runoff process, but two principal pioneers whose complementary work in contrasting environments has proved to be seminal, were Robert Horton and John Hewlett. It seems appropriate, therefore, to discuss runoff variations in the light of their individual contributions.

7.4.1 THE HORTON HYPOTHESIS

Horton (1933), working in semi-arid areas, proposed quite simply that the soil surface partitions falling rain so that one part goes rapidly as overland flow to the stream channels and the other part goes initially into the soil and thence either by gradual groundwater flow to the stream channel or through evaporation to the atmosphere. The partitioning device is the infiltration capacity of the soil surface, defined as 'the maximum rate at which rain can be absorbed by a given soil when in a given condition'. Depending on site conditions, typical flow velocities for overland flow may be of the order of ~100 m h^{-1}, in contrast with much lower values of ~0.01–1 m h^{-1} for subsurface flow.

During that part of a storm when the rainfall intensity is greater than the rate at which it can be absorbed by the ground surface, the excess precipitation will flow over the ground surface as overland flow (Q_o). When instead, the rainfall intensity is lower than the infiltration capacity then, the infiltration that takes place will first top up the soil-water reservoir until the so-called moisture capacity is attained, after which further infiltration through the ground surface will percolate to the groundwater reservoir thereby increasing the groundwater flow (Q_g) to the stream channel.

Horton (1933) suggested that infiltration capacity passes through a fairly definite cycle in each storm period. Starting with a maximum value at the onset of rain, it decreases rapidly at first, and then becomes stable or declines only very slowly for the remainder of the storm (see Section 6.4.2). The generation of infiltration-excess overland flow is most likely to occur in conditions where soil infiltration rates are exceptionally low, or rainfall intensities exceptionally high. Two environments for which such conditions are commonly believed to prevail are arid/semi-arid and tropical areas.

Ideal conditions for the generation of 'Horton overland flow' are found in sparsely vegetated arid and semi-arid slopes where bare soil is exposed to raindrop impact, such as those in the southwestern USA where Horton carried out his investigations (see Section 7.3.2), and especially where surface crusting occurs. Experimental evidence for this includes studies at the Walnut Gulch experimental area in Arizona, USA (Abrahams et al., 1994), a tropical semi-arid area of gently sloping terrain in Queensland, Australia (Bonell and Williams, 1986) and for many soil types in the Mediterranean area. In such areas infiltration rates drop rapidly during rainstorms and runoff contains virtually no subsurface or throughflow components. In addition, crust formation on predominantly sodic soils, causes a rapid drop in infiltration rate and results in hydraulic conductivity at the surface being several orders of magnitude lower than that in the subsurface horizons (e.g. Agassi et al., 1985). This was demonstrated for the Loess Plateau of China by Zhu et al. (1997) and for the Sahel region of Niger, where low infiltration surface crusts develop, even on cultivated deep sandy soils, resulting in significant amounts of overland flow (Rockstrom and Valentin, 1997).

Although it is usually assumed for convenience that the whole of a small basin will behave in the same way with overland flow occurring uniformly over the area it is clear that infiltration capacities may vary across a catchment, depending on the infiltration capacities of different soils, ground condition and vegetation cover.

Infiltration excess flow might also be expected to be common in tropical rainforests, due to the high rainfall intensities. However, widespread field evidence, succinctly reviewed by Bruijnzeel (1990) and Anderson and Spencer (1991), indicates that the infiltration capacities of forest soils are generally high, largely because of the presence of a litter layer. As a result, infiltration-excess overland flow is a rare phenomenon, only occurring where bare soil is exposed by litter removal, treefall, or landsliding, or where stemflow (see Chapter 3) concentrates the flow of water at the base of tree trunks.

In summary, the possible implications of the infiltration capacity cycle for short-term variations of runoff are as follows. Rain of high intensity may generate precipitation excess and therefore overland flow throughout a storm, although rarely over more than limited areas of the catchment; rain of moderate intensity will not generate overland flow until the initially high infiltration capacity has declined; and rain of low intensity may fail to generate overland flow at all. Furthermore, since infiltration capacity is likely to show a continued decrease through a sequence of closely spaced storms, we would expect a given amount of rainfall falling late in the storm sequence to generate more overland flow, and therefore a greater streamflow response, than the same amount of precipitation falling early in the storm sequence.

It is often stated in the literature that Horton perceived overland flow occurring uniformly *throughout* a catchment. In fact he was well aware that infiltration rates differed with soil type, land cover and management practice, and published measurements to show this (Horton, 1933). As a practicing consultant engineer, however, he recognised that such detailed extensive spatial measurements would not be taken up by his contemporaries and clients, and so adopted the pragmatic approach of assuming a uniform loss rate over a small basin and deriving a catchment average loss rate from more readily available rainfall and flow data. This linked neatly with the unit hydrograph approach advocated at about the same time. Some 30 years later Betson (1964) revisited this topic and his observations on the 'partial' areas of a catchment generating infiltration-excess overland flow was widely adopted as a way to explain the nonexistence of surface runoff in many areas during storms, without the need to invoke an alternative flow generation mechanism.

7.4.2 THE HEWLETT HYPOTHESIS

Field observations in the more humid and forested Coweeta catchment in South-eastern USA, showed that even during intense and prolonged precipitation, all of the precipitation infiltrated the soil surface (Figure 7.4a). Hewlett (Hewlett, 1961a; Hewlett and Hibbert, 1967) proposed an alternative conceptual model of streamflow generation in which infiltration and throughflow within the soil profile cause rising water tables to saturate the ground surface, first in the shallow water table areas immediately adjacent to stream channels, and subsequently in the lower valley slopes (Figure 7.4, B). In these surface-saturated areas the infiltration capacity is zero so that all precipitation falling on them, at whatever intensity, is excess precipitation or overland flow. This is termed **saturation excess overland flow** ($Q_o(s)$), in contrast to the infiltration-excess overland flow (Q_o) envisaged by

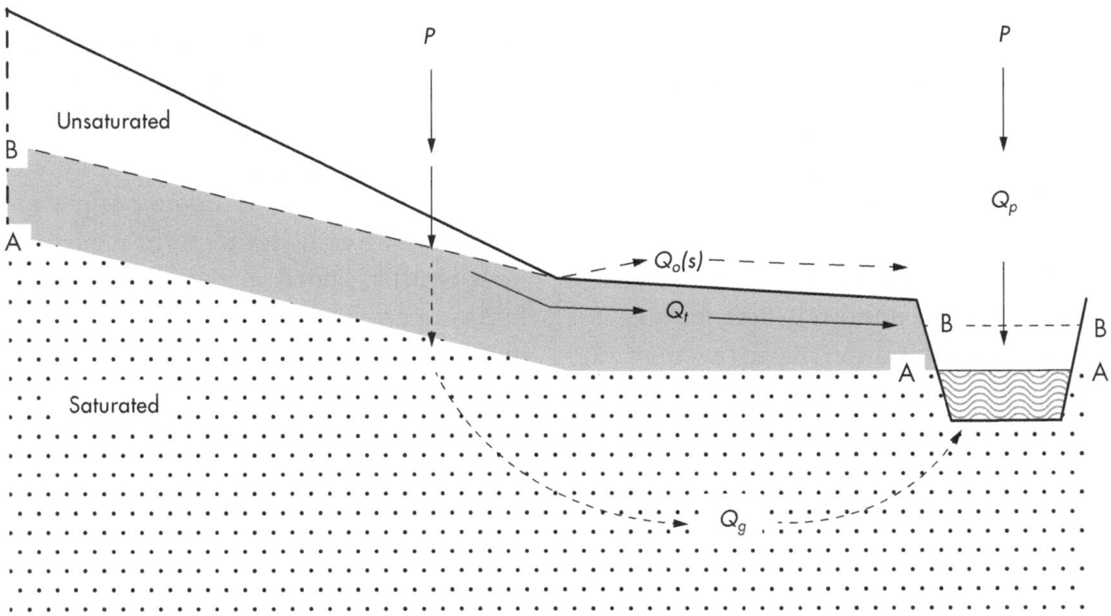

Fig 7.4: The Hewlett hypothesis response of streamflow to rainfall: (A) flowpaths in the initial stages of a storm, (B) flowpaths later in the storm. See text for details (from a diagram in Ward and Robinson, 2000).

Horton. According to Hewlett, only the saturated areas of the catchment act as a source of quickflow; all other areas of the catchment absorb the rain that falls and either store it or transfer it beneath the ground surface. He noted that the source area for quickflow is of variable size, growing as rainfall proceeds.

In one of those interesting coincidences which seem to characterise scientific progress, similar ideas were advanced independently in France by Cappus (1960). Additionally, Hewlett's hypothesis was contemporaneously supported by growing evidence of the pattern of subsurface water movement through valley slope profiles. This was generalised by Tóth (1962) in his mid-line model (see Figure 5.13) which shows infiltration and downward percolation in the upper slope, horizontal water movement through the middle slope material and upward movement near the base of the slope, reflecting the prevailing pattern of pore-water pressure. In such situations, the increase of pressure potential with depth in the lower slopes facilitates rapid saturation of the surface layers when even modest quantities of water are added to the soil profile by infiltration or shallow throughflow. The physical basis of this process was described in detail by Gillham (1984).

Hewlett's experiments with sloping soil models (e.g. Hewlett, 1961b; Hewlett and Hibbert, 1963) indicated water flow paths resembling those of a thatched roof, and showed no overland flow and no deep groundwater recharge, implying that all rainfall infiltrates, and is then transmitted through the soil profile. Note that the infiltrating rainfall contributes preferentially to storm flow, with upslope rainfall recharging the soil-water store in preparation for succeeding days and weeks of baseflow whereas downslope rainfall and channel precipitation provide most of the storm flow. The conclusion from this work appeared to be that throughflow was capable of producing runoff peaks in river hydrographs. Subsequent field studies have downplayed the role of *subsurface* stormflow as a major contributor of stormflow in humid areas and identified the role of individual areas of saturated soils as source areas for generating saturation overland flow, as will be seen below.

Disjunct source areas

Although Hewlett originally implied that variable source areas would be contiguous with the stream channels, later work showed that areas of saturation overland flow may also occur widely within a catchment area, often in locations far removed from the stream channels. This required a change in viewpoint from the 2-dimensional conceptual model of Hewlett to a 3-dimensional view of the drainage basin. These 'active' areas for generating runoff may be much larger than the area actually contributing to streamflow; to contribute quickflow to the stream channel these discontinuous or **disjunct** areas must have effective hydrological connections with the valley bottoms or lower slopes (Amboise, 2004). Apart from areas where crusting, compaction, sparse vegetation cover or thin, degraded soils (all frequently the result of human interference), result in infiltration-excess overland flow, the local source areas of quickflow often occur where **flow convergence** leads to surface saturation and saturation overland flow.

Three typical locations for flow convergence, shown in Figure 7.5, are:

a. Slope concavities in plan (hollows) where convergence leads to subsurface flow rates that may exceed the transmission capacity of the porous medium and lead, therefore, to the emergence of flow at the soil surface in the central areas of the concavities;

b. Slope concavities in section (breaks in slope) where, assuming uniform hydraulic conductivity throughout the section, subsurface flow rates will be directly proportional to the hydraulic gradient so that water will enter a concavity from upslope areas more rapidly than it can leave downslope; and

c. Areas of thinner soil whose water holding and transmitting capacity is low.

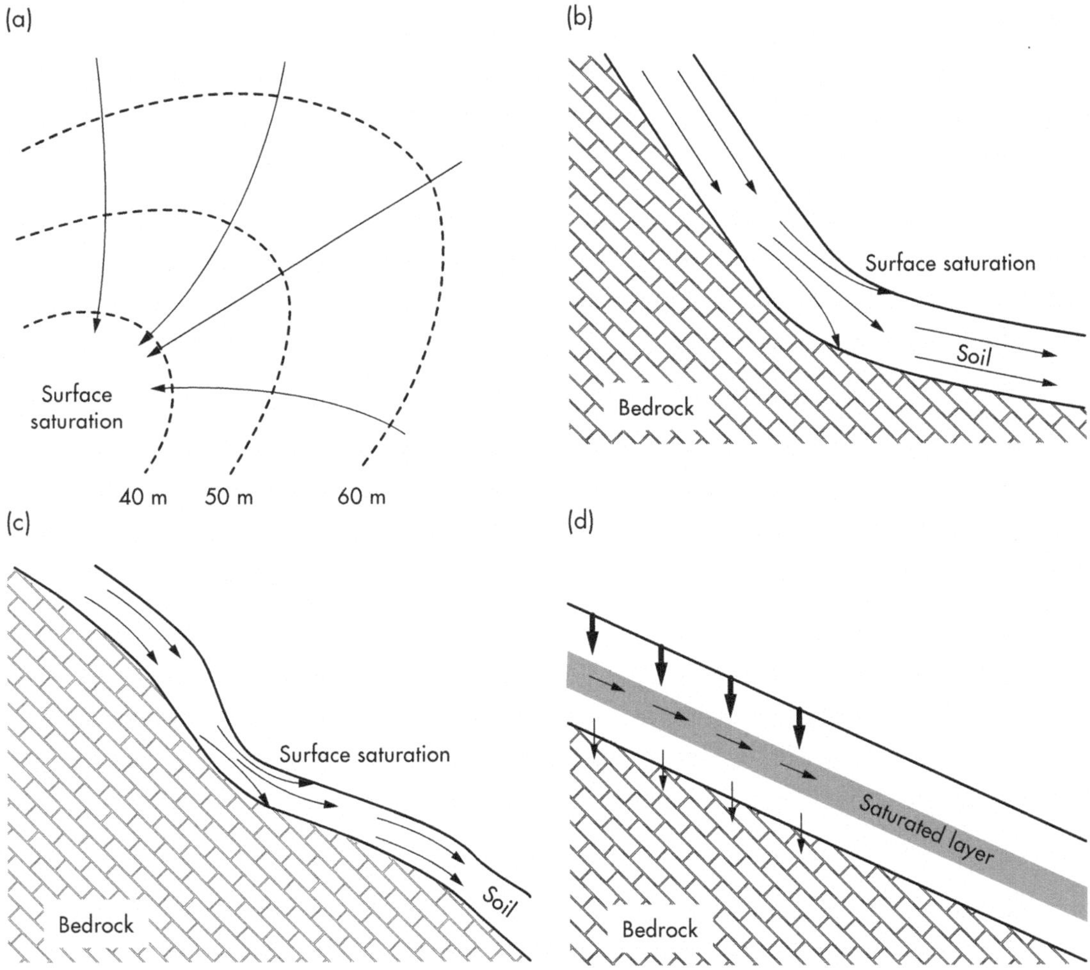

Fig 7.5: The principal locations of flow convergence in catchment areas. (a) Hollow, (b) Break in slope, (c) Thin soil, (d) Reduction in hydraulic conductivity. (From a diagram in Ward and Robinson, 2000)

Dunne and Black (1970) compared runoff producing mechanisms in three sections of a hillside - one concave (hollow), one was straight and the third was convex (spur), and found greater surface saturation and overland flow occurrence in the hollow, and least in the spur.

Slope concavity is more easily assessed than soil depth, both in the field and from maps and aerial photographs. Inevitably, therefore, much attention has been devoted to its effects on quickflow generation. Zaslavsky and Sinai (1981) presented field measurements of soil water concentrations in concave areas for European locations, including their own results for a site near Beer-Sheba, Israel, and O'Loughlin (1981) used computer models to demonstrate that the size of saturated zones on undulating hillslopes depends strongly on topographic convergence or divergence. Kirkby (1978) devised a simple, yet ingenious, topographic index, κ, to describe these areas of likely saturation:

$$\kappa = \alpha / \mathrm{Tan}\ \beta \qquad (7.1)$$

where α is the upslope area supplying water to the point, and β is the local slope (conducting water away). All points with the same value of κ are assumed to behave hydrologically in the same way. This topographic index has been used as the basis of the dynamic contributing area **top**ographic **model**, TOPMODEL (Beven and Kirkby, 1979; Beven, 1997). Rainfall-runoff models are discussed by Beven (2012).

TOPMODEL attempts to incorporateing these spatial patterns into a semi-distributed' conceptual model, with a limited number of parameters; it uses readily available topographic data and a limited amount of soil data. A fundamental assumption is that the water table follows the topography, allowing flow to be estimated using topography as a surrogate for hydraulic gradient. In this way, the drainage basin may be subdivided into several relatively homogeneous units, each modelled separately and with their simulated discharges being routed individually through the channel to the basin outflow. One of the reasons for the success of TOPMODEL is it can predict the geographical location of areas of saturated or near-saturated soil which can be easily checked in the field to test the validity of the hydrological concepts. Others have emphasized the relationship of dynamic quickflow-contributing areas to geomorphological structure, vegetation, field survey and remote sensing techniques.

TOPMODEL CODES AVAILABLE ONLINE

Topmodel and Dynamic Topmodel (Metcalfe et al, 2015) written in the programming language 'R' for data analysis and visualization are available from the CRAN network of ftp and web servers at:

https://cran.r-project.org/web/packages/topmodel/index.html and https://cran.r-project.org/web/packages/dynatopmodel/index.html

A fourth type of flow convergence, illustrated in Figure 7.5d, occurs as water percolates vertically through a soil profile. Partly because of the reduced hydraulic gradient as the flowpath of the percolating water lengthens, and partly because most soils, whether layered or not, exhibit a reduction of hydraulic conductivity with depth, rates of percolation decrease with depth, leading to the development of a layer, or layers, of temporary saturation. This accumulation of subsurface water normally moves downslope as throughflow before the build-up of saturation reaches the soil surface. In flat areas however, or in sloping areas having very high rainfall amounts and intensities, saturation overland flow will be produced. This is most likely where an impeding layer occurs at shallow depths in the soil profile, as in the pseudo-gley soils of central Europe, or with high-intensity tropical rainfall (e.g. Bonell and Gilmour, 1978).

Hydrological linkages

The existence of convergence-induced disjunct variable source areas effectively extends the Hewlett concept of runoff generation, provided that satisfactory **hydrological linkages** allow the rapid transmission of water between these areas and the stream channels. Various linkage mechanisms have been proposed, including an overland flow connection where disjunct upslope source areas are associated with the combination of heavy rainfall and shallow soils of variable depth.

Where soils are thicker, or where vegetation cover is denser, other linkage mechanisms are likely to prevail. Increasing recognition of the widespread occurrence of macropores and macrofissures has led to further consideration of their role (see Section 6.5.2). Early recognition that turbulent flow through large, quasi-cylindrical pathways, such as animal burrows or decayed root channels, could lead to subsurface stormflow moving rapidly through the slope material was provided by Hursh (1944) for the southern Appalachians, USA, and was confirmed much later for a forested catchment in Luxembourg by Bonell et al. (1984). However, biotic voids have been regarded by some hydrologists as 'pseudo-pipes' in contrast to the more widespread and hydrologically important pipes formed by hydraulic and hydrological processes. These latter types are found in a wide range of locations (see review by Jones, 2010) and particularly in blanket peats (Holden, 2005). In some cases they appear to be able to increase the quickflow contributing area to more than double that identified from surface contours (Jones, 1997).

The relatively high velocity of conduited, macropore subsurface flow suggests that the water arriving in the stream channel by this route will be 'new' water, i.e. water added by the current storm, rather than pre-storm 'old' water, already stored in the catchment. However, tracer experiments in a variety of environments confirm that the hydrological efficiency of macropore systems varies with factors such as changes in soil water status and soil structure which influence the exchange of water between the soil matrix pore space and the macropore systems. Accordingly, field results, even from the same locations, may seem inconsistent, though many appear to support Hewlett's view that 'old' water dominates the storm runoff hydrograph, even in areas where the existence of macropores is well established. This was demonstrated by Sklash et al. (1986), and Pearce et al. (1986) for catchments in New Zealand in which earlier analyses by Mosley (1979) had, in contrast, suggested that 'new' water dominated the storm runoff.

The role of throughflow

The preceding discussion of variable source areas and the ways in which these may be linked to the channel network must not be allowed to obscure the fact that, except in some arid and semi-arid areas, most of the water arriving at the stream channels (including some part of the quickflow component) has travelled below the ground surface as **throughflow** (Q_t) and **groundwater flow** (Q_g). The importance of these two flow-paths, once thought to be too slow and indirect to influence the short-term response of streamflow to precipitation, largely reflects the anisotropic nature of the hillslope soil profile and its effect on water movement.

The 'thatched roof' analogy of Zaslavsky and Sinai (1981) (see Figure 7.6) helps to explain the anisotropic hillslope soil profile and the way in which this encourages throughflow more or less parallel to the slope surface. No hydrologist, having measured the infiltration characteristics of bundles of straw, would recommend their use as a roofing material. And yet, even in the heaviest rain, the building remains dry, no water runs over the thatch as 'overland' flow, there is no 'groundwater' flow into the roof void and no evidence of zones of 'temporary

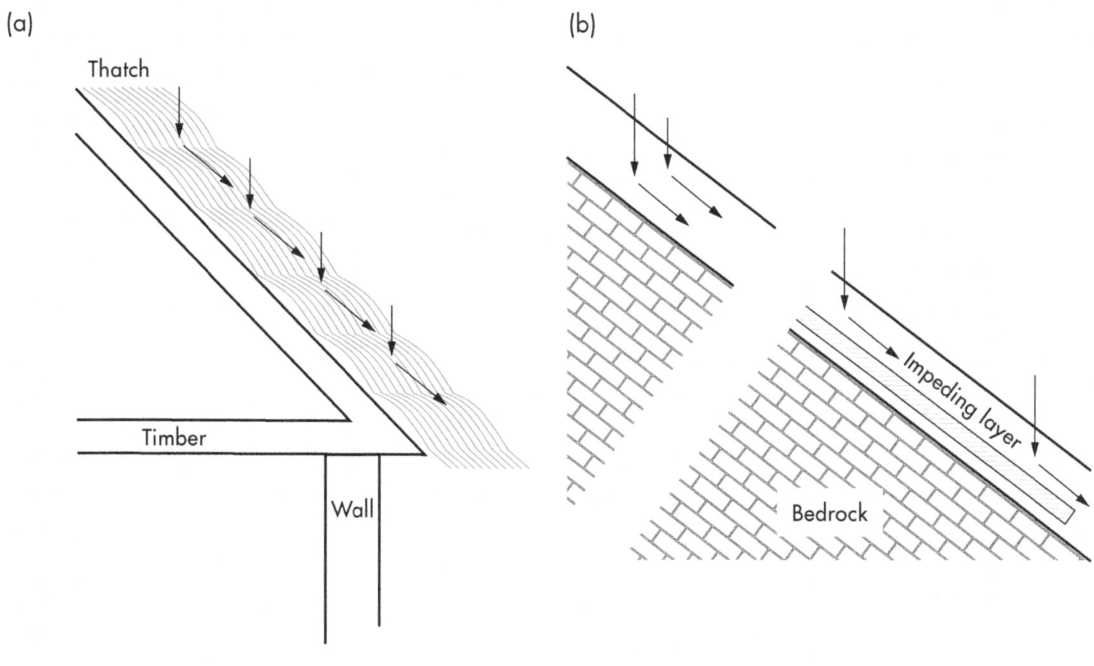

Fig 7.6: The thatched roof analogy of water movement through a hillslope soil profile, showing schematic flowpaths through: (a) a thatch, and (b) a sloping soil profile, with and without an impeding layer (From a diagram in Ward and Robinson, 2000).

saturation', i.e. all the rainfall is evacuated along the narrow layer of the thatch itself (Figure 7.6a). The thatched roof works because the alignment of the straw imparts a preferential permeability along the stems and because the roof has an angle of slope; it would not work if the straw bundles were placed vertically or if the roof were flat. In the case of the soil on a hillslope (Figure 7.6b), we know that, whether or not an impeding layer exists beneath the surface, there is normally a preferential hydraulic conductivity through the more open-textured upper layers parallel to the surface. As with the thatched roof, this may enable the sloping soil profile to dispose of rainfall without generating either overland flow or groundwater recharge.

Initially, Hewlett found it difficult to explain how throughflow, having a maximum rate of movement of about 5 m d^{-1}, could reach the stream channel quickly enough to contribute to the storm hydrograph. Two relevant factors are indicated in his theory which implies that: i) most of the throughflow contribution comes from the lower slopes, closest to the channel, and ii) as rainfall proceeds, expansion of the riparian area of surface saturation and ephemeral channels, shortens the effective flowpath for throughflow from more distant parts of the slope. Subsequently it was recognised that, in certain circumstances, soil water can move by a process of 'piston displacement' or 'translatory flow', whereby each new increment of rainfall displaces all preceding increments, causing the oldest to exit simultaneously from the bottom end of the hillslope profile.

This process had been confirmed by Horton and Hawkins (1965) in laboratory experiments with columns of sandy loam soil. Tritium-tagged water was added to the top of a column of moist soil which was then subjected to successive irrigations of simulated rainfall. Ignoring the effects of dispersion, each new rainfall caused a downward displacement of the tagged water and a corresponding outflow of untagged water from the bottom of the soil column. Eventually, after sufficient irrigations, the tagged water itself emerged. Regarding the soil profile as an inclined column receiving inputs of rainfall, Hewlett and Hibbert (1967) cited the displacement process to explain why each input of rainfall could be accompanied by a virtually instantaneous outflow of subsurface water at the slope foot. Subsequent confirmation of the important contribution of pre-event water to the storm hydrograph came from the gradual accumulation of field evidence across a wide range of catchment conditions (Buttle, 1994). In one example, detailed field measurements of soil water hydraulic properties and water chemistry formed the basis of successful attempts to model the displacement process on Mediterranean hillslopes (Taha et al., 1998).

A weakness of this explanation is that a given input will result in an equivalent output only if the available moisture storage capacity within the soil system is already filled or nearly full. In drier conditions rainfall inputs and/or displacements will be used to 'top up' the soil water store rather than to maintain the chain of displacements. This means that the mechanism will be most effective after a period of rain and/or on the lower (i.e. moister) slopes.

Hewlett and Hibbert (1963) sought further confirmation of the dominant role of throughflow from the sloping soil models referred to earlier. A sloping soil block was thoroughly wetted, covered to prevent evaporation and then allowed to drain, during which time the outflow was measured continuously. An outflow pipe established a free water table, which was used as the datum for all measurements, and below it was a saturated wedge of soil. The soil water content and outflow data were interpreted as showing that unsaturated drainage from the soil mantle was alone sufficient to account for the entire recession limb of the storm hydrograph in steep forested headwater catchments, and that the saturated wedge was not of itself a source but rather a conduit "through which slowly draining soil moisture passes to enter the stream" (Hewlett and Hibbert, 1963).

The role of groundwater

Although Hewlett and Hibbert's conclusions about the role of the saturated slope-foot wedge were correct for the conditions being investigated, the relationship between the saturated and unsaturated flow components of hillslope hydrology is partly a function of slope angle. In flat basins groundwater storage represents a large percentage of total storage, whereas in steep basins the soil moisture store is much the larger one. The steeper the soil body is inclined, the greater the contribution of unsaturated flow to sustained outflow.

In one sense this simply restates the long-established view that in highly permeable catchments, and in the flat lowland areas of larger drainage basins, groundwater is the major component of streamflow. However, in such areas the magnitude of the groundwater contribution often appears to diminish the response of rivers to precipitation. By contrast, the concern in this section is with the situation where groundwater may make a major contribution to the storm hydrograph in a wide range of hydrogeological and relief conditions, and where the response of rivers to precipitation is both rapid and pronounced. Hursh and Brater (1941) had advocated such a role for groundwater near the stream channels, although it was only much later that widespread field evidence, often based on tracer measurements, was used to show that groundwater can be a major and active component of storm runoff (e.g. Sklash and Farvolden, 1979).

Where shallow water tables are prevalent, groundwater and surface water are inevitably closely interlinked. Accordingly, in countries like the Netherlands, the important role of groundwater in runoff generation has long been recognised and runoff has been studied using combined groundwater and surface water models (e.g. Querner (1997). Interestingly, a groundwater-based concept of 'variable source areas' was introduced by De Zeeuw (1966), apparently unaware at the time of Hewlett's pioneering work. He argued that, in the Netherlands, the response of drain and ditch flow to precipitation depends on the number of drains and ditches that are deep enough to intercept the water table and that will, therefore, receive the more rapid local discharge compared with the slower regional groundwater seepage flow. As the water table levels rise, so more drains and ditches receive the quicker local flow.

In other circumstances, the ability of groundwater to contribute significantly to the storm hydrograph appears to reflect

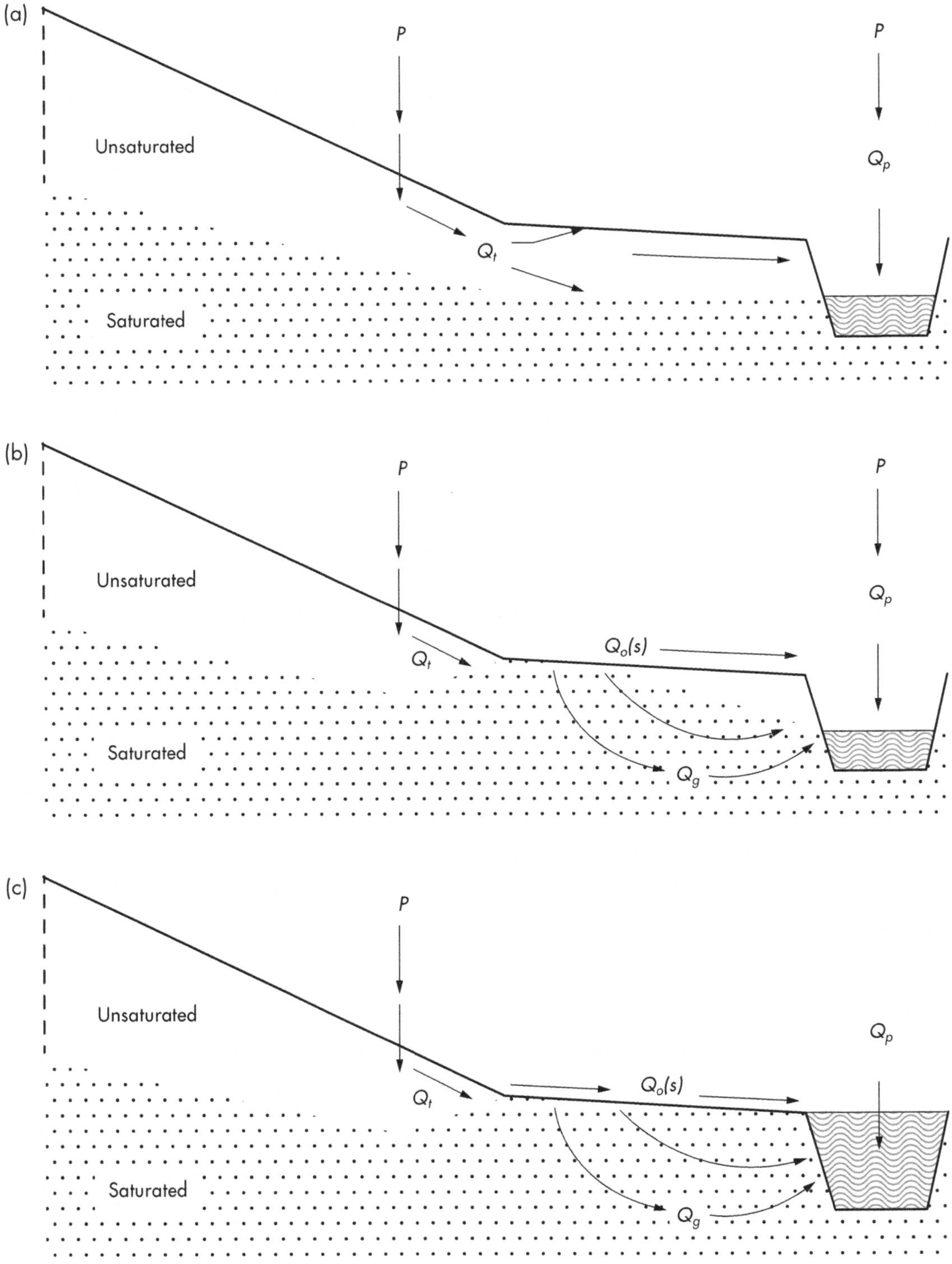

Fig 7.7: An integrated representation of the response of streamflow to precipitation. See text for details (from a diagram in Ward and Robinson, 2000).

the formation of a **groundwater ridge** adjacent to the stream channel, as illustrated in Figure 7.7b. Such a feature was identified by Ragan (1968) and Hewlett (1969) who referred to "an ephemeral rise in the groundwater table" near the stream channel which "helps produce the storm hydrograph"'. Later, Sklash and Farvolden (1979) used field evidence and computer simulation to show that the formation of a groundwater ridge, together with the resulting steepened hydraulic gradient and increased groundwater discharge area, was capable of producing large groundwater contributions to the stream channel.

Two factors encourage the formation of a groundwater ridge close to the stream channel. Firstly, the favourable gradient of moisture potential in the lower-slope areas, discussed in the opening paragraphs of Section 7.4.2, means that comparatively modest inputs of infiltration cause rapid increases of moisture potential in the surface layers (e.g. Abdul and Gillham, 1984). Secondly, the lower valley sides are often concave in profile and are therefore zones of convergence in which subsurface flow lines not only emerge at the ground surface, leading to surface saturation, but are also deflected downward, leading to concentrated groundwater recharge (Zaslavsky and Sinai, 1981).

Figure 7.7 summarises the Hewlett hypothesis of humid zone runoff formation. Water from rainfall (and snowmelt) infiltrates the slope surface and moves as throughflow (including macropore flow) in the slope mantle (Figure 7.7a). Convergence and infiltration in the lower-slope areas leads to surface saturation *and* groundwater recharge which will create both an overland flow and a groundwater contribution to the storm hydrograph (Figure 7.7b), with the groundwater ridge merging eventually in some locations into a wider riparian area of surface saturation (Figure 7.7c) which, with further rainfall, may extend onto the lower slope.

7.4.3 MULTIPLE PROCESSES

It is now appreciated that there are a number of streamflow generation models that may be operating in the same catchment, or individual hillslope, at different times and different places (Beven, 1989). These include infiltration excess overland flow from all (or part) of the catchment, subsurface flows below the main water table and in shallow layers near to the surface in perched water tables, as well as saturation excess overland flow from defined wetland areas and from individual areas defined by topography and subsurface soil characteristics.

Streamflow hydrology is thus potentially very complicated, with the addition of the role of preferential flow pathways, and evidence from geochemical separation of the hydrograph that in many basins the storm hydrograph is dominated by water stored in the basin prior to the storm event, rather than by storm rainfall (Beven, 2006). More generally, field evidence has confirmed that, as in temperate areas, there is a full range of runoff flowpaths on forested slopes in tropical areas and that the dominance of flow at different levels, on or below the surface, depends very much on the nature of the substrate. Indeed, Cassells et al. (1985) considered that the major significant difference between runoff processes in temperate and tropical forest catchments is that, in the

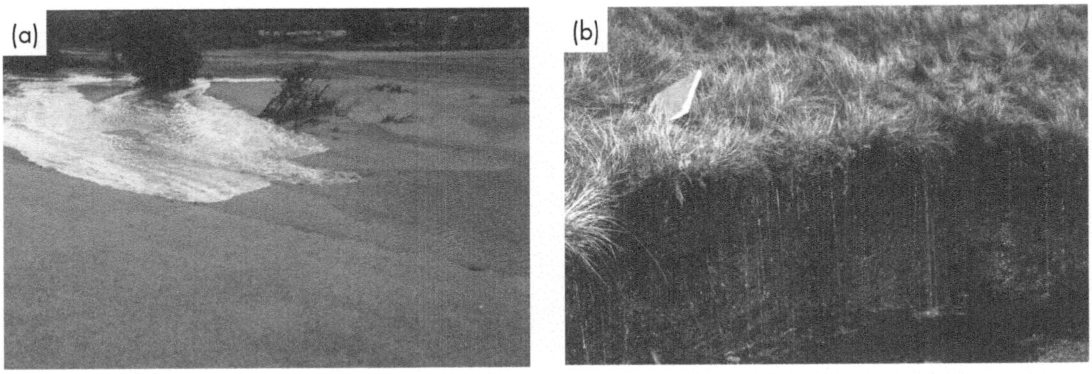

Photo 7.1: Different types of overland flow observed in the field: (a) Infiltration-excess overland flow generated by a severe thunderstorm flowed down to lower ground as a water front moving along a shallow depression. Niger about 40 km south of Niamey (dry desert climate). (Photo source: John Bromley). b) Saturation-excess overland flow running over a slip face following prolonged winter rain in mid-Wales near Aberystwyth (humid temperate climate) (Photo source: John Bell)

latter, wet areas are widespread throughout the catchments during storm events rather than being concentrated in riparian areas. Due to the high wet-season soil water contents, such areas can redevelop almost instantaneously with the onset of intense storms. As a result of surface outflows from these widespread areas, quickflow inevitably accounts for a large proportion of total streamflow. The concept of variable (or partial) source areas is an attempt to reconcile the absence of widespread overland flow with the spatial variability of channel flow and the rapid response of most streams to precipitation by postulating that over-the-surface movement of water is restricted to limited areas of a drainage basin.

Field experiments in a semi-arid area in Spain found that quickflow resulted, not from infiltration-excess overland flow but from saturation of the upper soil profile (Scoging and Thornes, 1979). Saturation overland flow also appears to be common in tropical rainforests because the infiltration capacities of forest soils are generally high, largely due to the presence of a litter layer (Bruijnzeel, 1990; Anderson and Spencer, 1991). In a small tropical catchment in Brazil, Nortcliffe and Thornes (1984) showed that quickflow is almost entirely the result of saturation overland flow from floodplain areas immediately adjacent to the stream channel. This result was later confirmed for a similar catchment by Hodnett et al. (1997), and in northern Queensland, Australia where highly transmissive surface soils are underlain at shallow depth by a relatively impermeable subsoil (Bonell et al., 1983). In the summer monsoon daily rainfalls commonly exceed 250 mm, causing a perched water table to rise to the surface, resulting in widespread saturation overland flow.

These findings confirm the intuitive conclusion that in general slope, slope material and slope vegetation are in equilibrium such that precipitation is normally able to infiltrate the ground surface. Only where one or more of these factors has been drastically modified, usually by human activity or during the course of 'catastrophic' meteorological events, is widespread overland flow generated. Were this not so, the entire land surface would be scarred by gullies.

7.4.3 RIVER CHANNELS AND NETWORKS

Typical speeds of hillslope flows are influenced by slope, and by soil properties such as surface roughness and hydraulic conductivity. Once water enters a stream channel there are a number of other factors to consider. The speed of travel (typically of about 1 ms^{-1}) is much faster than flow through the soil, or even over the soil surface, and will be controlled by factors including the channel roughness and the configuration of the stream channels, as flow is routed down the network to the catchment outlet at the point of interest.

The flow in river channels is often represented by the Manning equation, which under uniform and steady flow conditions relates flow velocity, V, at a river section to the channel characteristics by:

$$V = R^{2/3} S^{1/2}/n \qquad (7.2)$$

where R is the channel hydraulic radius (cross-section / wetted perimeter), S is the channel gradient and n is the Manning roughness coefficient, ranging from about 0.02 for a relatively smooth channel without weed growth, up to about 0.8 for a channel with dense weeds. Barnes (1967) provides a standard reference guide for estimating the roughness coefficients for natural channels with data and colour photographs for 50 different stream channels for which the roughness coefficient had been determined. This guidance was later extended to overbank flow on floodplains (Arcement and Schneider, 1989).

Channels may be artificially altered to enhance flood water conveyance by dredging and 'weed control. The impact on flow hydraulics was demonstrated in a study of the hydraulic effects of weed cutting to reduce flood risk in a lowland chalk stream (Old et al, 2014). Following the cutting and removal of about 40% of the in-stream weeds on three separate occasions the Manning roughness was decreased on average by over 40%, water levels dropped by 0.25 m (about 20% of the channel depth) and flow velocities increased by 45–55%. The weed cutting increased the bankfull conveyance capacity by up to 140%, thereby significantly reducing the risk of over-bank flooding of nearby properties.

The river hydrograph can be thought of as a wave that travels downstream, and the way in which the wave moves and may change shape as it travels, is known as **flood routing.** Although its detailed study is properly the province of a hydraulic engineer, it is an important consideration in catchment hydrology. As the wave travels along a channel reach the flood peak may be reduced and smoothed. The amount of this **attenuation** will depend principally on the ability of the channel to store the water, and the channel roughness acting to slow the water down. A full description of the propagation of a flood wave can be found in many engineering texts (e.g. Akan, 2006). Flood routing procedures may be classified as either *hydraulic* or *hydrological.*

The former are more accurate, but require a considerable amount of spatial data related to river geometry and morphology, and involve the numerical solutions of the convective diffusion equations along the channel. Hydrological or lumped methods such as the Muskingum routing method (e.g. Karahan, 2010) are much simpler, and use the continuity equation and flow/storage relationship, and estimate hydrologic parameters using past flow records at upstream and downstream sections of the river.

As upstream runoff peaks move down the channel into the lowland reaches of river basins, where floodplains are normally well developed, a significant proportion of the quickflow volume may be temporarily retained within the floodplain material. This **bank storage** helps to mitigate the height of the peak and extend the timebase of its hydrograph (e.g. Whiting and Pomeranets, 1997). With the arrival of the runoff peak, the water level in the channel increases, resulting in the flow of water from the channel into the adjacent floodplain material. This bank storage begins to drain back into the channel as the peak passes and the channel water level falls.

The drainage basin is the most commonly used unit for modelling hydrological processes, for water balance studies, for chemical budgets and for examining human impacts on hydrological systems. As the size of the catchment increases the channel storage and routing becomes increasingly important. The balance between hillslope and channel network affects the flow hydrograph. In small catchments, travel times are short with little storage, so hillslope hydrology determines the hydrograph. As catchment area increases, the range of channel network travel times to the outlet increases relative to the

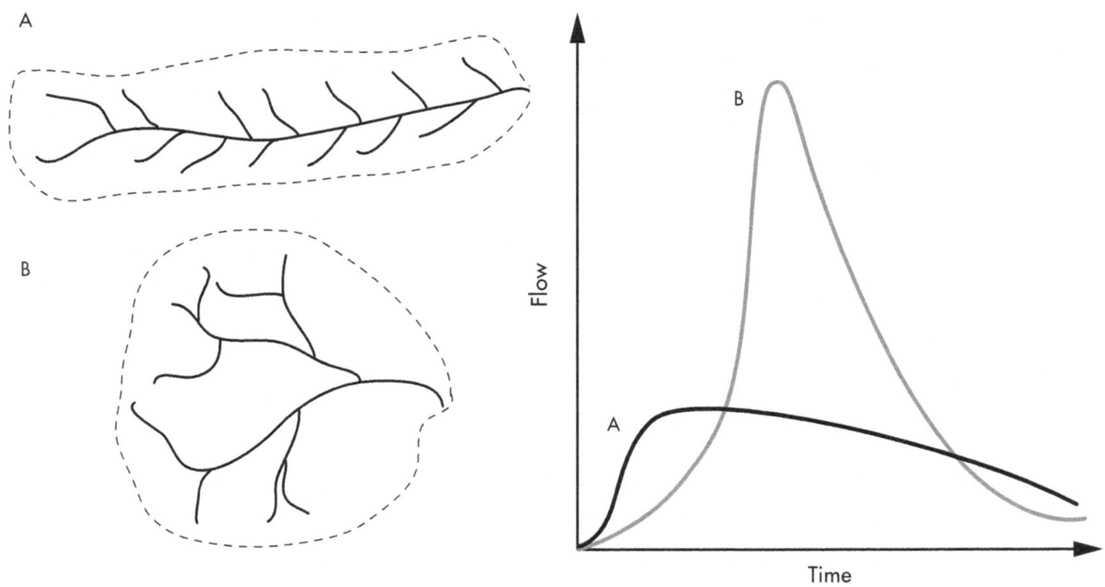

Fig 7.8: The geometry of the drainage network will have a strong influence on the shape of the catchment hydrograph.

Photo 7.2: Measurement of open water flow: (a) A v-notch weir is often used for small streams and small flows (typically less than 0.5 $m^3 s^{-1}$), (b) In a larger river a compound structure may be needed with a lower section to accurately measure low flows. (Photo source: Mark Robinson)

headwaters. The channel storage and routing, together with the effect of the channel network on flow synchronisation, then increasingly dominate the stream hydrograph. The basin shape and the pattern of the drainage network combine to influence the size and shape of flood peaks at the basin outlet, as shown in Figure 7.8.

This is important for understanding how changes within a basin, such as land cover, may (or may not) affect the pattern of flows at the drainage basin outlet. Thus, a local change may not be detected at a larger scale if channel processes dominate the river hydrograph further downstream. It used to be assumed by engineers (and is still accepted by many of the general public) that processes or activities that speed up flow on land will *always* worsen flood risk downstream, and processes that slow flow will *always* reduce flooding. Now, with computer modelling and GIS it is possible to consider location within a basin and changes to flow synchronisation. Thus, changing the speed of arrival of a flood wave (either faster or slower) from one sub-catchment may synchronise or desynchronise its arrival with the peak from another tributary where they meet.

7.4.4 MEASUREMENT OF RIVER FLOWS

River discharge is usually calculated by continuously measuring water level or **stage**, and combining these with measurements of river flow velocities, usually using a current meter, at different water levels and knowledge of the channel cross-section this can be used to compute volume of flow per unit time (usually m^3s^{-1}). River

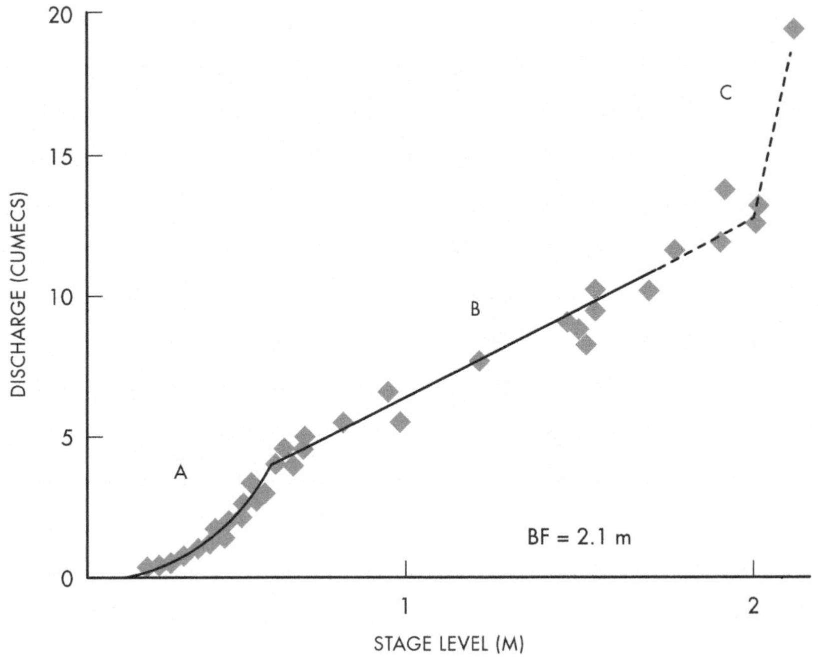

Fig 7.9: Stage discharge relationship for a complex stream section: (A) low flows are controlled by a flat vee weir, (B) rectangular main cross section, and (C) overbank flows. (Redrawn from data supplied by the NRFA, from measurements made by the Environment Agency)

flow measurement techniques can basically be divided into two groups: If the channel is stable, then a **flow rating curve** relationship may be established between the flow level and flow rate for a **rated section** of channel. If the channel is not stable (e.g. mobile alluvial bed), or more accurate flows are required, then a **gauging structure** may be built. This is an artificial channel section with a particular cross-section for which a stage-discharge relationship has been already established. Various types of structures, known as weirs and flumes have been calibrated. Details are given in works such as Herschy (2009), ISO (1999) and WMO (2010).

Rating curves whether for natural channels or gauging structures often use a power curve (similar in form to the Manning equation). The example here (Figure 7.9) shows the rating curve for a compound structure with the theoretical curves (solid lines) and **check gaugings** obtained using a current meter to measure flow velocity (and hence flow rate) through the channel at different stream water levels.

In addition to gauging structures and current meters, advanced technology including Acoustic Doppler Current Profilers (ADCPs) can map the depth and flow speed of rivers (e.g. Muste et al, 2004).

Ungauged basins

Many studies have attempted to derive relationships between flow properties and catchment characteristics, which can then be used to make flow estimates for ungauged basins (e.g. Thomas and Benson, 1975). One of the largest such studies was conducted as part of the Flood Studies Report (NERC, 1975). Over 1000 stream gauges were visited in the British Isles and their data quality was assessed including the state of the gauge and the accuracy of high flow measurements. The best 10% of the gauges were used and the most important parameters controlling the mean annual flood were found to be the catchment area, channel slope, soil type, drainage channel density, catchment wetness, and lake area. The derived relationships were included in the design flood estimates of the FSR, and some of the techniques developed were subsequently used in a study of European floods (Beran et al, 1984).

7.5 FLOW VARIATIONS – DAILY

Variations in runoff with time for catchments above about 100 km^2 are often studied using set time intervals (days, weeks, months, years). In the case of major continental rivers where the passage of flood peaks through the system may take several months, weekly flow values may be suitable, whilst for smaller basins, such as those of the British Isles), which respond rapidly to precipitation/melt events, hydrographs of daily mean flows may be more appropriate. The examples in Figure 7.10 illustrate the contrasting flow conditions of the flashy behaviour of the Dee, with its low permeability metamorphic and igneous geology and dominant quickflow component, with the subdued behaviour of the Lambourn, a highly permeability chalk catchment with a large delayed flow (or baseflow) component. The discharge scale

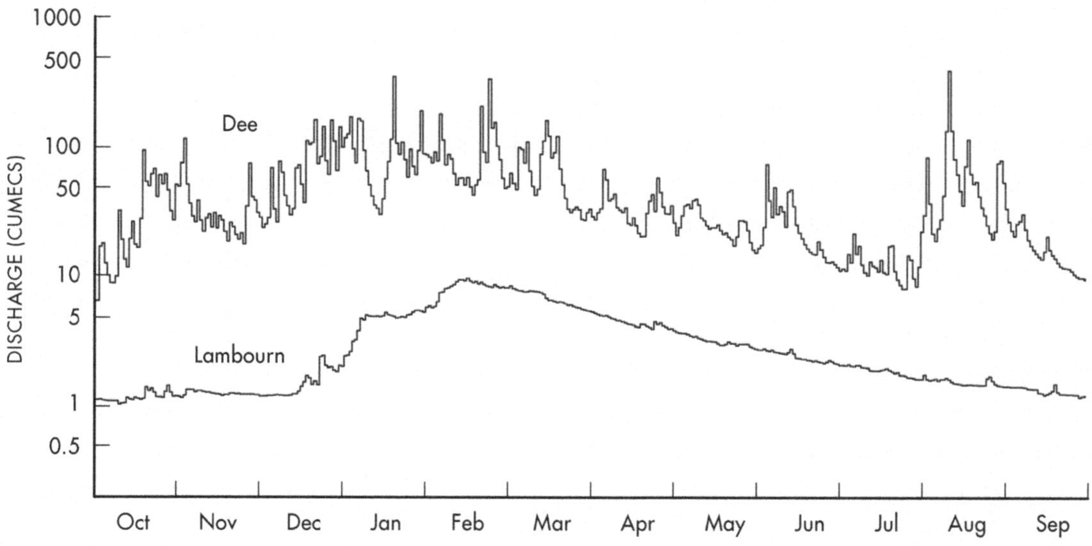

Fig 7.10: Hydrographs of daily mean flows (Oct 2013 – Sept 2014) for two British rivers: the Dee at Woodend (NE Scotland) and Lambourn at Shaw (SE England). (Data (m^{3}s^{-1}) provided by the UK National River Flow Archive, http://nrfa.ceh.ac.uk).

(m^3s^{-1}) is plotted as a logarithm to expand the low flow portion.

The long-term relationship between quickflow and delayed flow provides a basis for classifying streams as ephemeral, intermittent or perennial. **Ephemeral** streams consist solely of quickflow and therefore occur only during and immediately after a precipitation/melt event. There are usually no permanent or well-defined channels and the water table is always below the bed of the stream. Ephemeral streams are typical of arid and semi-arid areas, where they are characterized by large transmission losses. This means that runoff peaks, generated by storm rainfall, diminish rapidly downstream as they are absorbed by the dry stream beds, and under 10% of the runoff entering the channels actually leaves the catchment as streamflow. In some areas of inland drainage the percentage may diminish to zero for particular precipitation events. **Intermittent** streams, which flow during the wet season, and dry up during the season of drought, consist mainly of quickflow but delayed flow makes some contribution during the wet season, when the water table rises above the bed of the stream. A particular case occurs in high-latitude areas when flow ceases as subsurface water freezes during the winter. **Perennial** streams flow throughout the year because, even during the most prolonged dry spell, the water table is always above the bed of the stream, so that groundwater flow can make a continuous contribution to total runoff. Rarely is it possible to classify the entire length of a stream under only one of these three headings. A chalk bourne, for example, is normally intermittent in its upper reaches but perennial farther downstream; many other streams are ephemeral in their upper reaches but intermittent downstream.

The contrasting contributions of quickflow and delayed flow to total runoff, so clearly illustrated by the hydrographs of daily flow, reflect the integrated operation of a wide range of topographical, pedological, vegetational and geological factors that condition the runoff processes described earlier in this chapter. Thus the Dee with low permeability metamorphic and igneous geology has little baseflow, and in contrast the Lambourn is a high permeability chalk catchment with sustained high baseflow. The extremes of flow associated with a flashy stream and the more muted variations of a stream dominated by delayed flow may be quantified and compared more conveniently if the daily flow values are arranged according to their frequency of occurrence and plotted as a **flow-duration curve (FDC)**. This is a cumulative frequency curve showing the percentage of time that specified discharges were equalled or exceeded during a given year or period of years. Thus, for example the 95 percentile flow (Q95) is exceeded 95% of the time, or on average on all but 18 days of the year (i.e. 5% of 365). The FDC is often plotted on log-normal probability scale as this expands both ends of the flow-duration curve, since daily flows are often normally distributed and plot as a straight line. Streamflow data and flow-duration curves for many UK rivers are available from the National River Flow Archive (Dixon et al, 2013).

For ease of comparison a flow-duration curve may be plotted in dimensionless form (Figure 7.11), dividing the daily discharge values by the average daily discharge value

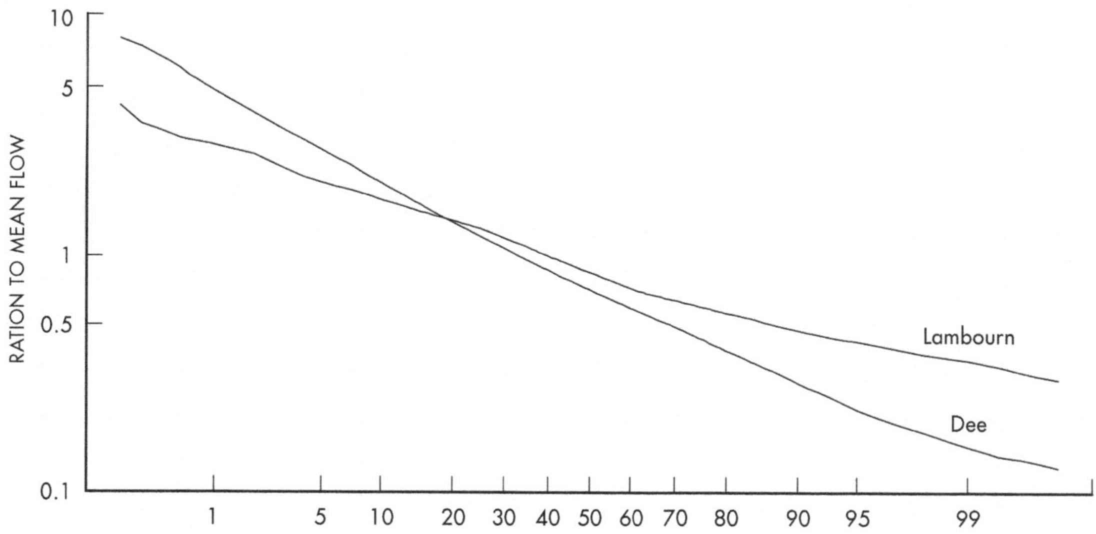

Fig 7.11: Dimensionless flow-duration curves of daily mean flows for the Dee at Woodend in NE Scotland and the Lambourn at Shaw in southern England (UK National River Flow Archive, http://nrfa.ceh.ac.uk).

for the period under study. The technique thereby combines in one curve the entire range of stream flows, and although not arranged chronologically, the shape and slope of the curve reflects the complex combination of hydrological and catchment factors that determine the range and variability of stream discharge.

Flow-duration curves that slope steeply throughout (e.g. Dee in Figure 7.11) denote highly variable flows with a large quickflow component, and gently sloping curves (e.g. Lambourn) indicate a large delayed flow component. In particular, the slope of the lower end of the flow-duration curve may reflect the perennial storage in the drainage basin, such that a small slope between 90% and 99% indicates a large amount of storage and significant groundwater contribution. The Q95 values i.e. the flow that is exceeded 95% of the time, as a proportion of the mean flow, are 0.22 for the Dee and 0.43 for the Lambourn. Inevitably, since low flows comprise baseflow, subsurface factors such as soil and geology are likely to play an important part. The strong control of geology is clearly identified, with groundwater discharge in the chalk catchment of the Lambourn sustaining low flows even in extreme droughts, whereas low flows from the mostly impermeable Dee catchment are at very low rates throughout the range of flow magnitudes.

7.6 FLOW VARIATIONS – SEASONAL

Over the course of the year many rivers demonstrate a broadly seasonal pattern that tends to recur each year, known as the **river regime**, when annual hydrographs are averaged over a decade or more. Natural river regimes are driven largely by climate, and modified by geology and vegetation. Many small rivers exhibit comparatively simple regimes having one period of high water and one period of low water each year. Comparisons between rivers of different sizes are facilitated if the data are plotted in dimensionless form, for example, as a ratio of the mean annual flow.

In temperate areas such as Britain, where rainfall is evenly distributed throughout the year, flow regimes are generally a reflection of the balance between rainfall and evaporation; low flow coincides with the peak of evaporation during the summer months and high runoff values occur during the winter months when evaporation is small. The River Thames at Kingston (Figure 7.12a) has a winter maximum partly due to higher winter precipitation, but mainly due to much higher summer evaporation. By contrast, in tropical areas evaporation tends to be high throughout the year, so that the rainfall distribution is the main determinant of river regimes, with high runoff occurring during the wet season. The Irrawaddy in Myanmar (Burma) experiences a tropical Asian monsoonal climate with a May to October flood season (Figure 7.12b) and the River Lobaye (a tributary of the Congo) in the Central African Republic has a tropical climate, with a wet season that lasts from June to October (Figure 7.12c).

In colder climates, flow regimes may result from the spring/summer melting of snowpacks or glaciers followed by a period of near-zero flow during the winter months when temperatures are low and icemelt is negligible; thus the low-flow period often coincides with the period of maximum precipitation and high flow is associated with the usually drier period of maximum melt. The Volga in Central Russia exhibits a spring snowmelt maximum (Figure 7.12d). Snowmelt floods are termed **nival**, as distinguished from **pluvial** or rain caused flows.

In some basins both snowmelt and rainfall may be important, giving rise to two low flow and two high flow periods. The Glama in Norway has a first high runoff period in spring resulting principally from snowmelt, followed by a period of low runoff. A second period of high flow may occur in the autumn as a result of rainfall exceeding evaporation and then a second period of low water levels in the winter due to precipitation being locked up as snow (Figure 7.12e). In Mediterranean areas a double peak may occur with a main peak due to winter rainfall and a second peak in some years as a result of intense summer convectional storms.

Large rivers may flow through several distinct relief and climatic regions, and receive the waters of large tributaries which themselves flow through varied terrain. As a result, the regimes of such rivers tend to change with distance downstream. The headwater reaches may be influenced by snow or glacier melt with a summer meltwater peak, whereas further downstream, inflows result from an excess of winter

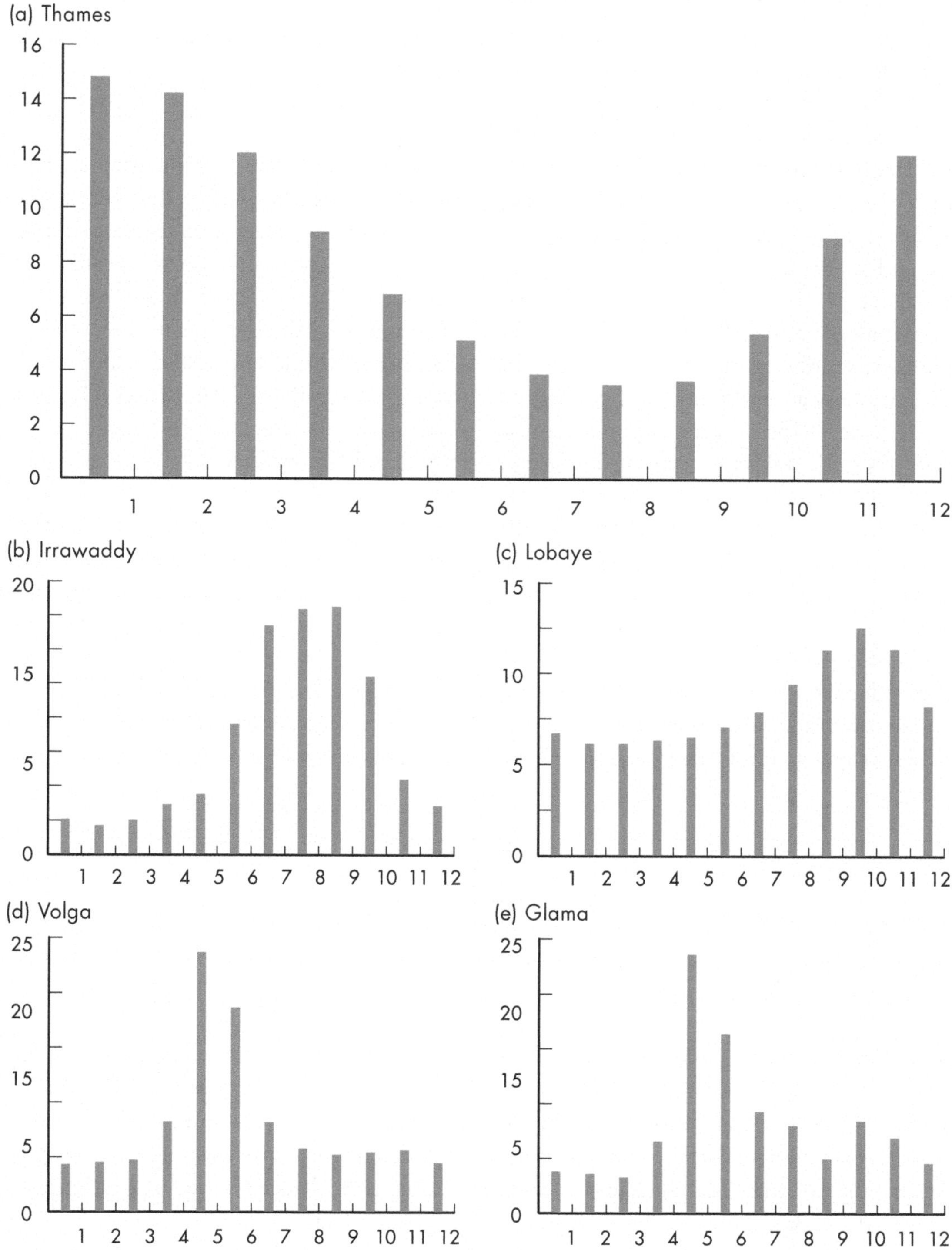

Fig 7.12: Examples of different river regimes based on long-term mean monthly discharges. 1 = January. Data provided by the Global Runoff Data Centre (GRDC) of the WMO, Federal Institute of Hydrology (BfG), Koblenz. Online retrieval: March-2016).

rainfall over evaporation. Thus, in the case of the Rhine, its headwater regime is dominated by meltwater flows with a summer maximum. Further downstream its flow is increasingly influenced by winter rainfall, and when it reaches the Dutch border, winter rainfall becomes the dominant factor and meltwater contributes only a secondary peak in the summer. Individual years may, of course, differ from these average patterns due to inter-year variations in rainfall, snow accumulation and melt.

Because British rivers are short, most have a 'simple' rainfall-evaporation regime with summer minima and winter maxima. The timing of these runoff extremes (both minimum and maximum) is progressively delayed towards the south and east as a result of a combination of geological and climatological factors. This spatial variation reflects both the pattern of increasing evaporation and also the greater water-holding capacity (and therefore later release of baseflow) from the large areas of sedimentary rocks in the south east, compared with the impermeable rocks of north and west Britain.

In the past, river regime graphs have been used in a rather simple, descriptive way. Increasingly, however, it is recognised that their dependence on climate means that they could also be used as an analytical tool for monitoring the changes in flow seasonality, both in time and space, which result from underlying environmental changes, particularly climate changes. Haines et al (1988) and Krasovskaia (1997), for example, used numerical approaches to obtain objective and reproducible grouping of monthly flow series to discriminate flow regime types, and Harris et al (2000) combined air temperature and discharge as the primary driving variables in riparian ecosystems. More recent studies (e.g. Moran-Tejeda et al, 2011) have looked at alterations in flow regimes over time due to environmental changes.

7.7 LONG-TERM VARIATIONS OF FLOW AND FLOW VARIABILITY

By definition river regimes are an expression of seasonal conditions averaged over many years, and so may imply a stability of long-term runoff which is misleading. According to the Global Runoff Data Centre (GRDC) database, there are about 450 sites globally with more than 100 years of flow data, but they are not evenly distributed (75% are in N America and 23% in Europe), or necessarily of high quality. One of the longest records, and one of the very few to have been 'naturalised' as far as possible to take account of net abstractions and discharges upstream of the gauging station, is the River Thames at Teddington. Its record from the 1880s is characterized by long periods of medium to high flow and shorter intervening periods of low flow. This bunching or grouping of wet and dry conditions has been referred to as 'persistence', and is an important complicating factor in the stochastic variation of precipitation driving the runoff process. It can make trend detection sensitive to the period of record studied

Records of UK river flows show that only 81 streamflow stations were in operation in 1953. There was then a steady increase to about 300 stations by 1960, followed by a rapid increase, as a matter of government policy, tripling to more than 1000 by 1975, and then numbers levelled out. The start of so many flow records in the same short time period from the 1960s is highly significant because that was a period of generally low flows across the country, and this artefact can bias the interpretation of flow trends over time (Marsh et al, 2015). In many areas of the world, flow records are even shorter than those for the UK, and the opportunities for identifying trends in the variation of runoff with time are therefore more limited and must be treated with caution.

7.8 EXTREMES OF RUNOFF

To some extent persistence accentuates the contrast between the extremes of flow. Flood conditions resulting from a given precipitation event may be more severe if that precipitation event occurs at the end of a long sequence of such events. Low flow or drought conditions will certainly intensify as the preceding dry period is prolonged. Thus, in many parts of England and Wales the extreme low flows recorded in August 1976 came at the end of the driest 17-month period so far recorded. Similar, less severe, low-flow conditions recurred between 1988 and 1992. However, the comparisons between extreme high and low flows should not be pursued too far, since although extreme low flow events are indeed dependent on antecedent conditions, extreme flood events are much more directly dependent on the severity of the causal precipitation/melt event. Severe floods can, and of course frequently do, occur in deserts and other persistently low rainfall areas.

Limitations of space mean that floods and droughts are discussed only briefly in the following paragraphs. For a more detailed treatment, the reader is referred to books on floods such as Smith (2013) and White and Watts (1994).

7.8.1 FLOOD FLOWS

Flood peaks are generated in river channels by a variety of causes (see Figure 7.13). Most river floods result directly or indirectly from climatological events such as excessively heavy and/or excessively prolonged rainfall. In cold winter areas, where snowfall accumulates, substantial flooding usually occurs during the melt season in spring and early summer, particularly when melt rates are high. Floods may also result from the effects of rain falling on an already decaying and melting snowpack. An additional cause of flooding in cold winter areas

Photo 7.3: Hydrological Extremes: a) River Piracicaba in flood at Piracicabain in Southern Brazil, during an extremely strong El Niño event in 1983 (Photo source: John Roberts). b) Thruscross Reservoir in the Washburn Valley, North Yorkshire during a drought, with exposed remains of the village of West End (Photo source: Ken Parkes, September 1995)

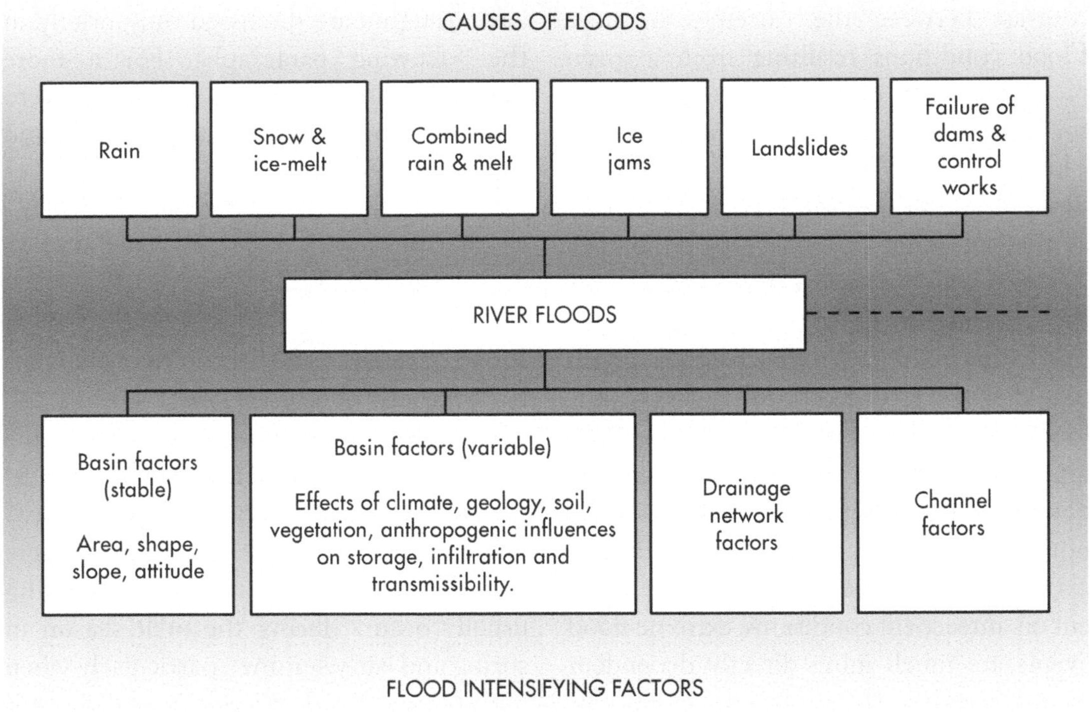

Fig 7.13: Causes of river floods, and flood intensifying factors. The dashed line indicates the interaction with estuarine and coastal floods (from a diagram in Ward and Robinson, 2000).

is the sudden collapse of ice jams, formed during the break-up of river ice

Some of the most devastating floods occur in tropical regions due to the intensity of the rainfall experienced there. One recent tragic example is the Pakistan flooding in 2010, when there were unusually intense monsoon rains attributed to La Niña (see Chapter 2). Heavy rainfalls of more than 200 mm were recorded in the Indus headwaters areas in the N of Pakistan during a 4-day period in July 2010. The resultant flooding left 20% of country under water; 2.6M ha of cultivated land were devastated and over 2000 people killed (see Smith, 2013). Temperate areas can suffer from extreme floods too, although generally with few fatalities. For example it is only just over a 100 years ago that the centre of Paris was flooded with extensive damage to property, but no loss of life (Jackson, 2010).

Hydrologists are concerned to know the rarity of such events, and often use **flood frequency** analysis of past data. There are many excellent engineering and statistical books detailing the frequency analyses of floods and droughts. Instead this section touches on aspects that the hydrologist may need to consider *interpreting* the results of such techniques. A flood **return period** or **flood interval** is the *probability* that a flood of a given size (or bigger) may occur in any year. Hydrologists may refer to a 100-year flood, which can be misleading to the layman. It does not mean it will occur once every 100 years, and certainly does not mean that after one has occurred there will be no further flooding of that magnitude for another 99 years. Rather, that it is the average time interval between years containing such a flood, i.e. there is 1% probability (p = reciprocal of the return period) that such a flood occurs in any one year. Following such a flood, the probability of a similar flood occurring in the following year remains 1%. Due to the common misunderstandings caused by time-dependent terms such as 'period' or 'interval', many hydrologists prefer the terms flood 'risk' or 'probability'.

Flood frequency analyses commonly make several assumptions that should be verified:

a. The data are **homogeneous**, i.e. the catchment has not undergone changes (such as land use), and neither has the climate. It is also assumed that the observed floods will come from the same 'population' as other potentially larger floods. If there are subsequent floods produced by entirely different mechanisms (for example tropical cyclones in an area where temperate storms are the norm), then using data from the normal events to predict flood risk in the future will miss these outliers,

b. The length of record and hence the **sample size** should be sufficiently large to encompass some major floods and not be restricted to a period of unusually flood rich or flood poor years,

c. It is dangerous to **extrapolate** too far beyond the record length – perhaps twice the length of the data set.

d. Estimation is only as good as the **quality of data** available; the measurements of extreme floods are of particular importance, but may be the least accurate.

Photo 7.4: Flood levels of the River Severn recorded on the south wall of the Water Gate to Worcester Cathedral. The highest recorded level was in 1770, and some other notable floods are shown in the inset. The bankfull level is 13.1 m. (Flood marks levelled to Ordnance Datum supplied by Environment Agency) (Photo source: Mark Robinson, May 2015).

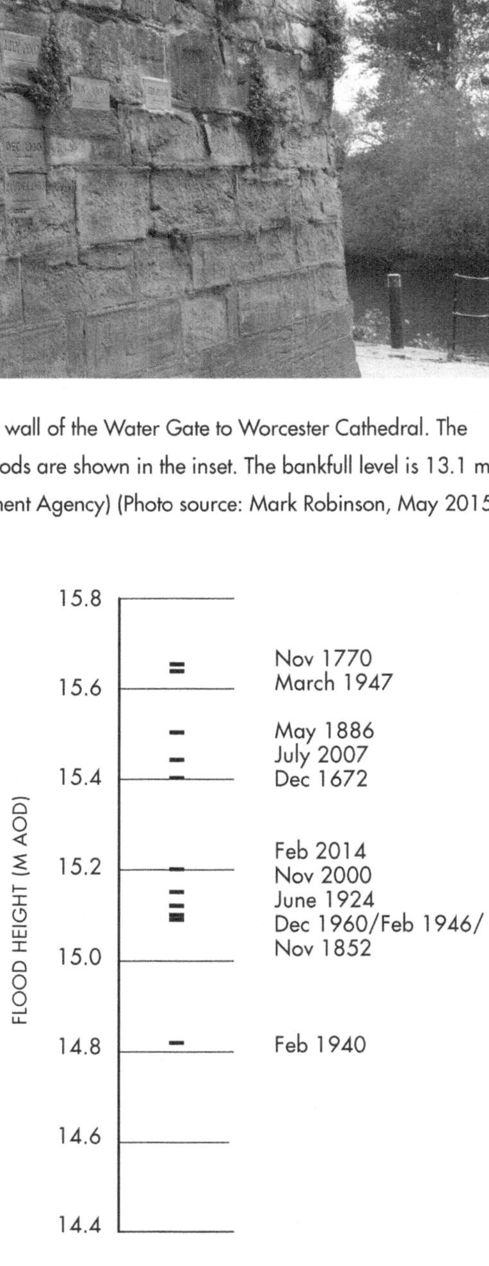

An excellent discussion of the potential pitfalls of flood frequency estimation is provided by Reed (2002).

Historic flood markers can provide valuable additional information about flood risk and a context for current conditions, but must be treated with caution. Channel dimensions may have changed, and obstacles such as bridges and mills may have been added or removed. There are also descriptive records of past floods (and droughts) from local community and church records as well as agricultural harvest records (e.g. Stratton, 1978).

Flood intensifying factors

As the lower part of Figure 7.13 shows, floods may be modified by a number of factors. These can operate either to ameliorate or to intensify flooding although, for the sake of brevity, only the latter function is considered in this discussion. For example, river floods may be intensified by factors associated either with the catchment itself or with the drainage network and stream channels. Most of these operate to increase the volume of quickflow and to speed up its movement. Few of these factors operate either uni-directionally or independently. Area, for example, is fundamentally important in the sense that the larger the catchment, the greater is the flood produced from a catchment-wide rainfall event. However, when a storm covers only part of the catchment, the attenuation of the resulting flood hydrograph, as it moves through the channel network to the outlet, is greater in a large catchment than in a small one. Again, basin shape and the pattern of the drainage network combine to influence the size and shape of flood peaks at the basin outlet as was shown in Figure 7.8. Some of the most complex relationships, those between the variable basin factors, have a significant influence on three important hydrological variables, i.e. water storage, infiltration, and transmissibility.

Water storage in the soil and deeper subsurface layers may affect both the timing and magnitude of flood response to precipitation, with low storage often resulting in rapid and intensified flooding. High infiltration values allow much of the precipitation to be absorbed by the soil surface and may thereby reduce catchment flood response, depending on the extent and growth of areas of saturation overland flow and on subsurface transmissibility; low infiltration values encourage infiltration-excess overland flow leading to rapid increases in channel discharge (see also Section 7.4).

Urbanisation will generally intensify flood peaks downstream due to the large areas of impermeable surfaces and also the urban storm sewer network removing runoff from these surfaces. Some rural activities may be similar to urban changes by replacing soil water storage and slower subsurface flows by overland flow. The most extreme cases are soil loss by erosion, which may result from the removal of protective vegetation, and soil crusting causing a change from subsurface flow to rapid overland flow.

Channel changes can take a number of forms; they may be farmland drainage to aid crops, which can unintentionally alter the rate at which water reaches the stream network and so influence flow regimes downstream (Robinson and Rycroft, 1999). Or they can be deliberate interventions such as channel dredging to increase the carrying capacity of a stream to reduce local overbank flooding. The removal of floodplain storage may simply move the problem further downstream, the extent of which depends on the design flood capacity of the scheme (Sear et al, 2000).

Spatial patterns of flooding

Although floods at any location in a river system are a function of the floods generated in the catchment upstream of that point, the relationship between flood behaviour in headwater catchments, and the flood behaviour of the entire river basin is often complex. The downstream flood hydrograph differs from the upstream hydrograph for

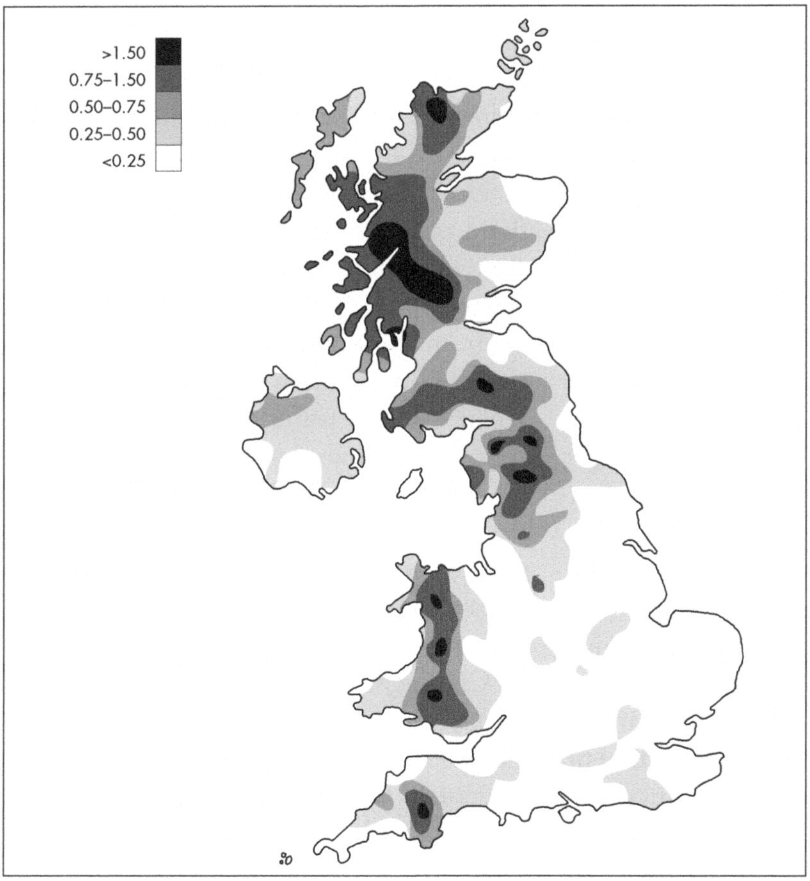

Fig 7.14: Map of the median annual maximum flood, QMED, ($m^3\ s^{-1}\ km^{-2}$) for gauging stations in the UK, based on the NRFA Peak Flow data set. (Data source: CEH)

the same event, partly because of lag and routing effects, partly because of the changing nature of the basin geology, physiography and climate from headwaters to outlet, and partly because of scale effects.

Scale effects are important in relation to both catchment conditions and precipitation inputs, and frequently restrict our ability to generalise from existing flood data and to predict flood occurrence and distribution. The fact that flood peak discharges tend to increase downstream when measured absolutely (i.e. m^3s^{-1}) but decrease downstream when expressed as specific discharge (i.e. $m^3s\ km^{-2}$) may in part reflect steeper slopes and higher rainfall in headwater areas, but there is also a mismatch between the scales of catchment and precipitation event. Large catchments normally have a lower specific discharge than small catchments partly because they may be only partly covered by a flood-producing storm, while smaller catchment areas may be completely covered (see Section 2.6.3), thereby generating high specific flood discharges.

From the preceding discussions of runoff processes and flooding, one might expect the flood-producing potential of each river basin and sub-catchment to be distinctively different. However there is some evidence of a spatial dimension to river flooding on a scale larger than that of a river basin. This can be illustrated for the UK, by an index of flood potential, the median annual flood (Figure 7.14).

The pattern of isopleths drawn through the QMED values shows a marked gradient from specific discharges exceeding 1.50 $m^3s^{-1}km^{-2}$ in the north and west to values well below 0.25 $m^3s^{-1}km^{-2}$ in the south and east. It should be noted that the comparatively low values of flood flow per unit area over south eastern England may be misleading in the sense that they are, to some extent, compensated by the large area of the catchments concerned. The highest instantaneous gauged discharge for the Thames at Kingston (9,950 km^2), for example, is about 900 m^3s^{-1} (Terry Marsh, pers comm., 2016), exceeding that of 663 m^3s^{-1} for the Tees at Broken Scar (820 km^2) and 830 m^3s^{-1} for the Clyde at Blairston (1,700 km^2).

The Flood Studies Report (NERC, 1975) generalised the relationship between mean annual flood (*QBAR*) and the flood of a given return period (*QT*) for defined geographic regions by a **Growth Factor**, by which the mean annual flood value is multiplied to obtain approximate values of floods having specified return periods. Subsequent analyses of European data (Beran et al, 1984) confirmed that, in areas having the same flood-producing mechanism, it is possible to 'pool' or average flood frequency curves to produce inter- as well as intra-national comparisons. The Flood Estimation Handbook (FEH), which succeeded the Flood Studies Report in the UK, adopts a more sophisticated approach in relating flood hydrology to catchment characteristics. This is achieved largely by using indicators such as stream length, stream density and slope gradient which can be derived from digital terrain models and other increasingly accessible digitised and gridded data which was not available at the time of the earlier report (Stewart et al, 2015).

In Britain, flood risk maps have been produced showing the 1% (1 in 100 years) and 0.1% (1 in 1000) chance of inundation each year for England and Wales (see http://watermaps.environment-agency.gov.uk/ and 0.5% (1 in 200 years) for Scotland (http://www.sepa.org.uk/environment/water/flooding/flood-maps).

In the USA the Federal Emergency Management Agency (FEMA) produces similar maps showing areas at flood risk (http://www.newfloodmap.com/).

7.8.2 LOW FLOWS

At the other extreme, the problems posed by low flows, although different, are equally varied and severe. Low flows not only reduce the amount of water available for supply but also lead to water quality degradation, as the diluting and reaerating capability of streams and rivers is reduced. This, in turn, leads to the aesthetic degradation of the affected channel reach. In contrast to the dramatic impact of flood flows, hydrologists were relatively slow to develop adequate methods for estimating low-flow characteristics, or even to standardise low-flow definitions. Some of the main analysis methods applied to low-flow data have been reviewed and discussed by Smakhtin (2001), WMO (2008) and Farquharson et al (2015).

Low-flow definitions

Low flows are a seasonal phenomenon, and an integral component of any river flow regime. Measures of low flows are needed for a variety of purposes. In addition to general hydrological descriptions, they are used for licensing abstractions, wastewater treatment design outflows, assessing the probability of severe drought, determining pollution dilution as well as assessing general river health for preserving the aquatic environment and the Water Framework Directive (WFD) requirements.

In broad terms low flows are determined by the balance between precipitation and evaporation and are therefore particularly susceptible to persistence when this results in the bunching of a sequence of dry years. Within a drainage basin experiencing essentially uniform climatological conditions, however, other more local or catchment controls play a significant role in determining the detailed pattern of low-flow variation.

There are a number of different ways of describing the low flow regime of a river flow and a drought event in terms of its frequency, duration and severity, and some of the main approaches are described below.

There is no low flow equivalent of the Probable Maximum Flood since the *minimum* flow that can be experienced is obviously zero. In most small catchments, and some larger ones in arid areas, zero flows can occur, and then it is then the *frequency* of zero flows that is the more useful index of low-flow conditions. In most rivers, however, the problems associated with low flows are manifest long before a flow of zero is attained, and are intensified as the duration of a given low-flow discharge is prolonged. Hydrologists are therefore concerned primarily with defining selected critical low-flow discharges and with identifying the *frequency* and *duration* of spells of low flow.

In contrast to high flows that can be identified in terms of instantaneous peak values, low flows are prolonged and a daily flow interval is less appropriate than, say, the mean annual minimum consecutive 10-day or 7-day flow. Annual minima can be derived from a daily flow series by selecting the lowest flow every year over different durations, and plotting low flow frequency curves, using e.g. a Weibull distribution. Mean annual minima for each of these durations may be calculated. Alternative flow frequency characteristics include the mean annual minimum 7-day flow, which is approximately the driest week in the average summer and has a return period of about 2.33 years, denoted 7Q2.33, while in the USA the 10-year return period 7-day flow (7Q10) is a widely used index of low flow (Smakhtin, 2001). Some analyses have been based on very extreme flows, including the 7-day 20-year flow (7Q20), but few gauging stations are designed to measure such extremely low flows and so the recorded discharges may be subject to large percentage errors.

Several flow conditions can be defined as 'low'. One is the lowest flow ever experienced, a condition that was closely approached over much of southern and central England in the late summer of 1976 and again on several occasions. A more usual measure, however, is the 95% exceedance flow (Q95), i.e. the flow that is exceeded 95% of the time, or on average on all but 18 days of the year, and this is the most common low-flow index used internationally (Farquharson et al, 2015). Thus in the UK's Low Flow Studies Report (IH, 1980) equations were derived for estimating the 95 percentile 10-day flow, denoted Q95(10) from values of BFI and either annual average precipitation or main stream length (IH, 1980).

A major advance for the estimation of low flows at ungauged sites was the development of an automated base flow separation procedure from river flow data to produce a **Base Flow Index (BFI)** that could be related to mapped geology (Beran and Gustard, 1977). The BFI is an index of hydrograph behaviour, so that a high BFI (close to 1) reflects a baseflow-dominated regime, whilst a low BFI indicates a 'flashy' regime dominated by quickflow. It was found that this index is closely related to catchment geology. Thus the Dee at Woodend which has low permeability metamorphic and igneous geology has a BFI=0.53; and the highly permeable Lambourn chalk catchment, a BFI = 0.97. The index is also sensitive to the storage effects of lakes and reservoirs. The BFI has been widely adopted as a general index of catchment response, and is routinely calculated for over 1,000 UK catchments (Marsh and Hannaford, 2008).

Gustard et al (1992) used HOST (Hydrology of Soil Types) maps instead of BFI, to facilitate flow estimates for ungauged basins. Relationships between soil type and low flow parameters were also discussed for the UK and continental Europe by Gustard et al. (1992) and Gustard and Irving (1994). Similar analyses have been conducted in New Zealand (NWSCA, 1984). Subsequently, software has been developed to facilitate the analyses, including LowFlows2000 and LowFlows Enterprise (Stewart el al, 2015).

Patterns of low flow in Britain

The most significant recorded periods of below average runoff for England and Wales, occurred for about 25 years commencing around 1885, and for a shorter duration in the 1930s and 1940s. Since then periods of low flow have been less sustained although sometimes quite dramatic in their impact. The variability, although modest in comparison with many semi-arid areas, nevertheless reflects a widespread pattern of alternation of lengthy periods of high and low flow.

The 1975–76 drought resulted in runoff values less than 40% of the long-term mean over much of southern England, and provided a major stimulus to low-flow studies (Rodda and Marsh, 2015). Another significant drought event occurred in the UK in 1984, despite rainfall in the 1980s being higher than for any decade since 1900. Then, during the period 1988–92, there occurred a drought that was, in places, as severe as the 1975–76 event. The 2-year period from July 1990 produced lower runoff totals in parts of lowland England than had previously been recorded (Marsh et al. 1994). The 1988–92 drought also affected large areas of continental Europe and by late 1990 more than 3000 km of rivers had dried up in southern France and low flows were causing irrigation problems from Hungary to Spain. A review of some notable British droughts is provided by Marsh et al (2007).

7.9 RUNOFF FROM SNOW-COVERED AREAS

Much of the land surface polewards of 40° N has a significant seasonal snow cover in most years, and there are also extensive areas of snow cover in high altitude areas. The problems of measuring snowfall, and estimating its water equivalent, were discussed in Chapter 2. Here we consider the amount and timing of snowmelt, which differs from rainfall-driven runoff in several important respects, as we have seen in Section 7.6. The storage of precipitation as snow and ice introduces potentially long delays in runoff, with in some cases low flows during the colder period of maximum precipitation, and high flows at warmer times of maximum melt, when there actually is little precipitation.

The prediction of snowmelt is important in areas of seasonal snow cover where winter snow comprises much of the total annual precipitation. It is needed to estimate seasonal flood risk and, in some arid areas to estimate the amount and timing of melt water for water-supply, and for irrigation and electric power generation. Mankin et al (2015) estimated that 2 billion people in the Northern Hemisphere rely on spring and summer snowmelt. They are mainly in a geographical zone between 25°N and 45°N, and include the Indus and Ganges rivers in Pakistan and India, the Huai in China and the Colorado and Rio Grande in N America.

In addition to difficulties in the estimation of snowpack water equivalent and the quantification of energy exchanges and rate of snowmelt, there remain uncertainties about the relationship between snowpack properties and water movement. The rate at which meltwater, generated near the surface of a snowpack, can move through the pack to the underlying ground surface and thence to the channel system is greatly affected by the structure and stratification of the snowpack. Most snowpacks develop layers of ice strata within the more permeable snow matrix due to the sequence of snow deposition, melting and refreezing. These layers divert the percolating meltwater so that complicated flow paths develop, which not only delay and diffuse the outflow of meltwater but also greatly increase the storage capacity of the snowpack (e.g. Singh et al., 1997).

7.9.1 SNOWMELT

The various approaches adopted to estimate the amount and timing of meltwater may be grouped broadly into empirical and conceptual models. The former use regression equations between snowmelt and usually a temperature index, and conceptual or physically based energy balance models attempt to explicitly represent the various hydrological processes.

Snowmelt results from many different processes involved in a net transfer of heat to the snowpack surface where meltwater is predominantly generated. The principal energy balance fluxes are solar radiation, long-wave radiation (especially at night), sensible heat transfer from the air to the snow by convection and conduction, and latent heat transfer by evaporation and condensation (and occasionally rain).

The relative importance of the energy balance components varies with time, both seasonally and diurnally, as well as with different weather conditions. Kuusisto (1986) summarized the findings of 20 studies of energy fluxes of melting snowpacks. On sunny days with little wind, net radiation may be the dominant source of energy gain. Its importance increases as the spring season advances due to increasing solar radiation as well as the decline in albedo with the ageing of the snowpack. Turbulent heat exchange (sensible heat) will dominate during the night and on cloudy days. It is greatest when there are strong, moist winds. Although dry air can cause some evaporation (sublimation) of the snow, humid air can have a much greater effect since water vapour condensation on snow releases sufficient latent heat to melt a much larger quantity of ice.

Snowmelt rates may differ with vegetation type, especially between forested and open sites. Forests dampen turbulent fluxes and shade direct solar radiation, although long–wave radiation is higher from the warmed tree branches. The role of tree canopy storage of snow is discussed in chapter 3. In general snowmelt in a forest is less rapid than in the open, largely as a result of the canopy shading the ground.

Before snowmelt can occur, the properties of a snowpack change: density increases, snow crystals become large–grained and the albedo reduces. With the onset of warmer temperatures, meltwater from the surface layer of the snow percolates down to lower, colder layers and refreezes. This freezing releases latent heat, warming the lower layers of the snowpack and, over time, tends to equalize temperatures at 0°C throughout the vertical profile of the snowpack. When this retention capacity of the snowpack is at its maximum (equivalent to 'field capacity' of a soil), the snowpack is described as being 'ripe'. Further addition of energy will result in meltwater runoff, as liquid water in excess of the retention capacity drains through the snowpack.

There are often insufficient meteorological data available to compute the energy budget of a snowpack, and so empirical methods are used to predict the magnitude and timing of snowmelt. Generally these are based on correlations between snowmelt and aspects of air temperature, although it is clear from the preceding discussion that due to the variation in the relative importance of the various heat transfer processes, no single index or method of estimating snowmelt will be applicable to all areas and for all weather conditions. In fact since the temperature of melting ice is 0°C, the overlying surface air temperature will be influenced by this, which undermines the temperature index approach.

More complex physically-based and spatially distributed models rely on many parameters, most of which are difficult to identify, and without calibration to observed data their predictions tend to be inaccurate however much 'physics' the models contain (Bloschl, 1999). Even when they are calibrated by streamflow or snow cover measurements there are substantial uncertainties in the parameter values, and so for operational systems the simpler temperature index approach is often used.

7.9.2 RUNOFF FROM GLACIERIZED AREAS

In high mountain areas the presence of glaciers and ice sheets contributes a further meltwater dimension to runoff variations. Glaciers form where perennial snow cover builds up and gradually it is compacted into ice. The ice, although solid, then flows slowly downslope under the influence of its own weight to lower altitudes where the glacier then melts (UNEP 2008). As in snow-covered areas, the hydrology of glacierized basins is largely thermally controlled. As well as snow and ice storage, large quantities of *liquid* water may be stored within glaciers and in marginal lakes alongside them. Meltwater will be routed over the glacier surface and either run off the glacier edges or enter the subglacial drainage system via 'sinkholes' or **moulins**. These subsurface conduits change morphology over the course of a melt season and together with the changing structure of the glacier ice and the dynamic interaction with the liquid water stored within it, means that the variations of runoff from glacierized basins is very complex. The runoff from such areas is also characterised by sudden outburst floods, releasing large quantities of water stored within, under or alongside glaciers.

Glacier runoff is thus characterized by two main components, a periodic, thermally driven meltwater regime, which produces distinctive diurnal and seasonal variations of flow, and an irregular component, which results from the occurrence of either extreme meteorological events or sudden releases of water from the glacial drainage system. The periodic pattern of an increase of discharge during the summer melt season broadly reflects the seasonal increase of available energy and progressive development of the glacier drainage system, together with a diurnal range that decreases in amplitude towards the end of the melt season.

Irregular variations of runoff from glacierized basins are exemplified by unusually high flood discharges resulting from (a) periods of very rapid melt over a week or more, which permit high rates of baseflow as well as of quickflow, (b) the occurrence of extreme high-intensity rainfall, especially late in the afternoon when meltwater runoff is at a maximum, or (c) sudden releases of water which has either been held in storage within the glacier, or as surface lakes on or adjacent to the ice, or has been dammed back by ice in tributary valleys. In some cases the flood outburst appears to be triggered when meltwater behind the glacial dam reaches a critical elevation.

Finally, long-term variations of glacier runoff, which result mainly from long-term changes of climate, are an amalgam of conflicting influences (Marty and Blanchet, 2012). Periods of persistently warmer summer weather result in increased ablation and high runoff values, whereas a series of cool summers favours increased storage and low runoff. However, continued ice removal causes shrinkage of the lower sections of glaciers and since these are the zones of highest melt rate, their progressive disappearance results in a corresponding loss of potential for meltwater yield.

UNITS OF RUNOFF

Runoff is normally expressed as a volume per unit of time. The cumec, i.e. one cubic metre per second (m^3s^{-1}), and cumecs per square kilometre ($m^3s^{-1}km^{-2}$) are commonly used units. Runoff may also be expressed as a depth equivalent over a catchment, i.e. millimetres per day or month or year. This is a particularly useful unit for comparing precipitation and runoff rates and totals since precipitation is almost invariably expressed in this way.

FLOW TERMINOLOGY

It should be noted that the term 'flood' is used in this chapter and widely by hydrologists to mean a hydrograph peak, and does not imply overbank flow.

Similarly the term 'runoff' does not necessarily imply a restriction to flow over the surface.

REVIEW PROBLEMS AND EXERCISES

7.1 Describe the various processes which generate streamflow and the circumstances in which one or another of them may become dominant.

7.2 Explain the difference between 'infiltration-excess' overland flow and 'saturation' overland flow.

7.3 Outline the role of the following in the generation of quickflow: convergence; macropores; throughflow; groundwater ridge.

7.4 Discuss whether hydrograph separation is a valid exercise.

7.5 To what extent are there distinctive runoff-producing conditions in semi-arid and tropical rainforest areas?

7.6 Define the flow duration curve and explain what information it may yield on drainage basin characteristics.

7.7 In the light of growing information about climate variability and climate change, discuss the following concepts: river regimes; mean annual flow; the 100-year flood.

7.8 Define the following terms: annual maximum series; partial duration series; return period; dry weather flow; naturalised flows.

7.9 Outline the main causes of river floods.

7.10 Discuss the influences of geology and climate on the generation of low flow conditions.

7.11 Compare the movement of water through a snowpack and through a soil profile.

7.12 Explain what is meant by the following: water equivalent of a snowpack; snowpack energy balance; a 'ripe' snowpack; glacier runoff.

WEBSITES

Environment Agency (England and Wales)
Flood Risk Maps
http://maps.environment-agency.gov.uk/wiyby/
Flood warnings
https://fwd.environment-agency.gov.uk/app/olr/home
Scottish Environment Protection Agency
http://www.sepa.org.uk/environment/water/
Environmental Protection Agency (Ireland)
http://www.epa.ie/irelandsenvironment/water/
Office of Public Works (Ireland)
http://www.opw.ie/en/flood-risk-management/
Global Runoff Data Centre
http://www.bafg.de/GRDC/EN/Home/
National River Flow Archive (UK)*
http://nrfa.ceh.ac.uk/
HiFlows
http://nrfa.ceh.ac.uk/peak-flow-data
Hydrological Outlook UK
http://www.hydoutuk.net/
National Water Information System (USA)
http://waterdata.usgs.gov/nwis/sw
Water Survey of Canada
http://wateroffice.ec.gc.ca/index_e.html#

*River flow records for around 1,500 UK gauging stations totaling 59,000 years' of daily data, are available to download directly from the NRFA website.

8

WATER QUALITY

'If you could tomorrow morning make water clean in the world, you would have done, in one fell swoop, the best thing for improving human health by improving environmental quality'

WILLIAM C. CLARK , PROFESSOR OF INTERNATIONAL SCIENCE, PUBLIC POLICY & HUMAN DEVELOPMENT, HARVARD UNIVERSITY

8.1 INTRODUCTION AND DEFINITIONS

The preceding chapters in this book dealt with the storages and fluxes of water through the hydrological cycle, but made little mention of the nature of the water itself except in specific areas where soil erosion was severe, or where salinization occurred. In fact hydrologists recognise the crucial importance of **water quality**, both its *chemical* characteristics due to dissolved material, and *physical* characteristics such as temperature, taste and suspended solids. Water quality research is one of the most rapidly developing aspects of hydrology. Given the increasing usage and pollution of water sources by human activities, and the continuing development of new chemicals it may be argued that the challenges of water quality are often even more difficult and demanding than those of water quantity, and in many areas of the world the use of water is limited by its quality rather than by its quantity.

Information on water quality is of great importance for a wide range of purposes, including water supply and public health, agricultural and industrial uses. Water quality is also important to preserve aquatic and terrestrial habitats. There is thus a pressing need to monitor and control the ever-increasing human impact on water chemistry through various forms of pollution.

In principle, evaporation provides a continuous recycled source of pure (distilled) water for precipitation. This then becomes increasingly concentrated with dissolved material as it moves through the atmosphere and in subsequent stages of the hydrological cycle as it comes into contact with organic matter, soil and rock material, and is used by humans for carrying away waste products. In fact no waters are free from human influences – even Arctic precipitation contains constituents discharged into the atmosphere. Much remains still to be learned about the chemistry of these natural processes, as well as the mechanisms and pathways of solute fluxes. The hydrologist studying the rates and pathways of water movement is thus placed in a unique position to help find solutions to these problems.

This chapter provides a brief overview of general principles and discusses a number of selected aspects of this important area of research; further details can be found in specialist books (e.g. Stumm and Morgan, 1996). Before turning to the water quality processes at different stages of the hydrological cycle, the properties of water and the nature of chemical reactions are reviewed.

8.1.1 PROPERTIES OF WATER

Water occupies a central role in the transport of chemicals around the surface of our planet. Although its appearance is bland and pure water is almost colourless, tasteless and odourless, it has certain properties that make it unique (see Section 1.1.1). Water is a **chemical compound** of two commonly occurring **elements**, hydrogen (H) and oxygen (O), but differs in behaviour from most other compounds to such an extent that it has been called a 'maverick' compound. Many of its physical and chemical properties are unusual, but of most importance for water quality studies is the fact that virtually all substances are soluble to some extent in water. The water **molecule** (H_2O) is strongly attracted to most inorganic substances (including itself). It is, in fact, practically impossible to produce and store absolutely pure water (Lamb, 1985).

The unusual properties of water can be accounted for by its molecular structure. Each water molecule comprises two **atoms** of hydrogen attached to one oxygen atom by a very strong and stable mechanism, involving sharing a pair of electrons, known as a **covalent bond**. The two hydrogen atoms are not on diametrically opposite sides of the oxygen atom but at an angle of 105° apart (Figure 8.1a). This produces a bipolar molecule, equivalent in effect to a bar magnet, with an unbalanced distribution of **electrical charge**: the oxygen atom on one side of the molecule has a negative charge while the side with the two hydrogen atoms has a positive charge. As a result of this **electrostatic** effect, adjacent water molecules tend to interact by a process known as **ionic** or **hydrogen bonding** (Figure 8.1b). It is the combined strength of these two types of bond that accounts for the unusually large latent and specific heat capacity of water. It is also responsible for water's cohesive nature and large surface tension which enables it to 'wet' surfaces and move through materials such as soils and plant stems by capillarity.

These properties account for the solubility of many materials in water. The atoms in many substances are held together not by strong covalent bonds but by weaker electrostatic attraction. These bonds may

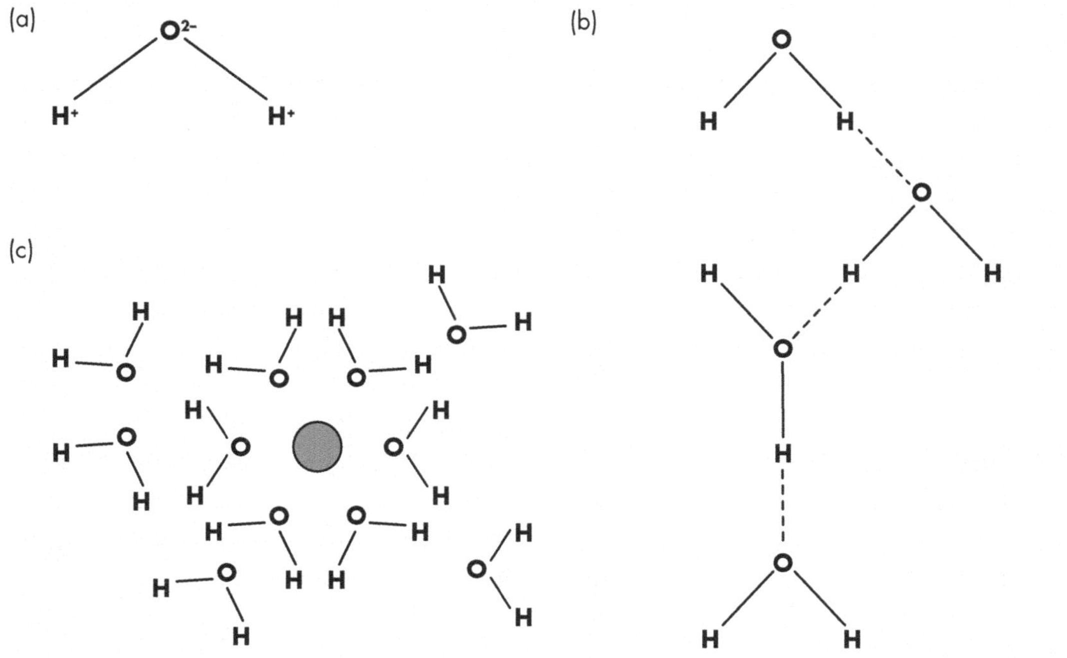

Fig 8.1: Structure of water: (a) water molecule comprising one oxygen atom and two hydrogen atoms (protons), (b) hydrogen bonding between water molecules, (c) hydration 'shell' of water molecules surrounding a cation in solution.

be weakened further by the bipolar water molecules which act to cancel out some of the electrostatic attraction and enable the atoms to move apart as separate electrostatically charged atoms or groups called **ions** – positively charged cations and negatively charged anions. When these ions become surrounded by water molecules and have little direct influence on each other, the substance is said to be **dissolved** (Figure 8.1c). The liquid (in this case, water) is called the **solvent** and the dissolved solid is called the **solute**. The mixture of solvent and solute is called a **solution**. This chapter is concerned with **aqueous solutions**, in which water is the solvent. Most inorganic compounds dissociate into ions when they dissolve in water, although there may be some interaction between oppositely charged ions to form **complex ions**. Organic compounds occur in solution as unchanged molecules (Hem, 1985). The powerful solvent action of water is vital for plant and animal life as it provides the medium for the transport of chemicals and nutrients. In that sense it may be considered to be the original 'elixir of life' (Lamb, 1985). However, this same solvent action also works to transport harmful pollutants and toxic substances through the environment.

8.1.2 WATER QUALITY CHARACTERISTICS

The notion of 'water quality' encompasses many different factors. Commonly quoted **determinands** include physical characteristics such as colour, temperature, taste and odour, as well as chemical characteristics such as acidity, hardness, and the concentrations of various constituents including nitrates, sulphates and dissolved oxygen and man-made pollutants including pesticides and herbicides. There is no simple single measure of the purity of water, and the term 'quality' only has meaning when related to some specific use of water. Thus, the concentration of total dissolved material in raw sewage is similar to that in many groundwater supplies used for drinking water – both are about 99.9% pure water, but they are obviously very different in other respects!

There are many texts that detail analytical procedures for the evaluation of different aspects of water quality (e.g. Drinan and Spellman, 2012; Viessman et al., 2013). It is outside the scope of this chapter to deal with them in detail, but certain aspects will be outlined which are of importance for discussions in subsequent sections in which the data and results from the completed analyses are discussed. For many purposes it is not the total amount of a particular element that may be of interest but the chemical form in which it occurs. Thus, for example, nitrogen may occur in a number of **species**, including organic nitrogen, ammonia (NH_3), nitrite (NO_2) and nitrate (NO_3), and these may have very different effects on the suitability of the water for different uses.

Methods of chemical analysis are available to identify and measure the concentrations of many elements and compounds in water. The most commonly used unit for expressing the concentration of dissolved constituents is as the weight of solute per unit volume of water, e.g. milligram per litre (mg l^{-1}). For some purposes the weight of solute per unit weight of solution is used, measured most often as parts per million (p.p.m.). For most practical purposes the two systems yield the same numbers; however, for highly mineralized water with solute concentrations greater than 7,000 mg l^{-1} a density correction should be used to convert between the two. For the calculation of the masses of substances involved in chemical reactions the concentrations may be expressed in **moles** per litre, where a mole of a substance is its atomic or molecular weight in grams. For thermodynamic calculations, described later, **chemical activities** rather than concentrations are used. A correction factor, or activity coefficient, usually represented as γ_i, is applied to the concentration values to allow for non-ideal behaviour of ions in solution. Its value is unity for ideal conditions; for dilute solutions (less than 50 mg l^{-1} of dissolved ions) the coefficient is generally >0.95, but at high concentrations (e.g. 500 mg l^{-1}) for ions with a large charge, or valency, it may be as low as 0.7.

For many purposes it is useful to express chemical species by their **equivalent weight**. This is the molecular weight of the ion dissolved in water, divided by its ionic charge. Concentration are often expressed in units of milliequivalents per litre (meq l^{-1}) or as mg l^{-1}. Table 8.1 gives the chemical formulae of many of the common ions in solution, which are discussed in this chapter, and conversion factors between the different methods of expressing concentration. In any solution the overall number of positive and negative electrical charges must be equal to maintain electrical neutrality, i.e. the total meq l^{-1} of cations must equal the total meq l^{-1} of anions. This requirement of **electroneutrality** is useful for checking the accuracy of the determination of ionic concentrations and for ensuring that all of the significant ionic species in a solution have been accounted for. Some ions, however, such as silica (SiO_2), do not have a charge, and therefore an equivalent weight cannot be computed.

An aspect of water quality which is of great importance, since it affects many chemical reactions, is the **acidity** of the water (Drever, 1997). Whether or not solutes are present in water, some of the water molecules will dissociate into hydrogen (H^+) and hydroxyl (OH^-) ions. Since the resulting concentrations of H^+ ions are very low, they are expressed in terms of the pH, or negative $\log_{10}$ of the H^+ ion activity, i.e.

$$pH = \log_{10} (1 / [H^+]) \qquad (8.1)$$

The square brackets denote chemical activities in moles per litre. Values of pH less than 7 (10^{-7} moles/litre of H^+ ions) are said to be acidic, while those above 7 are **alkaline**. A pH of 7 at 25°C is said to be **neutral** but as hydrogen ion behaviour is temperature dependent, this value decreases somewhat with increasing temperature. Natural waters, that are uninfluenced by pollution, generally have pH values of between 6 and 8.5 (Hem, 1985). This may appear a small variation but it should be remembered that since pH has a logarithmic scale, a change of one unit corresponds to a ten-fold change in H^+ ion concentration.

The term 'acidity' applied to aqueous solutions may also be defined as the ability to react with OH^- ions, and this may be determined by titration with an alkali (Stumm and Morgan, 1996). It is a function of a number of solute species (including, for example, iron) and is not simply related to the H^+ concentration. In contrast, the 'alkalinity' of water (i.e. its ability to react with H^+ ions) can usually be identified with the concentration of CO_3^{2-} and HCO_3^- ions. The 'strength' of an acid refers to the extent to which it dissociates in solution.

To understand the chemical processes in natural waters that affect the composition of water and to make quantitative statements about them requires the application of certain fundamental concepts, of which some of the most useful are the principles of chemical thermodynamics.

NAME	SPECIES	FORMULA WEIGHT (APPROX.)	MEQ $L^{-1} \rightarrow$ MG L^{-1}
Aluminium	Al^{3+}	26.9	8.994
Ammonium	NH^{4+}	18.0	18.037
Bicarbonate	HCO_3^-	61.0	61.013
Calcium	Ca^{2+}	40.1	20.040
Carbonate	CO_3^{2-}	60.0	30.003
Chloride	Cl^-	35.4	35.448
Hydrogen	H^+	1.0	1.008
Hydroxide	OH^-	17.0	17.007
Iron (ferrous)	Fe^{2+}	55.8	27.925
(ferric)	Fe^{3+}	55.8	18.615
Magnesium	Mg^{2+}	24.3	12.152
Nitrate	NO_3^-	62.0	61.996
Nitrite	NO_2^-	46.0	45.998
Phosphate	PO_4^{3-}	95.0	31.656
Orthophosphate	HPO_4^{2-}	96.0	47.985
	$H_2PO_4^-$	91.0	96.993
Potassium	K^+	39.1	39.093
Silica	SiO_2	60.1	-
Sodium	Na^+	23.0	22.988
Sulphate	SO_4^{2-}	96.1	48.031

Table 8.1: Names and formulae of some common chemical species showing their electrical charge (valency), formula weight and conversion factor to equivalent weight units. The concentration (mg l^{-1}) divided by the final column equivalent weight (formula weight/valency) gives milliequivalents per litre (based on data in Hem, 1985).

8.2 PROCESSES CONTROLLING THE CHEMICAL COMPOSITION OF WATER

Chemical processes in natural waters are primarily concerned with reactions in relatively dilute aqueous solutions; these are usually **heterogeneous** systems comprising a liquid phase with either or both a solid and a gaseous phase. Due to the great complexity of natural water systems it is usual to employ simplified models to illustrate the principal regulatory factors controlling the chemical composition of natural waters. Many reactions are **reversible**, being able to proceed in both directions, and in practice a dynamic equilibrium will be established between the two opposing reactions. The behaviour of such reversible reactions may be studied using the principles of **chemical thermodynamics**. This enables the likely direction of a reaction over time to be determined and the final equilibrium solute concentrations in the water to be predicted (Sposito, 1994; Stumm and Morgan, 1996). The final products of an **irreversible** reaction will be determined by the quantities of the reactants available.

The solution of gaseous carbon dioxide in water is a reversible reaction producing carbonic acid (H_2CO_3), and may also form the ions HCO_3^- and CO_3^{2-}:

$$CO_2(g) + H_2O \leftrightarrow H_2CO_3\,(aq) \quad (8.2)$$

$$\updownarrow$$

$$HCO_3^- + H^+ \quad (8.3)$$

$$\updownarrow$$

$$CO_3^{2-} + 2H^+ \quad (8.4)$$

The second and third steps produce hydrogen ions (H^+) and will lower the pH of the solution. Letters within brackets in equations in this chapter indicate the physical state of the substance: g = gaseous, aq = aqueous species occurring in solution, and c = crystalline solid.

Similarly, a solid may dissolve in water; an example of this is calcite ($CaCO_3$) which occurs in many carbonate rocks:

$$\underset{\text{Calcite}}{CaCO_3\,(c)} + H^+ \leftrightarrow \underset{\text{bicarbonate or hydrogen carbonate}}{HCO_3^-} + Ca^{2+} \quad (8.5)$$

Depending upon the pH of the water there may be subsequent interactions between the dissolved carbonate species, i.e.

$$HCO_3^- \leftrightarrow \underset{\text{carbonate}}{CO_3^{2-}} + H^+ \quad (8.6)$$

or

$$HCO_3^- + H^+ \leftrightarrow \underset{\text{Carbonic acid}}{H_2CO_3} \quad (8.7)$$

The **equilibrium constant** (K) of a reversible reaction has a constant value for a given combination of reactants and products at a given temperature. Experimentally obtained values at standard temperature (usually 25°C) are available in the chemical literature. Alternatively, the equilibrium constant of a reaction may be calculated from the Gibbs free energy (Drever, 1997).

The solution of $CaCO_3$ in water may be used to illustrate the use of these principles to give the final equilibrium values of a set of reactants. The equilibrium constant is calculated from the ratio of the activities of the products divided by the activities of the reactants, i.e. for the reaction given in Eq. (8.5),

$$K = [Ca2^{+}]\,[HCO_3^{-}]\,/\,[CaCO_3(c)][H^{+}] \quad (8.8)$$

where K for this reaction has a published value (Jacobson and Langmuir, 1974) of 81, and the activity of a solid (here $CaCO_3$) is taken as unity, so the equation becomes:

$$81 = [Ca^{2+}][HCO_3^{-}]\,/\,[H^{+}] \quad (8.9)$$

Therefore, given measurements of the pH, solution temperature and concentrations of calcium (Ca) and bicarbonate (HCO_3^-) it is possible to say whether the system is in equilibrium. If the quotient is $<K$, the water may dissolve more calcite (assuming it is present); if it equals K the water is at equilibrium; or if it is $>K$ the solution is supersaturated and could precipitate calcite. The fact that the final equilibrium condition depends upon the amounts of the reactants and products is known as the law of **mass action** (Schnoor, 1996). It does not provide quantitative information on the *rate* of a reaction, although in general the further it is from equilibrium the faster it may be.

For a chemical in gaseous form the **partial pressure** is used in such calculations. This is the proportion (by volume) of the particular gas, multiplied by the total pressure (measured in atmospheres). More complex reactions may be dealt with by combining several equilibrium equations, For example, the dissolution of CO_2 in water produces H^+ and HCO_3^- ions, which are a reactant and product respectively of the dissolution of calcite. Adding the equations for these two reactions (Drever, 1997) enables the solubility of calcite to be expressed as a function of the partial pressure of CO_2.

In practice there are many limitations to the application of thermodynamic procedures in real world situations since, outside of the chemistry laboratory, there are likely to be exchanges of energy and reactants with the surrounding environment and equilibrium may not be attained. Nevertheless, the principles have proved very useful for indicating the direction and the maximum extent of reactions and are widely adopted. A number of computer programs are available to facilitate the calculations of equilibrium conditions, of which one of the best known is PHREEQC (Parkhurst and Appelo, 2013).

The rate of different chemical reactions can vary enormously. It is more likely to be attained in deep aquifer systems, where movement is slow and residence times are long, than for rapid near–surface flows. While thermodynamics deals with equilibrium states, **chemical kinetics** is concerned with the mechanism and rate of operation of chemical changes and with the factors controlling the reaction rate (Stone and Morgan, 1990; Schnoor, 1996). For a reversible reaction the ratio of the forward and backward reaction rates equals the equilibrium constant. Many reversible reactions involve a sequence of intermediate steps, some rapid, some slow. Kinetics can identify the slowest or limiting change that determines the overall rate of the reaction (Drever, 1997; Stumm and Morgan, 1996).

In addition to these chemical considerations of the thermodynamics and kinetics of reactions, hydrology plays an important role in determining solute composition and concentrations. Apart from snow and ice, and with the exception of some deep groundwater systems, water is generally in continual movement. Its velocity, and hence residence time, will strongly influence whether or not the water attains a chemical equilibrium for a particular reaction. Many reactions are **diffusion controlled**, i.e. their rate constants are controlled by the physical speed at which reactants can diffuse together, rather than by the rate of chemical reaction at a given point. The dynamic nature of flow, particularly in the soil zone, means that in periods between storms pore water solute concentrations may increase as minerals are dissolved, but are then flushed out by new waters in the next storm. The flushing frequency and interstorm period may be important variables for river chemistry.

Succeeding sections of this chapter deal in turn with water quality behaviour and changes as the water passes through different components of the hydrological cycle. These begin with the composition of precipitation and then the soils and groundwater, ending with the mixture of chemicals found in rivers and lakes. Both natural and man–made sources are considered, as it may often be difficult or impossible to separate the two. Finally, the relation between water quality and the characteristics of the region or catchment area is discussed.

8.3 ATMOSPHERIC SOLUTES

At the instant that a droplet is formed in the atmosphere the water is very pure, but its chemistry will alter rapidly both within the cloud and in its fall through the atmosphere to the Earth's surface. Particulate material may act as nuclei for raindrop formation (Section 2.1), determining its initial chemical composition, and as the precipitation moves through the atmosphere it will accumulate further particulates by entrainment and various gases in the atmosphere will dissolve in the droplets. The particulates in the atmosphere originate from a wide variety of sources including ash from volcanoes and power stations and wind–blown dust. Of particular importance as cloud condensation nuclei are **aerosols**. These are very small particles (less than 1 mm) which may be liquid or solid material and originate from the land or sea, or from chemical reactions in the atmosphere.

This natural 'scrubbing' of the atmosphere by precipitation, removing gases and particulates, is a major means by which the air is purged of materials that might otherwise accumulate to reach dangerous concentrations. There is often a period of much improved visibility following the removal of particulates from the atmosphere by heavy rain or snowfall.

The oceans comprise about 70% of the Earth's surface, and sea salts are a major source of dissolved material in precipitation. Seawater droplets become entrained

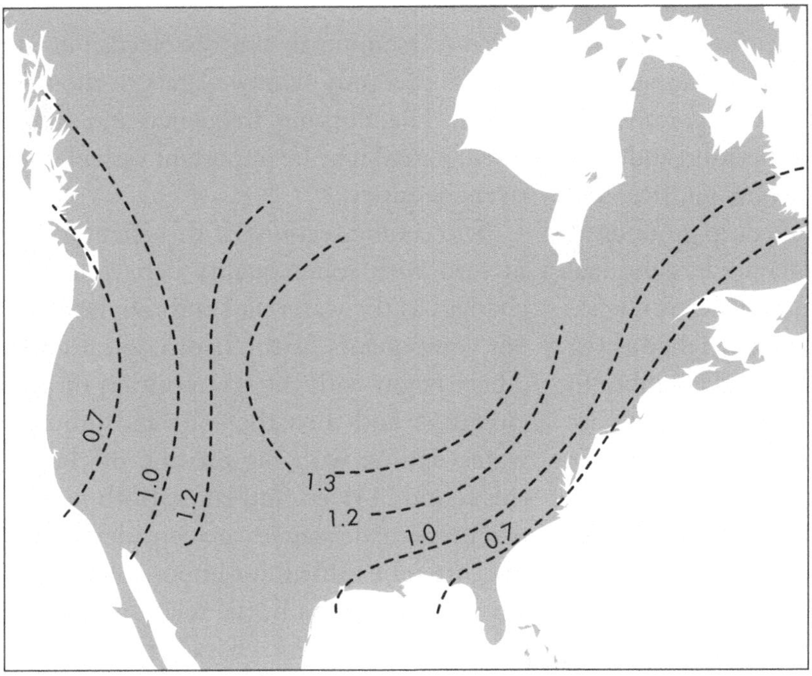

Fig 8.2: Ratio of Na/Cl concentrations (mg l^{-1}) in precipitation over the conterminous USA. Complied from data in Munger and Eisenreich (1983).

in the atmosphere when waves break and are carried upwards by turbulence, becoming increasingly concentrated as their water evaporates. This may continue until just a solid particle is left, carried in the wind until it is dissolved in rain. The supply of sea salt to the atmosphere varies with meteorological conditions and the state of the sea surface. A number of studies have mapped the concentrations of the dissolved elements in precipitation (e.g. Munger and Eisenreich, 1983) and have demonstrated a decline with distance inland from the coast in those elements, including Na^+, Cl^-, Mg^{2+} and K^+, which are derived from marine sources. In contrast, the solutes in precipitation falling over inland areas are derived predominantly from terrestrial sources, and include Ca^{2+}, NH^{4+}, SO_4^{2-}, HCO_3^- and NO_3^-. These comprise substances from natural sources such as gases from plants and soils and wind–blown dust and, in addition, oxides of sulphur and nitrogen produced by the burning of fossil fuels and from industrial and vehicle emissions. As a consequence there are differences in the relative amounts of these two groups of ions, e.g. the ratio Na^+/Cl^- (mg l^{-1}) which is about 0.56 in sea water increases with distance from the ocean due to Na enhancement from terrestrial sources, e.g. dust (Figure 8.2).

The relative concentrations of some marine–derived ions in precipitation may differ from sea water due to **fractionation** and enrichment. Rising bubbles tend to retain ions with larger charge/mass ratios, ejecting them into the atmosphere on bursting at the surface. Sodium, chloride and sulphate, in contrast, occur naturally in similar proportions in precipitation as in the oceans, and the ratio of Cl^- to SO_4^{2-} in rainfall may be compared to that in the oceans to determine the 'excess' input of sulphate to the atmosphere (i.e. above that from natural marine sources), assuming that all the Cl^- in a rainwater sample is of the marine origin.

There has been considerable concern about the possible effects of the many polluting substances emitted into the atmosphere by human activities. These **anthropogenic** materials comprise both particulates and gases, and while many are naturally occurring constituents even in 'unpolluted' atmospheres, some of the substances may be harmful if present in high enough concentrations. Appreciable amounts of pollution in precipitation began in the nineteenth century during the Industrial Revolution. Peat soil profiles near industrial towns in northern England contain widespread accumulations of soot and heavy metals within the layers formed in the last two hundred years. This pollution and the associated gases (which are not retained in the peat record) are held responsible for major changes in natural vegetation species at that time over extensive areas of the Pennine uplands of Britain (Ferguson and Lee, 1983). Not surprisingly, early concern about atmospheric pollution was centred on the clearly visible depositions of particulates, rather than the dissolved material in precipitation, leading many countries to introduce smoke abatement legislation such as Clean Air Acts. The term **acid rain** was probably used for the first time in the 1870s (Smith, 1872). In the 1950s Gorham (1958) showed evidence of links between atmospheric pollution and the acidity of precipitation and surface waters in small pools, although the significance of this work was not recognized by other workers at that time.

It was not until the late 1960s that a direct link between 'acid rain' and damage to the environment was first identified with declining fish populations in Scandinavia as a consequence of the acidification of freshwater rivers and lakes (Havas, 1986). Since that time acidification of fresh waters

Photo 8.1: Forest in Jizera Mountains, Czech Republic, 1995, severely damaged by 'acid rain'. (Photo source: Mark Robinson).

was observed in other parts of the world, particularly in the more industrialised Northern Hemisphere and it has been found to also affect forests, crops and soils.

The primary chemical pollutants are sulphur dioxide (SO_2) and nitrogen oxides NO and NO_2 (often referred to together as NO_x) which are produced by the burning of fossil fuels. These undergo oxidation in the atmosphere, to sulphuric acid (H_2SO_4) and nitric acid (HNO_3), in a number of complex reactions, involving sunlight, moisture, oxidants and catalysts. Many of the reactions involve photochemical oxidants, including ozone (O_3), the hydroxyl radical (OH^{-}) and hydrogen peroxide (H_2O_2).

The reactions in very simplified form may be viewed as:

$$NO + O_3 \rightarrow NO_2 + O_2 \quad (8.10a)$$

$$NO_2 + OH^- \rightarrow HNO_3 \rightarrow NO_3^- + H^+ \quad (8.10b)$$

$$SO_2 + SO_2 \rightarrow SO_3 \xrightarrow{H_2O} H_2SO_4 \rightarrow SO_4^{2-} + 2H^+ \quad (8.10c)$$

Sulphuric acid is the main cause of acid deposition because sulphur is emitted in much larger quantities than nitrogen, and the sulphuric acid molecule in solution releases two H^+ ions whereas nitric acid releases one.

Although the term 'acid rain' has been widely used it is a misleading name, and acid deposition is more accurate. Due to the dissolution of atmospheric CO_2 even pure water has an acidic pH of about 5.6 (UNEP, 1995) i.e. well below the neutral pH 7. This value has been used as the reference level for distinguishing 'natural' from polluted 'acid' rain. However, other natural materials, including dissolved aerosols and volcanic gases, can also influence the pH, resulting in still lower values. Furthermore, while rain is usually the most important mechanism by which atmospheric water transfers pollutants to the ground it is not the only one. In some areas snow is an important component of precipitation, while in upland and coastal areas frequent cloud or mist can provide a significant contribution to acidification. Although the volumes of fine water droplets of mist and cloud are much lower than rainfall they generally contain much higher concentrations of acid than the larger drops of rain and may cause a proportionately much more important chemical input than the quantity of water deposited would suggest. This so–called **occult** deposition occurs much more efficiently on aerodynamically rough surfaces such as tall vegetation such as trees, in exactly the same way that such surfaces have greater interception losses due to the enhanced air turbulence.

The forms of **wet deposition** (rain, snow and mist) are efficient processes for removing material from the atmosphere, but are restricted to the times when condensation and precipitation occur. Wet deposition also depends upon the source of the air mass and the time it has recently spent over areas of high emissions. Appreciable acid deposition also takes place by means of the **dry deposition** of gases and aerosol particles onto the surfaces of soils, plants and water bodies. The main process is by the absorption of gases, such as SO_2 and NO_2, rather than particles, and will depend upon the chemical and physical affinity of each gas for a particular surface (Fowler, 1984). In contrast to the episodic nature of wet deposition, this is a continuous process, although the rate of deposition may fall as the collecting surface approaches 'saturation' (Fowler and Cape, 1984). Subsequent oxidation to SO_4^{2-} and NO_3^- takes place on the soil and vegetation surfaces when they are wetted by rain or dew. In addition, gases may also pass into the plant stomata and be metabolized.

The relative importance of dry and wet deposition varies with factors such as geographical location and season (due to differences in the amounts of rainfall and of artificial emissions). In general, dry deposition usually occurs within two or three days and dominates close to emission sources and wet deposition is more important at greater distances as there is greater opportunity of being oxidized to sulphuric and nitric acids which are then dissolved in precipitation. The ratio of dry/wet deposition declines systematically from about 10 close to pollution sources to <1 for areas over 300 km away (Fowler, 1984). The pattern of acid deposition is, however, not simply related to the distribution of sources. In Britain, for example, while the largest inputs by dry deposition are in the industrial parts of the country, the largest loads of wet deposition are in fact upwind of the emission sources. These are the remote uplands of the north and west of the country where, despite low concentrations of H^+ ions in the rain, the total input is greatest due to the high rainfall amounts.

Acid deposition causes the acidification of soils and waters if the neutralisation of the acids by weathering is too slow (Stumm

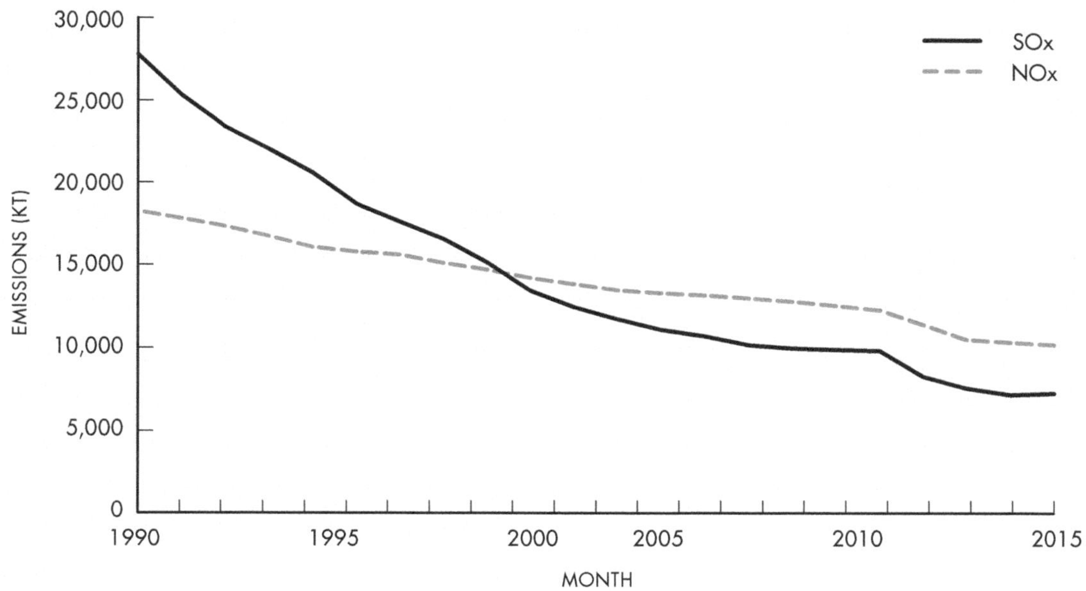

Fig 8.3: Declining annual emission loads across Europe, 1990–2011. Sulphur oxides (SOx) decreased by about 75% and nitrogen oxides (NOx) by over 40%. Data from European Environment Agency (www.eea.eu/data-and-maps). Downloaded June 2016.

and Morgan, 1996). The sensitivity of a particular catchment to acid deposition depends on a variety of factors, including the geology, soils and land use; this is discussed further in Sections 8.5 and 8.6, particularly in terms of the concept of a 'critical load'. The areas where damage is most severe are not necessarily those receiving the greatest deposition of acidity.

Following concern about damage to ecosystems from acid deposition the Convention on Long Range Transboundary Air Pollution (CLRTAP) was established under the United Nations Economic Commission for Europe (UNECE) and has been key in developing cross-border air pollution control strategies. Emissions of sulphur oxides decreased by about three quarters between 1990 and 2011 due to measures including switching to low sulphur fuels and fitting flue gas desulphurisation technology on power stations, while emissions of nitrogen oxides fell by nearly half primarily due to fitting catalytic converters on petrol fuel vehicles, although reductions in NO_x for diesel vehicles were smaller than expected (Figure 8.3). These reductions were in large part achieved by measures proposed by science, aided by separate developments such as the collapse of the old Soviet bloc in Eastern Europe and the closure of their highly polluting industries (Reis et al, 2012). Corresponding reductions for the UK are about 90% for SO_2 and 72% for NOx since 1990 (Battarbee, et al, 2014).

Globally, acid deposition is still an issue. China is now the world's largest emitter of air pollutants, in large part due to the manufacture of goods for foreign consumption as North America and Europe outsourced their manufacturing to China (Lin, et al, 2014). The Chinese have an ambitious program of installing flue gas desulphurisation to their coal-fired power stations, which has resulted in a significant decline in their energy sector emissions (Klimont et al, 2013). This is mostly driven by human health concerns, because large amounts of calcareous dust blowing from the deserts of central Asia have been able to largely neutralise the environmental impact of their emissions on surface acidification (Skeffington, pers. comm. 2016). Comparable reduction strategies are not yet in place in several other Asian countries, most notably India where emissions are still rapidly increasing (Klimont et al, 2013). Comparable reductions in NO_x have been smaller, in part due to the increased popularity of diesel vehicles which have higher real-world emissions than expected.

At present surface water acidity is declining across Europe and North America and biological systems are recovering in many areas, although the pattern is quite complex (Battarbee, et al, 2014). So it is with some justification that the work to counter acidification can be claimed as one of the greatest success stories for environmental science (Williams et al, 2015).

8.4 INTERCEPTION AND EVAPORATION

Different land uses affect the water budget of an area, including much higher evaporation losses from forests than from grassland due to the greater interception losses of the taller and aerodynamically rougher trees (Chapter 3). This will increase the solute concentrations of the remaining water reaching the forest floor, although it will not affect the total solute loads. Vegetation can directly influence the total amount of solutes reaching the ground in several ways. Trees provide efficient collecting surfaces for the deposition of fine mist or fog droplets, which may have much higher concentrations of solutes than rainfall, and they may also receive deposition of particulate materials and absorb gases into their leaves by stomatal uptake. In a study of water chemistry changes due to pollution in the canopy of a coniferous forest in northern Britain, Cape et al. (1987) found that the sulphate loads in the rainfall above the trees amounted to only 30% of that reaching the ground via throughfall and stemflow. They concluded that the bulk of this gain was due to leaching of SO_4^{2-} from the foliage, and that this material originated from gaseous SO_2 taken up by stomata and from particles containing SO_4^{2-} which had been deposited externally on the vegetation.

Several studies of throughfall under different tree species have found greater acidity under conifers than under hardwoods, which may be due to more efficient 'filtering' of atmospheric pollutants by the former, and the greater cation exchange capacity of the latter (Joslin et al., 1987). As rainwater passed through the canopy of a deciduous forest in the north–eastern USA, there was a large increase in solute concentrations and much of the acidity was neutralized (Bormann and Likens, 1994). Such neutralization of acid inputs will, nevertheless, still result in acidification of the overall soil–plant system as these cations are ultimately derived from the root zone and will be lost in drainage water.

The pattern of chemical input to the ground surface under trees is likely to be very spatially variable. If the canopy is discontinuous there will be direct incident precipitation between the tree crowns and enhanced input of leached material under the canopy. Stemflow may provide high solute concentrations in water to a very localized zone immediately surrounding the base of each tree trunk. The role of vegetation in water chemistry does not end when the water reaches the ground due to the intimate role of vegetation in soil chemical systems, involving organic matter and nutrient cycling, as well as in physical processes including soil structure that affect soil water movement.

8.5 SOIL WATER AND GROUNDWATER

Before discussing the water quality processes operating in these subsurface zones it is necessary to give some attention to the media in which the water resides and through which it passes, since they may provide important sources of, and sinks for, solutes. The relative abundance of the elements in the surface layers of the Earth is determined by the composition of the Earth's crust; the materials in the rocks and soils derive directly or indirectly from rock minerals formed originally under conditions of extreme heat and temperature, and which are found in igneous and some metamorphic rocks. Cooling magma formed **primary minerals** such as feldspars, quartz and micas. Apart from quartz, which is very resistant, these are unstable at the Earth's surface, and are prone to chemical alteration to more stable **secondary minerals**, such as clays and iron oxides. In addition, the operation of biochemical processes forms new minerals such as calcite. About 75% per cent of the land surface of the Globe comprises these reworked sedimentary rocks which are much more important for holding and transmitting water than the relatively impermeable and low–porosity igneous and metamorphic rocks.

8.5.1 WEATHERING OF ROCKS

Approximately 99% by weight of the Earth's crust comprises just eight elements: 47% oxygen (O), 28% silica (Si), 8% aluminium (Al), 5% iron (Fe), 3.5% calcium (Ca), 3% sodium (Na), 2.5% potassium (K) and 2% magnesium (Mg). The chemicals are combined into minerals, which have a definite chemical composition. The concern of the hydrologist centres on the weathering of these minerals to make substances available to go into solution and the behaviour of the soil and rock systems in retaining, cycling and leaching these chemicals.

Interest in rock weathering processes was given a tremendous impetus by the development of the **critical load** concept, which has come to dominate European legislation on air pollution due to concerns about acidification. This may be defined as "the maximum deposition of a given compound which will not cause long–term harmful effects on ecosystem structure and function, according to present knowledge". In other words it is a threshold deposition rate that ecosystems can tolerate without long–term damage. This means that acid inputs should not exceed within-soil alkalinity production – essentially the production of base cations by mineral weathering. Sverdrup and De Vries (1994) describe one approach to calculating critical loads. In practice there are many difficulties in applying the concept (Schnoor, 1996). For example, which pollutants are the most critical, and how do they interact to cause ecological damage? Which elements of the environment are we trying to protect – the most sensitive lake in a region, or a lesser target so that some resources may be lost?

The most effective mechanism of chemical weathering is the action of rainwater, containing dissolved acids, on rock minerals. The main source of natural acidity in the environment is provided by the solution of CO_2 in water to form carbonic acid (H_2CO_3). This dissociates in water to form bicarbonate, and to a lesser extent carbonate, ions and generates H^+ ions (Eqs. 8.2 to 8.4).

The CO_2 is dissolved from the atmosphere but, due to plant root respiration and the decay of organic matter, the concentration of CO_2 in the air in soil pores may be 100 times greater than that in the atmosphere. This results in a much higher concentration of carbonic acid in the soil water than is found in surface water such as rivers and lakes. In most humid areas other, stronger acids may also be important, including very dilute H_2SO4 and HNO_3, as well as organic acids formed from decaying vegetation.

The major mechanism of mineral weathering is by **acid hydrolysis**, whereby H^+ ions replace cations in the mineral, leading to an expansion and decomposition of its silicate structure. An example of such chemical action is the weathering of the mineral orthoclase feldspar, found in igneous rocks, to the clay mineral, kaolinite (see Eq. 8.11).

Although chemically a reversible reaction, in practice it is essentially irreversible because the feldspar cannot be reconstituted to any significant extent without imposing very great temperatures and pressures. The silica and potassium are removed in solution to groundwater and to streams, pushing the reaction to the right, and kaolinite clay accumulates as part of the soil mantle. Such secondary minerals may, under suitable conditions, undergo further chemical weathering to even more stable chemical forms. Another example of mineral weathering is given by Eq. 8.5 which describes the chemical solution of calcite (a major constituent of limestones). The rate of weathering, and the resulting concentrations of solutes in streams and underground waters for a given reaction, will depend on factors including the temperature, and the flux of water removing products in solution and bringing new water into contact with the minerals. In general, reaction rates speed up with increasing temperature, and rates of weathering and leaching in the tropics are several times greater than in temperate areas. Tropical soils have much higher clay contents (often 60% or more) than temperate areas (e.g. 35% clay is considered high in Britain), and this much stronger weathering history may explain why the solute loads of streams in the tropics are much lower than those in temperate regions.

The rate of supply of minerals by chemical weathering of rocks is fairly slow, and if there was no mechanism for retaining the substances that are in solution they would be quickly washed out in drainage water. Such mechanisms do exist, however, and are closely related to the operation of biological processes and have an important control over short–term solute dynamics.

$$\underset{\substack{\text{Feldspar}\\\text{(primary mineral)}}}{2KAlSi_3O_8\,(c)} + \underset{\text{Acid}}{2H^+\,(aq)} + 9H_2O \leftrightarrow \underset{\substack{\text{Kaolinite}\\\text{(secondary mineral)}}}{Al_2Si_2O_5(OH)_4\,(c)} + \underset{\text{(In solution)}}{4H_4SiO_4\,(aq) + 2K^+(aq)} \quad (8.11)$$

8.5.2 ADSORPTION AND EXCHANGE REACTIONS

When plants colonize the weathered rock debris they have a direct physical effect by controlling the removal of particulate weathering products by erosion, and they also result in a number of chemical changes. Plants take atmospheric gases into their foliage and dissolved minerals into their root systems and return chemicals to the soil as crown leaching (see Section 8.4) or as leaf litter and other partially decomposed plant remains, known as **humus**. The humus combines with the soil clays to form colloidal complexes which have extremely large areas per unit weight. Their surfaces have electrical charges which enable them to attract and **adsorb** a 'swarm' of dissolved ions (Figure 8.4). This electrostatic attachment of adsorbed ions is sufficiently weak for them to be easily exchanged by other ions in solution. The exchange of ions between the soil exchange surfaces and the soil solution is a continuous process. The rate of exchange is generally rapid and, following a change in the composition of the soil solution, requires only a few minutes for a new equilibrium to be established between the adsorbed ions and those in solution. The amount of a given ion that is adsorbed depends upon the abundance of the different ions in solutions, the ion exchange capacity of the clay or humus and the relative strength of adsorption of the various ions.

With the exception of kaolinite (which has the fewest surface charges), the clay and humus surfaces have more negative than positive charges and consequently attract more cations than anions. The **cation exchange capacity (CEC)** measured in meq per 100 g or centimole/kg (cmol kg^{-1}) varies from about 10 for kaolinite to 100–150 for montmorillonite, and organic colloids have capacities of 200 or even more. Values of CEC for topsoils typically range from about 5 cmol kg^{-1} for sandy soils up to 50 cmol kg^{-1} for a heavy clay with a high organic matter content. The adsorption strength of the different ionic species increases with their charge

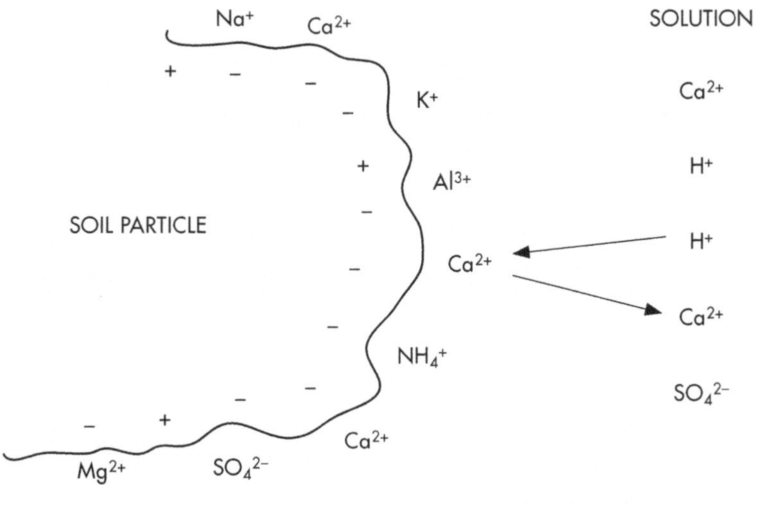

Fig 8.4: Ion exchange between cations in solution and cations adsorbed on soil particles

and decreases with their hydrated ion radius (i.e. the charged ion surrounded by a 'shell' of polar water molecules). Thus for equal concentrations of ions in solution (in equivalent weight units) the relative strength of adsorption of cations is:

$$Al^{3+} > H^+ > Ca^{2+} > Mg^{2+} > K^+ > NH^{4+} > Na^+$$

The capacity of a given exchange surface for holding cations varies with the pH of the solution. As acidity increases, e.g. due to the accumulation of organic matter, there is an increase in the number of H^+ ions in solution. These H^+ ions are strongly adsorbed onto the exchange surfaces and consequently will displace some of the previously adsorbed cations – excepting Al which is too strongly held. Increasing acidity also increases the rate of Al supply to solution from mineral weathering. Since there is a dynamic equilibrium between the adsorbed Al and that in solution some of this Al is adsorbed on the surfaces as Al^{3+} or Al–hydroxy ($Al(OH)_x$) ions. The H and the Al ions tend to dominate acid soils, and both contribute to H^+ concentrations in solution – the H ions directly and the Al ions indirectly by hydrolysis, releasing H^+ ions (see Eq. 8.12):

Under low pH conditions (Eq 8.12a) Al becomes soluble in the form of Al^{3+} and Al hydroxy cations which are very strongly adsorbed on exchange surfaces. At higher pH conditions these Al ions react with OH^- ions to form insoluble $Al(OH)_3$, making the exchange surfaces available to other cations. In contrast to Al and H these **base cations**, which principally comprise Ca, Mg, K and Na, act to neutralize acidity and dominate the CEC in neutral and alkaline soils. In acid soils, Al and H tend to be the dominant cations due to their greater adsorption by the soil, while Ca, Mg, K and Na are leached out in solution. The percentage of the CEC accounted for by base cations is known as the **base saturation**, and acid soils which are therefore poor in these plant nutrients have a low base saturation. The leaching of these cations depends both upon the equilibrium between the cations in solution and on exchange surfaces, and also on the presence of a **mobile anion**, such as $SO4^-$, or NO_3^- which is not itself readily retained on soil surfaces or taken up by plants and can transport the released cations from the soil in solution (e.g. as $CaSO_4$, or $MgSO_4$) in drainage water. Anions may also be adsorbed on soil particles, but to a much lesser extent than cations.

Increasing pH

$$Al^{3+}\,(aq) + H_2O \leftrightarrow Al(OH)^{2+}\,(aq) + H^+ \qquad (8.12a)$$

$$Al(OH)^{2+}\,(aq) + H_2O \leftrightarrow Al(OH)_2^+\,(aq) + H^+ \qquad (8.12b)$$

$$Al(OH)_2^+\,(aq) + H_2O \leftrightarrow \underset{\text{Gibbsite}}{Al(OH)_3\,(c)} + H^+ \qquad (8.12c)$$

The ions adsorbed on the exchange surfaces are, like those in solution, largely available to plants and microorganisms. Plants can take up, or absorb, ions selectively, and different plant species have different nutrient requirements. The uptake of these ions must not disturb the overall charge neutrality within the plant, however. Thus the roots must excrete H^+ ions if there is an excess cation absorption, or HCO_3^- or OH^- ions if anion absorption predominates. Plants can thus directly influence the pH and the ionic composition of the soil water around their roots. Important ions for plant nutrition include the cations K, Ca, Mg and Na, and the anions phosphorus (P), nitrogen (N) and sulphur (S). The concentration of these ions in the soil solution consequently varies systematically over a growing season, becoming depleted as the plants take up nutrients, and then increasing again over the winter period, or when there is no crop.

From the preceding discussion it will be evident that the pH of the soil solution has a very important effect on the way in which substances are gained (by mineral weathering), retained (by ion adsorption or plant uptake) or lost from the soil (dissolved in drainage water). The soil solution 'acidity', defined in terms of its ability to neutralize OH– ions, is related to the concentration of H and Al ions. This can be considered to comprise two forms: an **active acidity** due to the H^+ (and Al) ions in solution and a **reserve** or **exchange acidity** comprising the H and Al ions adsorbed on exchange surfaces. The reserve acidity is very much larger than the active form, being about 10^3 times greater in sandy soils and about 10^5 times greater for a clay soil with a high organic matter content. Since the active and reserve acidities are in a dynamic equilibrium any change in the concentration of the H^+ ions in solution (e.g. an increase caused by acid deposition from the atmosphere or a reduction due to adding lime to farmland) will tend to be balanced by the adsorption or release of adsorbed H^+ ions. Thus any pH change in the soil solution following the addition of an acid or base will be negligibly small until there has been a significant change in the (much larger) reserve of adsorbed ions. Most natural waters exhibit this resistance to change of the pH which is known as **buffering** (Stumm and Morgan, 1996). It is therefore important to distinguish between *intensity* factors (pH) and *capacity* factors, i.e. the total acid or base neutralizing capacity. The buffering capacity of a soil is related to its cation exchange capacity and is important in determining the effect on a soil of external inputs of water and solutes. Thus, for example, the soils and surface waters of the north and west Britain, which have developed on igneous rocks, low in cations, were more susceptible to acidification by atmospheric deposition than the soils of lowland, eastern Britain where the soils have a much higher content of exchangeable cations. Different buffering mechanisms correspond to broad soil pH ranges. In neutral and slightly acid soils $CaCO_3$ is an efficient buffer (Eqs. 8.5 to 8.7). Aluminium is a major source of buffering in the pH range 4–5 (Eqs. 8.12a to 8.12c) while in very acid soils, pH <3.5, iron oxides control the pH.

Once in solution the movement of solutes may be influenced by a number of processes, both physical and chemical, and these are discussed below.

8.5.3 SOLUTE MOVEMENT IN SOILS AND GROUNDWATER

Dissolved materials will be transported by the flow of water, but they will rarely travel at exactly the same rate. Movement of a solute species depends on three mechanisms: advection, dispersion and reaction (Freeze and Cherry, 1979). The mass flow of water will carry solutes by **advection** (sometimes also called **convection**) which, in the absence of other processes, results in chemicals being transported at the same rate as the macroscopic velocity of the water. The velocity of solutes in practice will, however, vary from this rate due to **hydrodynamic dispersion**, leading to a range of velocities. This results from two processes: mechanical dispersion and molecular diffusion, which are similar in effect, and in consequence are generally considered together. Mechanical dispersion is a consequence of the complexities of the pore system of the medium through which the water is moving. Flows are faster through large pores than through small pores, and across the middle of pores than near the pore walls. Flows also vary due to the tortuosity of pore networks, with flows in some pores at an angle to the mean direction of water flow (Figure 8.5). Molecular diffusion is much smaller and results from the random, thermal–kinetic motion of molecules, and occurs regardless of whether or not there is net water movement. Both mechanical dispersion and molecular diffusion cause a movement of solute from areas of high to low concentration, and this response to concentration gradients makes the solute species become more diffuse with time. The resulting intermixing of chemicals between the moving 'mobile' water in the larger pores and the largely 'immobile' water held in finer pores of the matrix can influence the overall rate of transport of solutes).

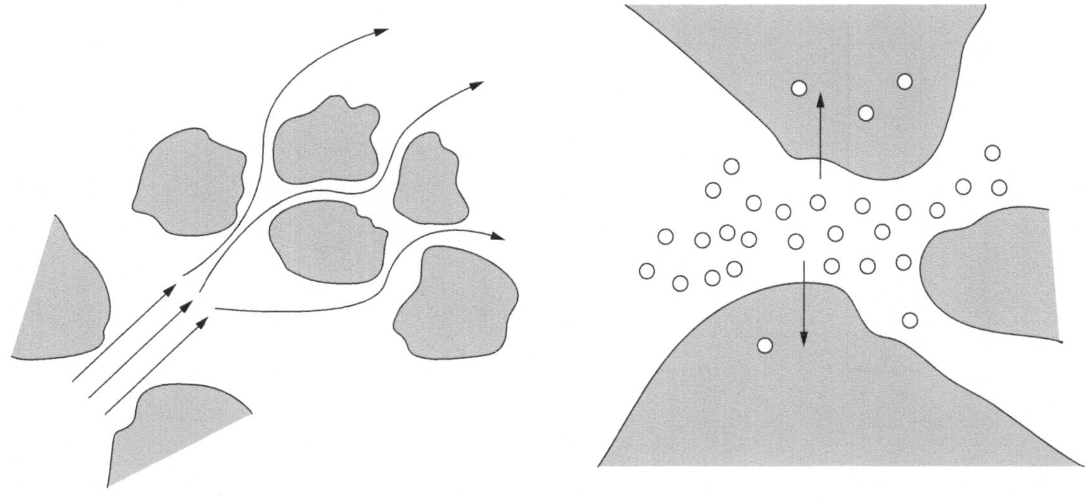

Fig 8.5: Dispersion of a solute in porous media: (a) mechanical dispersion, (b) molecular diffusion.

Changes in solute concentrations may also take place due to **chemical reactions**. These may be with the solid matrix of the medium or between dissolved substances or, in the unsaturated zone, between the solution and the gas phase. Many different chemical and biochemical reactions can occur, including oxidation, reduction, solution and precipitation, but of the most general importance is the ion exchange process of **adsorption**.

The effects of these processes on the movement of a solute species through a porous medium may be presented in graphical form as a **breakthrough curve (BTC)**. Figure 8.6 shows such curves for the simplest case of a non–reactive tracer (of concentration C_0) continuously added to steady, saturated, downward flow through a column of a porous, uniform medium. As the tracer replaces an existing solution (tracer concentration zero), there is an increase in the concentration of the tracer in the outflow (C). Advection alone results in a simple 'piston flow' displacement of the existing solution by the new solution with a sudden, step–like change (line A) in the proportion of new solution in the outflow C_e (= C / C_0) (Figure 8.6). This moves at the average linear velocity of the water. Hydrodynamic dispersion (line B) results in some of the tracer moving faster, and some slower, than this velocity. With increasing flow distance from the tracer input this spread of the concentration profile due to dispersion becomes greater.

For groundwater systems the adsorption reactions are normally very rapid relative to the flow velocity and an equilibrium state is usually assumed between the solute species in solution and adsorbed on solid particles (Freeze and Cherry, 1979). In this situation the relation between the concentration of a solute species adsorbed per unit weight of a

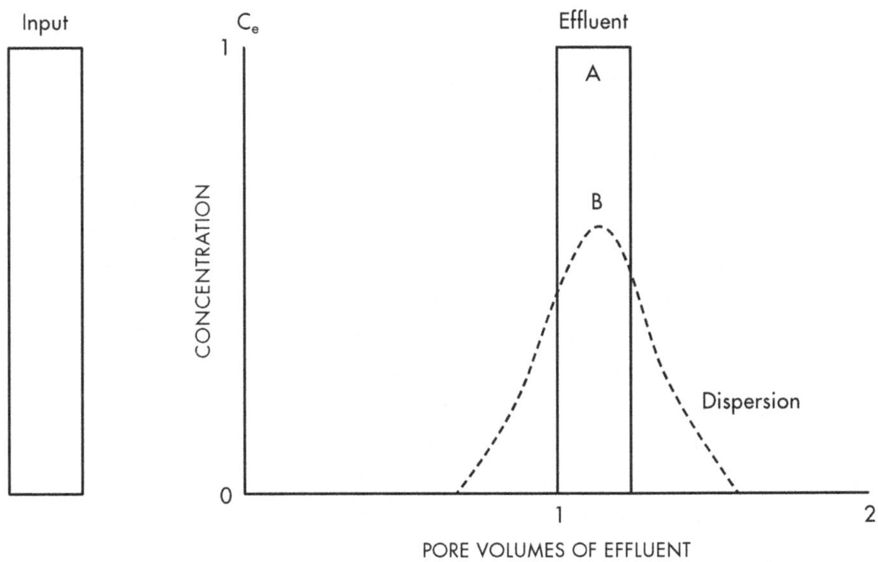

Fig 8.6: Schematic breakthrough curves resulting from piston type displacement of a pulse input of solute (A) and the effect of dispersion (B)..

particular medium and the concentration in solution at a given temperature can be described by the **adsorption isotherm** (Stumm and Morgan, 1996). From this 'partitioning' of the chemical between solid and liquid phases a retardation factor can be calculated to describe the delay in solute movement due to adsorption (Smettem, 1986). The effect of greater adsorption is to alter the time distribution of a pulse of tracer, reducing the average rate of movement (shifting the BTC to the right) and making the concentration curve more asymmetrical (Figure 8.7).

The displacement process may be described by the convective–dispersion equation for one–dimensional, steady–state, uniform flow in a homogeneous saturated medium as:

$$\underbrace{\delta C/\delta t = D\,\delta^2 C/\delta Z^2}_{\text{Dispersion}} - \underbrace{V\,\delta C/\delta Z}_{\text{Convection}} + \underbrace{[\rho \delta S/n\,\delta t]}_{\text{Sorption}} \quad (8.13)$$

where the third term is an extension to account for the gain or loss of the solute by reaction. C is the concentration of the solute species under consideration, Z is the distance along the flow line, t is the time, D is the coefficient of hydrodynamic dispersion and V is the average flow velocity; ρ and n are the bulk density and porosity of the medium and S is the mass of the chemical adsorbed per unit mass of solid. If the adsorption rate is slow compared to the flow rate, $\delta S/\delta t$ can be estimated from reaction kinetics.

In some soils and rocks macrovoids can act as conduits for the rapid movement of

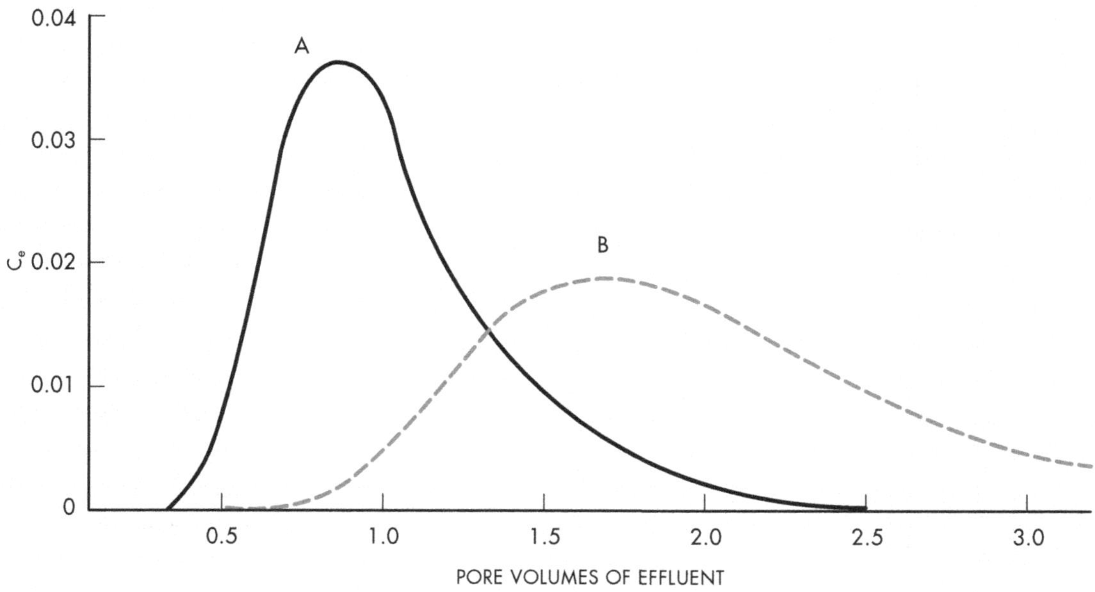

Fig 8.7: Schematic breakthrough curves, showing the effect increasing adsorption of the solute by the medium (curves A to B).

water and solutes. Such channels may include shrinkage cracks and root channels in soils (see Section 6.5.2) and fracture lines and jointing in rocks (see Section 5.7), and, under certain circumstances, enable flows to effectively bypass the soil or rock matrix.

Schematic breakthrough curves are shown in Figure 8.8 for media with different ranges of pore sizes. For simplicity, adsorption and dispersion effects are ignored. The rate of inflow is sufficiently great for preferential flow to be generated in the larger pores. This pattern may be contrasted with the classical infiltration equations developed by soil physicists (Section 6.4.2) which deal with ideal situations of a homogeneous medium (uniform pore size) and a uniform initial moisture content. In the case of preferential or bypass flow, the water does not come into chemical equilibrium with the soil as it can only react with the soil mass by lateral diffusion. The obvious consequence is that greater amounts of dissolved chemicals will pass

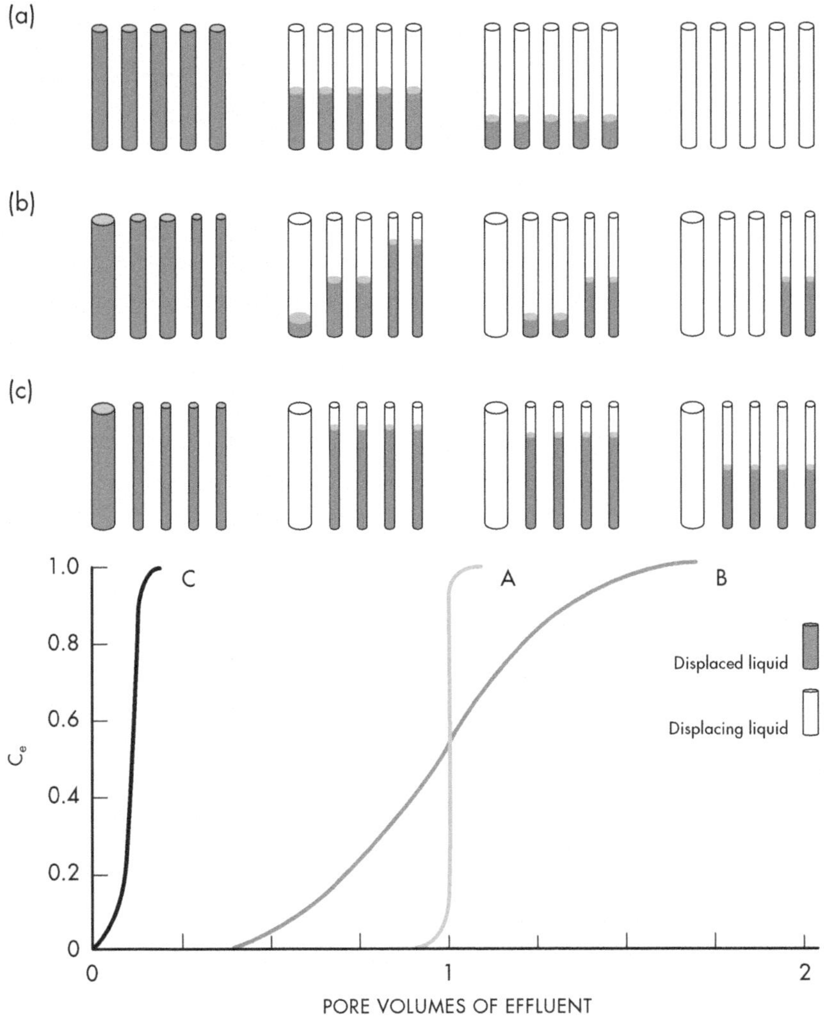

Fig 8.8: Breakthrough curves for a continuous input of solute: (a) piston-type displacement, (b) a medium with a limited range of pore sizes, (c) a medium with bypass flow through interconnected macropores (from an original diagram by Bouma, 1981).

unchanged through the medium to underlying layers and to receiving water bodies. Macropores may be important in the contamination of groundwater since, under high surface inputs of water, they enable solutes to bypass the natural purification actions in the soil matrix and penetrate beyond the reach of plant roots.

In a study of nitrate losses under arable cultivation Smettem et al. (1983) observed very high stream nitrate losses in runoff immediately after the application of nitrogen fertilizer. They attributed this to bypass flow with nitrate moving in the larger, water–filled pores. Only a month later the nitrate had relocated in the immobile water in small pores in the soil peds and was 'protected' from leaching by subsequent bypass flow.

Macropore flow results in an early rise in the BTC while adsorption on the solid phase and solute diffusion from mobile to immobile water zones will cause a delay. Subsequent desorption and diffusion back to the mobile flow zone will both give rise to a long tail. A number of studies have found that the pore water velocities may be approximated by a log–normal frequency distribution.

In field situations it is often unlikely that an equilibrium will be reached between the concentration of the solute and the soil matrix due to the sorption rate being much slower than the rate of change of the solute concentrations. This will be the case for situations with (a) rapid and varying soil water fluxes, (b) varying solute concentrations and (c) slow adsorption rates. The relative importance of these factors will vary between sites where different hydrological processes are operating as well as between different chemicals at the same site.

Subsurface flows will have a much greater opportunity for reactions with the solid phase than flow over the ground surface. Consequently, subsurface flows may carry much higher concentrations of solutes than overland flow. Overland flow will, however, take up organic chemicals from leaf litter and vegetation matter, and on poorly vegetated sloping sites the erosion of soil particles (and any adsorbed chemicals) may occur. The distinction between subsurface and surface flow chemistry is not necessarily clear cut, however, since infiltrating water must pass through the surface organic layers, and subsurface lateral flow may subsequently reemerge at the ground surface further downslope as 'return flow'. The chemical and physical properties of different soil horizons may also give rise to different solute concentrations in the water following different flow paths. Whitehead et al. (1986), used a simple conceptual two–compartment soil model, with an upper layer representing a thin acidic organic soil overlying a more alkaline mineral subsoil, to investigate the effect of changes in flow paths on stream chemistry. Stream storm runoff was very acid, whereas the baseflow in dry weather periods was much less acid and had lower Al levels since a greater proportion of the flow was moving through the subsoil and undergoing acid buffering.

Knowledge of soil processes, including adsorption, can be used to limit the loss of agricultural chemicals into the streams by discouraging their application close to streams - the intervening 'buffer' strips act to filter out the chemicals as long as they are not breached by ditches or field drains. Similarly, by applying liquid chemicals at low intensities it will be less likely that bypass flow in macropores will be generated, so that chemicals will pass slowly through the soil matrix and have a greater opportunity to be adsorbed. There may, however, be problems with farm sludge where it is stored or spread over the ground in large quantities near to watercourses including ditches and field drains.

8.5.4 TRACERS

The direction and speed of water movement may be studied by 'labelling' the water by adding a tracer. This should move as part of the water flow and it is therefore important to select one which is unreactive. A wide range of substances has been used as tracers, including salts, fluorescent dyes and **isotopes**, which are atoms of a given element that have a different atomic mass but the same chemical properties, and may be used to 'label' chemical species in solution. Davies et al (2014) used the ratio of oxygen isotopes ^{16}O and ^{18}O to determine the source of phosphorous in rivers – sewage works, leaking water mains or fertiliser. A detailed discussion of their relative advantages and disadvantages has been given by various authors (see Cook and Herczeg, 2000; Kendall and McDonnell, 1998).

Tracers can be used in a number of ways. Some of the earliest work, for example, used dyes to study the connectivity of conduit flow paths through limestone terrain. Tracers have also been used to study the relative importance of intergranular seepage through the matrix and fissure flow in the unsaturated zone of chalk aquifers. Radioisotopes with a known rate of decay have been used to date groundwater bodies to determine whether their water was derived under present conditions (and so may be sustainably used) or else derived from recharge in a much rainier period.

8.5.5 CHEMICAL EVOLUTION OF GROUNDWATER

Since the kinetics of mineral weathering are often slow, it is unlikely that thermodynamic equilibrium with the flowing water will be reached in the soil zone, whereas in large groundwater systems residence times are much longer, and equilibrium may be progressively established as water moves through the medium. Where groundwater recharge occurs there is a net transfer of mineral matter from the soil zone to the underlying saturated zone. As this groundwater moves along flow lines to discharge areas, it will normally dissolve further material from the matrix rocks and the concentration of total dissolved solids will increase. Some materials may be precipitated out due to changes in temperature and pressure affecting their solubility, or to reactions with ions dissolved from the matrix forming new insoluble compounds.

In a classic study of the chemical changes in groundwater as it moves from areas of recharge to areas of discharge, Chebotarev (1955) studied over 10,000 Australian groundwater samples and found a progressive evolution in the dominant anion species with increasing travel distance and age:

$$HCO_3^- \rightarrow SO_4^{2-} \rightarrow Cl^-$$

This sequence is determined by mineral availability and solubility (Freeze and Cherry, 1979). In broad terms, minerals which dissolve to release Cl^- (e.g. halite) are more readily soluble than those releasing SO_4^{2-} (e.g. gypsum and anhydrite), which are in turn more soluble than those releasing HCO_3^- (e.g. calcite and dolomite). Thus Cl^- and, to a lesser extent, SO_4^{2-} may have been largely leached from the recharge zone of a groundwater system. Furthermore, HCO_3^- enters the groundwater system in recharge zones from the dissolution of CO_2 from the atmosphere and the soil air.

While this chemical sequence is a useful conceptual model, these changes cannot be defined quantitatively to particular ages or distances, and it is rarely observed in its entirety due to the often dominant effects of the local physical environment. Thus some waters may not evolve past the CO_3^- or SO_4^{2-} stage, while if the water comes into contact with a highly soluble mineral such as halite it may evolve directly to the

Fig 8.9: Trilinear diagram for presenting the chemical composition of groundwater. From an original diagram by Davis and De Wiest (1960).

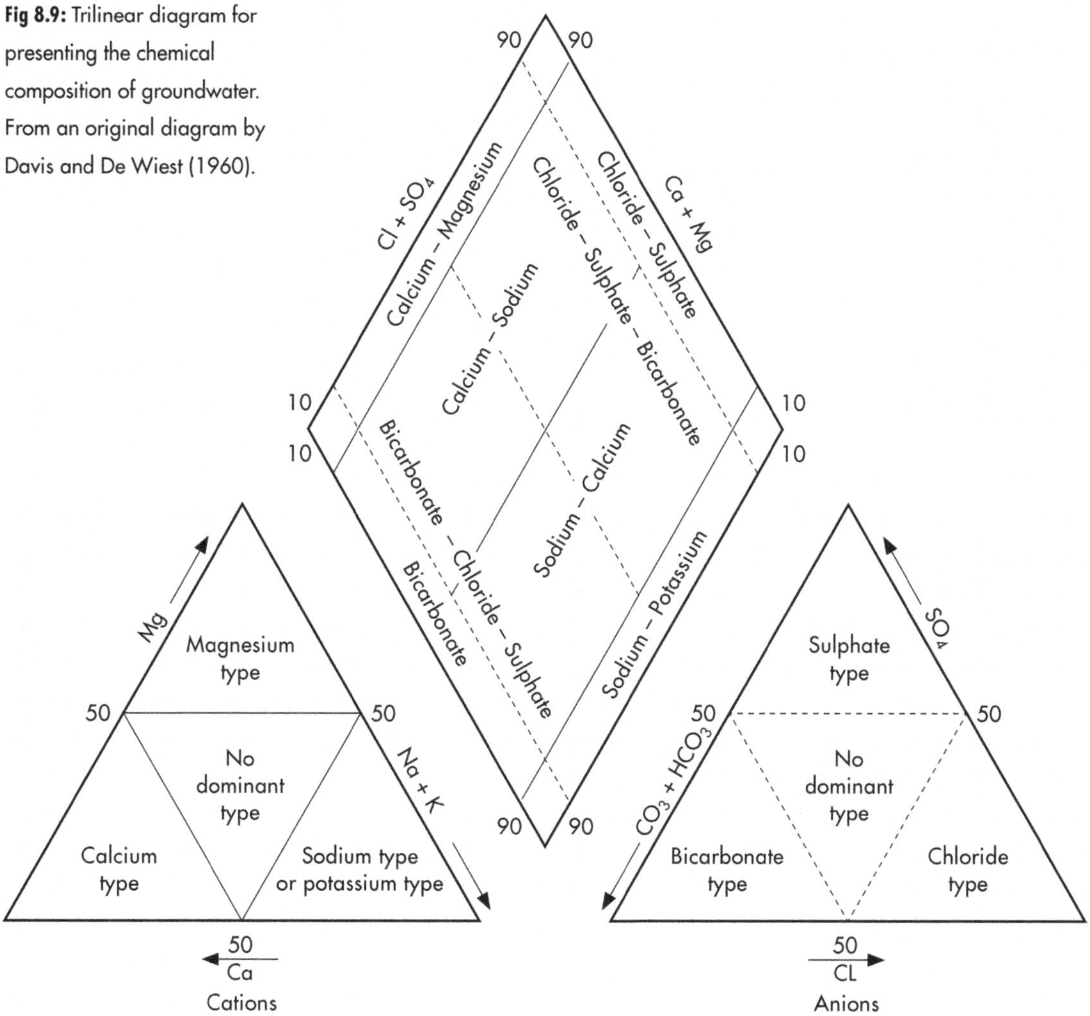

Cl^- stage. Many sedimentary deposits are not homogeneous, but comprise assemblages of minerals, and one of the most important factors controlling the chemistry of their groundwater is the order in which the minerals are encountered by the flowing water. Furthermore, in a heterogeneous aquifer system, water flow will occur predominantly in the more permeable strata, and the mineral composition of these layers, rather than the composition of the whole aquifer, will be the main factor influencing water chemistry. The pattern is much less clear for cations than that outlined by Chebotarev for the anions, but as a broad generalization Ca and Mg are replaced by Na. Groundwater evolution along an 800 km flow line in a hydraulically continuous aquifer in North Africa using chemical and isotopic indicators was studied by Edmunds et al (2003).

8.5.6 PRESENTATION OF WATER CHEMISTRY DATA

It is often useful to summarize and display data on the chemical composition of groundwater, for example to study changes along a flow line, or to detect mixing of waters of different compositions. There are many different ways of presenting the data, and some of the most important and commonly used were summarized by Hem (1985). Perhaps the most widely adopted system is the trilinear diagram attributed to Piper (1944) and used for the major ion composition. Groundwater is characterized by three cation constituents: Ca^{2+}, Mg^{2+} and Na^+ plus K^+, and three anion constituents: CO_3^-, SO_4^- and Cl^-. The trilinear diagram (Figure 8.9) combines three plotting fields. The cation and anion compositions are plotted in the left– and right–hand triangles respectively (expressed in equivalents per litre). The intersection of the rays projected from the cation and anion plotting points onto the central field represents the overall major–ion character of the groundwater. The values given are the relative rather than absolute concentrations, and since the latter are important in many situations, it is common to plot in the central field not a point but rather a circle centred on that point whose area is proportional to the total dissolved concentration of the water (Hem, 1985). This type of presentation is useful for describing differences in chemistry of water samples and it can be extended to give identifiable categories or **facies**, based on the concept of the composition of groundwater tending towards equilibrium with the matrix rocks through which it is flowing. The hydrochemical facies represent the observed spatial pattern of solute concentrations in the groundwater which results from the chemical processes operating, the rock types and the flow paths in the region.

8.6 RUNOFF

Surface waters, comprising rivers, streams and lakes, are almost always a mixture of waters from different sources. They receive varying inputs from overland flow, throughflow and baseflow, which due to their very different flow paths and residence times, have different solute and sediment loads. These waters are then mixed together within the channel and routed downstream. In addition to such **non-point** sources, there are **point sources** of pollution entering the river system as industrial waste discharges and effluent from sewage treatment works.

The following sections deal with the water quality changes and processes occurring within stream channels and lakes and then look at the water quality of a water course in the context of the overall catchment, in terms of its precipitation chemistry, mineral weathering, vegetation and land use.

8.6.1 PROCESSES IN STREAM CHANNELS

A number of changes may occur when water from the soil zone or groundwater enters a stream channel. These include physical, chemical and biological processes which result, perhaps most obviously, in changes in water temperature, dissolved oxygen and the ability to transport solid particles. Solute concentrations may increase due to the solution of materials in the stream bed and banks. In semi–arid areas, in particular, solute concentrations increase as stream water volume is depleted by evaporation.

The solute composition of water flowing into stream channels will be related to the flow paths and residence times of that water in the soil and groundwater systems. Thus baseflow in streams during periods of dry weather is likely to comprise water that has been in contact with mineral material for some time and so has a higher solute concentration than flows in storm periods which have had a much shorter period of contact with the vegetation and soil material. As a consequence, solute concentrations in river water are generally found to be inversely related to flow rates. In fact some studies have used this to develop chemical means of apportioning a stream flow hydrograph into 'baseflow' and 'stormflow' components.

There may also be interactions between the solutes entering the channels and sediments in the channel. Pollutants may be adsorbed by stream sediments and then carried with the sediments as they move through the river system, or may be deposited for a considerable time on the beds of lakes and rivers, or on flood plains during times of overbank flooding. In some cases these reactions with the sediments may be irreversible, e.g. cesium–137 becomes fixed between the lattice plates of illite, while in other cases the effect is simply to introduce a time delay before the adsorbed materials are released back into solution. The study of sediment movement and storage is important, both in its own right, in relation to problems of land erosion and sedimentation of channels and lakes, and also for the fate of adsorbed hazardous or toxic substances.

A crude, but convenient, classification of the sediment movement in a river is into **suspended load**, comprising the finer particles which are held in the flow by turbulence, and **bed load**, comprising the coarser particles which move by sliding, hopping or rolling along the stream bed. In practice there is no clear division between the two modes of transport, and particles that may move as bed load in times of low flow may be carried as suspended sediment in times of high flow.

The mechanism of sediment movement and the transporting capacity of given flow conditions are discussed in various texts on hydraulics and fluvial geomorphology (e.g. Robert, 2003; Richards, 2004). Rating curves between flow and sediment discharge are commonly used in sediment yield studies as a means to interpolate between the often infrequent sediment measurements. Hydraulic factors control the sediment transport capacity of a stream and so are most appropriate for **alluvial rivers**, which have beds and banks formed of river deposited material that can be transported by high flows. Many rivers are non–alluvial and the discharge of sediment loads, and in particular that of the finer suspended load, is controlled more by the quantity and timing of the supply of sediment into the stream than by the capacity of the flow to transport it. In such rivers, although there may be a positive relation between flow and sediment load (the higher flows can move larger particles and storm erosion supplies new sediment to the channels), the relationship between flow and sediment discharge may be very poorly defined (Figure 8.10). Improved estimates of sediment loads may be obtained by the use of separate rating curves for different seasons and for rising and falling river levels. Ideally, however, sediment sampling should be sufficiently frequent to define the variations over time. Turbidity meters provide one means of achieving a detailed picture of temporal variations without excessive expense and sampling frequency, but are not suitable for all situations.

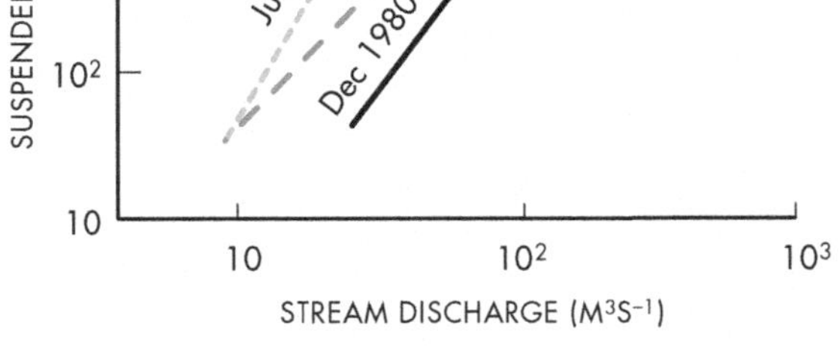

Fig 8.10: Changes in the suspended sediment-discharge relations for the Toutle River, Washington, following the Mount St Helens volcanic eruption in May 1980 that deposited huge quantities of easily eroded ash. Redrawn from an original diagram by Nordin (1985)

Syvitski et al (2005) estimated that human activities through soil erosion probably increased global sediment transport in rivers by about 2.5 billion metric tons per year compared to pre-human sediment loads. However globally the flux of sediment reaching the oceans has probably reduced by 1.5 billion metric tons per year due to trapping in reservoirs, although there are large regional variations.

The pattern of sediment and solute behaviour in a river is complicated by the fact that solute and sediment travel at different speeds and both move more slowly than the rate of propagation of a flood wave, and the distributed nature of river networks results in a complex mixing of inputs from various tributaries. Consequently, they may therefore be best understood in terms of the characteristics of, and processes operating in, the drainage basin as a whole. This is discussed in Section 8.6.3, but attention must be paid first to the study of point source inputs into a river network which may be viewed in terms of the processes operating within the channels. Point inputs to stream channels may have some important consequences; an example, which is of fundamental biological importance, is the role of certain types of organic waste, such as sewage and animal wastes, in consuming dissolved oxygen. Oxygen is critical for aquatic life, and when large amounts of biodegradable materials enter a stream or lake, their chemical and microbial breakdown consumes appreciable amounts of oxygen, hence reducing the concentrations available to aquatic life. In a well aerated turbulent stream dissolved oxygen concentrations may be close to saturation, about 10 mg l^{-1}. Low concentrations (<3 mg l^{-1}) may be harmful to fish (UNEP, 1995) and, if the dissolved oxygen becomes exhausted, further decomposition will occur by means of anaerobic processes, which generally produce noxious odours.

The pollutant 'strength' of an effluent, i.e. its potential for removing dissolved oxygen, is measured in terms of its biological or biochemical oxygen demand (BOD). This is a laboratory–derived measure of the consumption of oxygen under standard conditions, usually at 20°C over a period of 5 days. While 5 days is too short for some resistant chemicals, such as those found in wood pulp wastes, for which a 20–day BOD is often used, it is sufficiently long for the majority of the oxygen demand for domestic sewage and many industrial wastes. Even so, the necessity of at least a 5–day delay may be unacceptable for some purposes, such as monitoring the performance of a treatment plant, and the quicker measure of **chemical oxygen demand (COD)** can provide a value in a couple of hours, which can then be related by an empirical relation for the site to BOD (Lamb, 1985). Figure 8.11 illustrates a typical dissolved oxygen 'sag' curve due to the deoxygenation and subsequent recovery after the addition of organic effluent to a stream from a single outfall point. This curve represents the balance between the rate of consumption of oxygen (which progressively declines as the waste is broken down) and the replenishment of oxygen, mainly by solution from the atmosphere. The rate of reoxygenation increases with the severity of oxygen depletion and the turbulence of the water and reduces with increasing water temperatures. Since reoxygenation is much slower than depletion, the minimum levels of dissolved oxygen may occur some distance, perhaps many kilometres, downstream of the effluent outfall. The magnitude of the oxygen sag will depend on the flow in the river, to dilute the effluent, so that any problems of low oxygen levels will be particularly critical at times of low flow in warm weather.

8.6.2 LAKES

The discussion of water quality changes in open water bodies has, so far, centred on flowing water in channels. The properties of lakes and reservoirs are also important. Firstly, they generally tend to reduce the wide variations in water quality of the rivers entering them. Secondly, due to their reduced water velocity and turbulence, they may cause important differences in water quality. Much of the sediment and organic matter entering with river water may settle out, reducing turbidity, and allowing sunlight to penetrate further, which in turn allows more photosynthesis and a richer growth of plants and algae. Under natural processes this gradual accumulation of organic matter will eventually result in the infilling of the lake and ultimately the creation of a peat bog. The rate of this aquatic plant growth may be limited by a shortage of dissolved nutrients, particularly phosphorus and nitrogen. Increased levels of these nutrients are released into streams from agricultural fertilizers and domestic sewage; and by removing the natural limits to aquatic plant production this may result in excessive growth of algae and/or other aquatic plants, termed **eutrophication**. This can cause problems for water supply abstractions by affecting the taste and smell

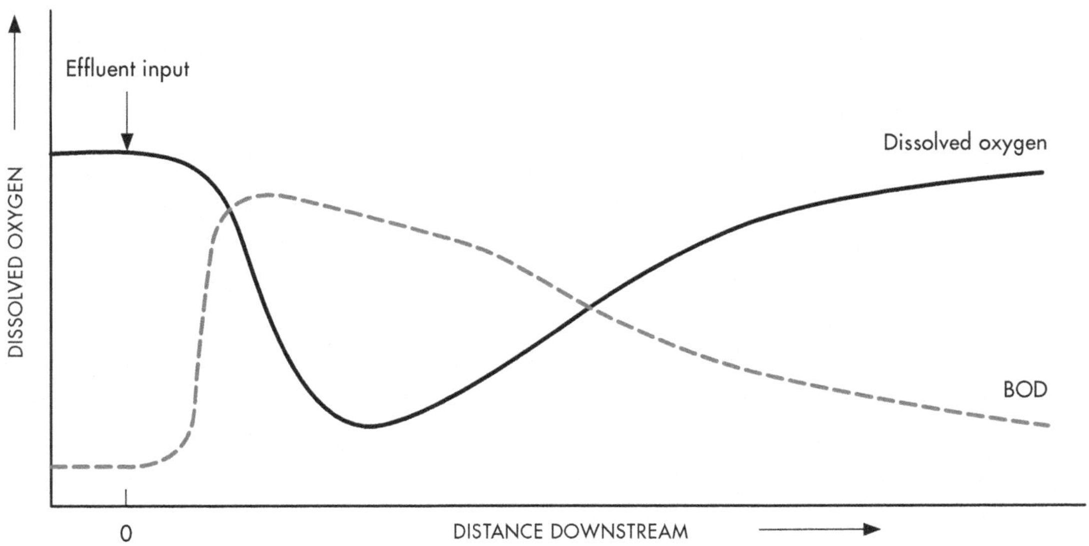

Fig 8.11: Schematic dissolved oxygen 'sag' curve and biochemical oxygen demand (BOD) in a river following the input of an organic effluent.

and cause filtration problems. In some cases it may result in severe oxygen depletion when the plants die, and can lead to the death of fish and other aquatic life.

An important aspect of some lakes is **thermal stratification**, which results from water being at its maximum density at 4°C. In deep lakes or reservoirs there may be little mixing of the surface layer of water with that below during most of the year. In summer the surface waters are warmer and less dense than the water below and in winter the surface water may be colder and denser. As a result, the surface layer (or epilimnion), which may be some 5–8 metres deep, has a high dissolved oxygen content, abundant sunlight and a rich plant life, while in contrast the deeper water (or hypolimnion) is insulated from direct contact with atmospheric oxygen and so may have little or no dissolved oxygen and receives less sunlight. Under these conditions the accumulation of decaying organic matter, sinking down from the surface, leads to stagnant conditions. In the deep water of lakes, phosphate and silica may be released into solution, and ammonia and other gases produced from the biological decay. These two layers of the lake water may remain distinct until with the onset of cold winter weather the surface water cools and becomes denser, allowing rapid mixing of the layers by wind action. This 'overturn' of the lake water may occur over only a few hours, and the rise of the bottom stagnant water may cause serious reductions in the dissolved oxygen levels both within the lake and in streams flowing from it. The sudden upwelling of nutrients to the surface can also result in algal blooms.

8.6.3 CATCHMENTS

The foregoing sections briefly discuss the main processes controlling the water chemistry of the different components of the hydrological cycle. The following sections give a few examples of how the water quality at the outlet of a catchment can be interpreted as the integration of these processes. These discuss the influence on water quality of geology, climate and human activities, including the use of chemicals such as fertilizers and pesticides, and detail the main water quality characteristics of a number of British rivers.

Geology and Climate

Since weathering reactions and their soluble products will be largely determined by the minerals available, it is to be expected that the natural composition of stream solutes will be primarily determined by the soil type and the underlying geology. On a global scale, about 60% of the total natural dissolved load of rivers is derived from rock weathering (Walling and Webb, 1986). The rest comes mainly from the dissolution of atmospheric CO_2 (to form HCO^{3-}) and from sea salts.

A number of studies have shown the importance of lithology on stream chemistry. In a global study of annual flow and solute data for nearly 500 largely unpolluted catchments Walling and Webb (1983) found that loads increased with annual runoff, although at a slower rate than the increase in flows, so solute *concentrations* actually decreased. There was much scatter in the solute/runoff relationships, due largely to differences in catchment lithologies. They could not clearly distinguish the climatic controls on solute loads due to differences in the predominant rock types. Temperate regions, for example, have a greater abundance of sedimentary rocks while tropical areas have predominantly crystalline (igneous and metamorphic) rocks. Peters (1984) studied the major ions and total dissolved solids in 56 single bedrock–type catchments across the USA, and concluded that lithology was a major factor determining dissolved loads, with the highest values in streams draining limestone basins. Bluth and Kump (1994) noted that the dissolved load of rivers depended not only on lithology, but also on the climate and the rate of physical weathering. The latter is necessary to expose fresh rock material to attack (Drever, 1997). Thus, solute loads may be higher from steep or poorly vegetated areas where physical erosion continually removes the solid weathering products. In many situations the subsurface waters pass through several different rock strata, so the controls on the resulting chemical composition of the surface waters may be very complex.

The water quality of streams depends on a number of interrelated environmental factors, but in general terms the most important natural factor, along with geology, is climate. The most important control on the speed of chemical weathering is the availability of liquid water. Temperature is less crucial, and while weathering reactions are faster at higher temperatures this may be offset by less rapid exposure of bedrock by physical erosion due to a denser vegetation cover

Table 8.2 shows typical stream solute concentrations that have been reported in the literature, arranged into broad classes by rock type. Igneous rocks are generally rather impervious and their minerals are resistant to weathering; they generally have the lowest solute concentrations while sedimentary rocks commonly give higher values (but this is very dependent upon their mineral composition). Streams on metamorphic rocks tend to have solute concentrations that are intermediate between igneous and sedimentary rocks, depending on the original material and the degree of alteration. Precipitates such as limestone (mainly $CaCO_3$) and dolomite ($CaMg(CO_3)_x$) generally have much higher concentrations, while evaporates such as halite (NaCl) and anhydrite ($CaSO_4$), which are derived from soluble minerals deposited by evaporation of water, may produce the highest concentrations of dissolved solids. In areas where rock

ROCK TYPE	TOTAL DISSOLVED SOLIDS (MG L^{-1})	PRINCIPAL IONS
Igneous and metamorphic	< 100	Na, Ca, HCO_3
Sedimentary (detrital)	50 – 250	Variable
Limestone and dolomite	100 – 500	Ca, Mg, HCO_3
Evaporites	< 10 000	Na, Ca, SO_4, Cl

Table 8.2: Solute concentrations in streams in catchments underlain by different rock types.

weathering is very slow the stream solutes may be strongly dependent upon atmospheric inputs (Bormann and Likens, 1994).

Stream sediment yields in Western Europe are generally low (<10 tonnes km^{-2} y^{-1}), with the highest yields observed in south east Asia (>500 tonnes km^{-2} y^{-1}) due to a combination of heavy rainfall, steep slopes and readily erodible soils (UNEP, 1995).

Human impacts

Human activities have had an increasing effect on the water quality of all the components of the hydrological cycle. This has been both through the addition of chemicals (whether into the atmosphere, onto the land or directly into watercourses) and by the alteration of catchments by land–use changes. The following discussion is by no means comprehensive, since limitations of space prevent a full treatment of all the many influences and interactions. It is intended, therefore, to indicate some of the processes involved and some of the management options available, by reference to the examples of chemical fertilizers and pesticides on farmland and to the planting of some agricultural land for commercial forestry.

Fertilizers are applied to the land to correct nutrient deficiencies limiting plant growth. They are particularly important for arable crops since nutrients are removed when the crops are harvested. The main fertilizers in use supply nitrogen (N), phosphorus (P), potassium (K) and sulphur (S) (Haygarth and Jarvis, 2002).

There has been a great deal of concern regarding the possible links between nitrates in water and health risks to humans, including stomach cancer and 'blue baby' syndrome (WHO, 2011). The World Health Organisation / European Union standards specify a maximum admissible NO_3 concentration of 50 mg l^{-1} (= 11.3 mg l^{-1} N as NO_3) in drinking water. Nitrate (NO_3) is the dominant source of nitrogen in the rivers and largely comes from agriculture (Jarvie et al., 1997). Nitrate concentrations have increased substantially over the last 50 years in response to increases in fertilizer inputs during the first half of the twentieth century. They result from the long-term release of contamination because within-catchment attenuation and aquifer storage result in long water residence times. In terms of environmental strategies for reducing nitrate pollution, recent reductions in fertiliser application will take a very long time to translate into major reductions in NO_3 within the rivers (Howden et al., 2011).

Soils naturally contain a large amount of nitrogen, most of which is held in the humus rich surface horizons. This is, however, largely unavailable to plants (and to leaching) and only slowly becomes available by a two–stage process comprising decomposition (mineralization) to ammonium (NH_4^+) and then oxidation (nitrification) to NO_3^- by soil bacteria. As noted earlier, NO_3 is a 'mobile' anion which is not adsorbed by the soil and will move with the water flux, but in the upper soil layers it is taken up by plants. Nitrogen fertilizer is generally applied to crops in spring or early summer when plant growth is rapid, and the crops are subsequently harvested in late summer. This is reflected in the seasonal pattern of nitrogen concentrations in the stream water draining an

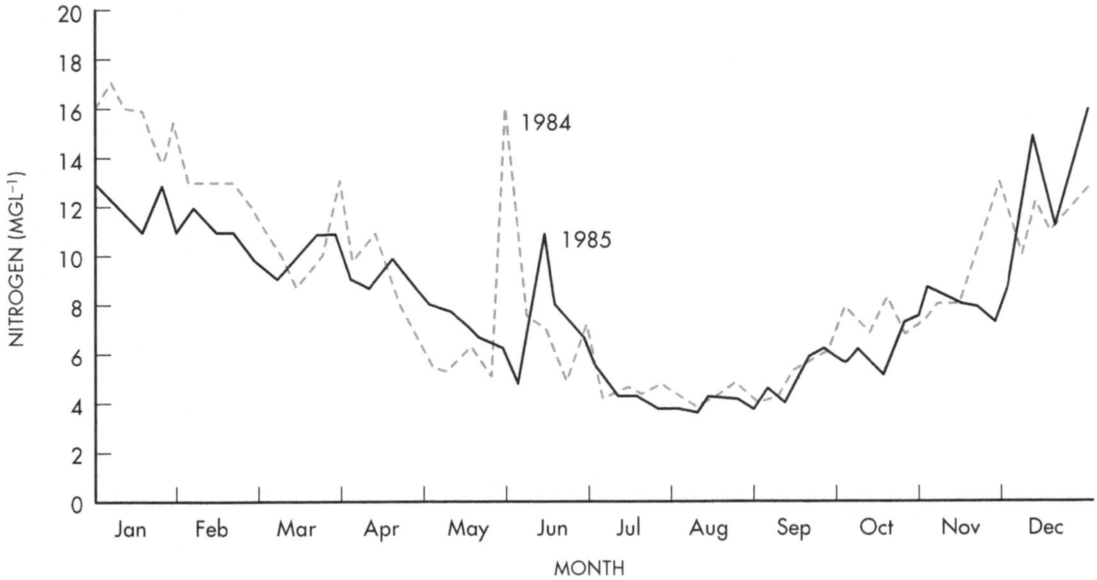

Fig 8.12: Seasonal changes in nitrate concentration (mg l^{-1}) in the River Stour at Langham, SE England. Data supplied by the Environment Agency.

intensively farmed arable catchment (Figure 8.12). Nitrate leaching losses are greatest in the winter when plant uptake is minimal, and lowest in the summer when the plants are growing rapidly. The late spring peak values correspond to the 'flushing' of nitrogen by the first big storm after the fertilizer application.

In catchments where the dominant source of nitrate is sewage effluent the seasonal pattern of concentrations is reversed, with the fairly constant input of effluent undergoing greater dilution by the generally larger flows of river water in winter. The Land Ocean Interaction Study (LOIS) provided a large scale multidisciplinary study of UK river inputs to the North Sea. This included major ions, nutrients, trace metals, pH, alkalinity and suspended solids (Leeks et al., 1997). Variations in water quality were shown to be related to regional differences in geology, climate, land use and population density (Robson and Neal, 1997).

Agricultural practices will also affect the rate of loss of nitrate from farmland. Ploughing leads to aeration of the soil and greater mineralization and nitrification, and in the case of ploughing up of old grassland it can result in an increase in the release of nitrogen over a number of years. Similarly, artificial drainage will increase nitrate leaching and will also alter the flow paths by diverting some of the leachate from downward movement towards the groundwater to lateral flow in pipes discharging into streams.

There has also been concern regarding the increasing concentrations of nitrate found in groundwater supplies. Foster and Young (1981) studied pore water nitrate levels in the unsaturated zone of the chalk aquifer in Britain. Cores were drilled at 60 locations and the interstitial water was extracted by centrifuge. Nitrate concentrations were found to be closely related to the history of agricultural practice on the overlying land. Thus, all sites under arable farming had $NO_3^- - N$ levels >11.3 mg l^{-1} and many had over double this figure, while under permanent unfertilized vegetation, such as rough grassland and woodland, concentrations were much lower, i.e. generally less than 5 mg l^{-1} and often less than 1 mg l^{-1}. Values beneath fertilized grassland were intermediate between these land use categories, i.e. generally in the range 5–10 mg l^{-1}. Under arable crops water in the upper unsaturated zone was 20–30 mg l^{-1}, and generally 10–15 mg l^{-1} in the underlying saturated zone. This suggests that as this water moves downwards to the water table nitrate concentrations in the groundwater will continue to increase for some years, even if no further fertilizer is applied.

Phosphorus (P) is a common constituent of agricultural fertilizers, manure, and organic wastes in sewage and industrial effluent. It is an essential element for plant life, but in high concentrations, it can result in eutrophication, causing excessive plant and algal growth and low night-time dissolved oxygen concentrations. Losses from agricultural land and effluents from sewage treatment works (STW) have both been implicated. There have been great efforts to reduce losses from both sources. Phosphorus occurs in rivers in both dissolved and particulate forms. Dissolved inorganic phosphorus, known as soluble reactive phosphorus (SRP) typically dominates in UK lowland rivers and generally correlates with effluent inputs and population density, indicating the relatively greater importance of sewage sources (Davies and Neal, 2007; Jarvie et al., 2006). For many UK river basins there has been a major removal of SRP in effluents from the main sewage treatment works since the late 1990s and this has led to corresponding reductions in SRP concentrations within the lowland rivers (Bowes et al., 2010). However, the initial reductions can be less than expected due to a net release of SRP from contaminated in-stream sediments.

Physical effects such as temperature and flow may also affect biological functioning, and most UK lowlands rivers have been greatly affected by factors such as water abstractions for water supply and river management such as river straightening that have changed the flow regime. Other human influences include bank-side clearance of trees that increase light levels and temperature, and impoundments/sluices which reduce flow velocities, promoting algal growth. These all affect the functioning of the river ecosystem and can impact on algal development so it is inappropriate to target SRP as the sole cause of biological decline or the solution to the problem if SRP levels are to be reduced. Thus, despite four decades or more of remediation measures to reduce P inputs, in many cases there has not yet been overwhelming success in reducing nuisance algal growth (Jarvie et al., 2013b). Lags in response to remediation and release of both P and N from 'legacy' stores in catchments and water bodies pose a major challenge in terms of meeting water quality targets. Indeed, nutrient legacies from past land management may continue to impair water quality over timescales of decades, perhaps longer (Jarvie et al., 2013a; Sharpley et al., 2013). With the

Water Framework Directive goal of good ecological status, better understanding of the role of changing nutrient concentrations in relation to other factors is vital in designing appropriate P concentration standards to protect and improve river water quality.

While much attention has been given to the loss of fertilisers from agricultural land, much less is known about the leaching of **pesticides** (literally meaning 'killer of pests'). In temperate climate areas the most commonly used pesticides are **herbicides** to kill weeds while in tropical areas **insecticides** have revolutionized health and life expectancy by controlling insects which are the principal vectors of disease. However, while pesticides have saved millions of lives in developing countries widespread concern has arisen, especially in developed countries, regarding their environmental effects. Pesticides may accumulate in the food chain and so pose a threat to fish, birds and mammals, leading to a long-standing debate about whether they are lifesaving chemicals or the 'elixirs of death'.

Pesticides comprise an enormous range of both naturally derived and synthetic chemical compounds, designed to be toxic and persistent (stable), to protect against pests, disease and weeds. There was a rapid increase in the use of pesticides from the 1940s, reaching a peak in the 1990s, followed by a declining trend in Europe and N. America (Parris, 2011). Due to their potential environmental damage there has been a change from the persistent organochlorine pesticides to less persistent chemicals which will degrade, by chemical hydrolysis or bacterial oxidation, over a few months rather than years. These newer compounds are, however, generally more soluble in water, posing the risk of contamination of groundwater and surface water in the first few weeks after application, before processes such as degradation, adsorption and volatilization make the residues unavailable. The fate of an applied chemical in the soil depends upon its partitioning between the soil particles and the soil solution.

At present, the European Union has adopted an arbitrary standard for pesticides in drinking water of 0.1 $\mu g\ l^{-1}$ for each pesticide, regardless of toxicity. In Britain the most commonly used herbicide is *isoproturon* which is used to control weeds in cereal crops. Its high usage and predominantly autumn application results in its contaminating surface and groundwaters. Concentrations >10 $\mu g\ l^{-1}$ have been recorded in streams during storms that occurred within a few days of its application (Harris et al., 1991). Williams et al. (1995) noted that pesticide transport to streams was increased by bypass flow through structural cracks in a heavy clay soil. Garmouma et al (1998) related the varying pattern of herbicide concentrations in the streams of four catchments in northern France to differences in applications rates and weather conditions, land use and soil types.

Johnson et al (2001) and Haria et al (2003) studied pesticide penetration of herbicides *isoproturon* and *chlorotoluron* to an unconfined chalk aquifer by soil sampling, shallow coring and groundwater monitoring. Where the water table was deep recharge only occurred through the chalk matrix and little herbicide was detected at depth; in contrast where the water table was shallow the capillary fringe ensured that the profile was close to saturation with limited capacity for accommodating incoming rainfall, resulting in rapid preferential flow down fractures and the potential for herbicides to contaminate groundwater.

A major new concern is the pesticide *Metaldehyde* which is contained in slug pellets and used particularly in oil seed rape crops. It is very soluble and applied as a liquid and may wash out in next storm. It does not break down and enters water courses (Kay and Grayson, 2014). There is currently no way to treat it as it is not amenable to the treatment processes normally used to reduce pesticide concentrations in drinking water, and it is not broken down by other processes, including chlorination and ozonation.

There are many different types of land-use change and disturbance to catchments resulting from man's activities which may alter the water quality of streams. The planting of forestry on former agricultural land may benefit the environment, for example by locking up carbon in the plant material and so helping to reduce 'global warming', and by providing a land use that is less dependent upon the application of chemicals. However, as was shown earlier forests may alter water quality both directly through the increased 'capture' of atmospheric pollutants and indirectly due to increased evaporation losses and hence an increase in the solute concentrations. Forestry management practices will also affect water quality. The drying of organic soils by drainage and by evaporation losses from forestry increases aeration and may result in substantial losses of carbon and nutrients including nitrate and ammonium. If the land is of poor quality, fertilizers such as phosphorus and potassium may be used and herbicides are sometimes necessary to control competing weeds when the young trees are first becoming established. Felling of the forest for timber will also affect stream water quality, with an increase in nitrates in streams draining felled areas (Bormann and Likens, 1994) due to the decay of the organic debris, soil disturbance and the lack of any plants to take up the released nutrients.

Emerging Pollutants

In recent years, there has been increasing concern over the environmental risks of the so-called 'emerging pollutants or contaminants' that originate from a wide variety of products including human and veterinary medicines, nanomaterials, personal care products (such as deodorants), paints and coatings. They are not necessarily *new* chemicals; they may have been present in the environment for a long time, but the unintended consequences of their presence and significance is only now being recognised. Emerging pollutants detected in surface waters and groundwater, include antibiotics, insect repellents, cancer drugs and oestrogens (Ashton et al, 2004; Boxall, 2012).

The potential problems to the aquatic environment caused by **endocrine disrupters** in sewage and industrial effluents are perhaps the best known example. These include the natural female oestrogen hormones and the main active component of the oral contraceptive pill, and can interfere with the endocrine (hormonal) system of plants and animals which controls a wide range of physical processes including metabolism, growth and reproduction. Effects include a high incidence of intersexuality (hermaphrodite) condition in male fish downstream of STWs (Jobling et al., 1998). Given the increasing number of substances with endocrine disrupting properties (ranging from antifouling paint to some detergents) there is a large potential for ecological damage. As it would have been extremely difficult and expensive to catch and examine thousands of fish and to measure the very low concentrations in sewage effluent and rivers an alternative approach was needed. One innovative approach is to use the human population as a proxy to estimate the likely loading to a particular STW; and then knowing the type of treatment at that works and its likely efficiency in removing the compounds to predict the effluent oestrogen concentrations (Johnson and Williams, 2004). This was then combined with Low Flows 2000, a GIS based water resources model (see Section 7.8.2) and the location and type of STWs, and the chemical's degradation rate to predict concentrations of oestrogens throughout Britain's rivers (Williams et al., 2009).This approach has also been adopted to predict the likely

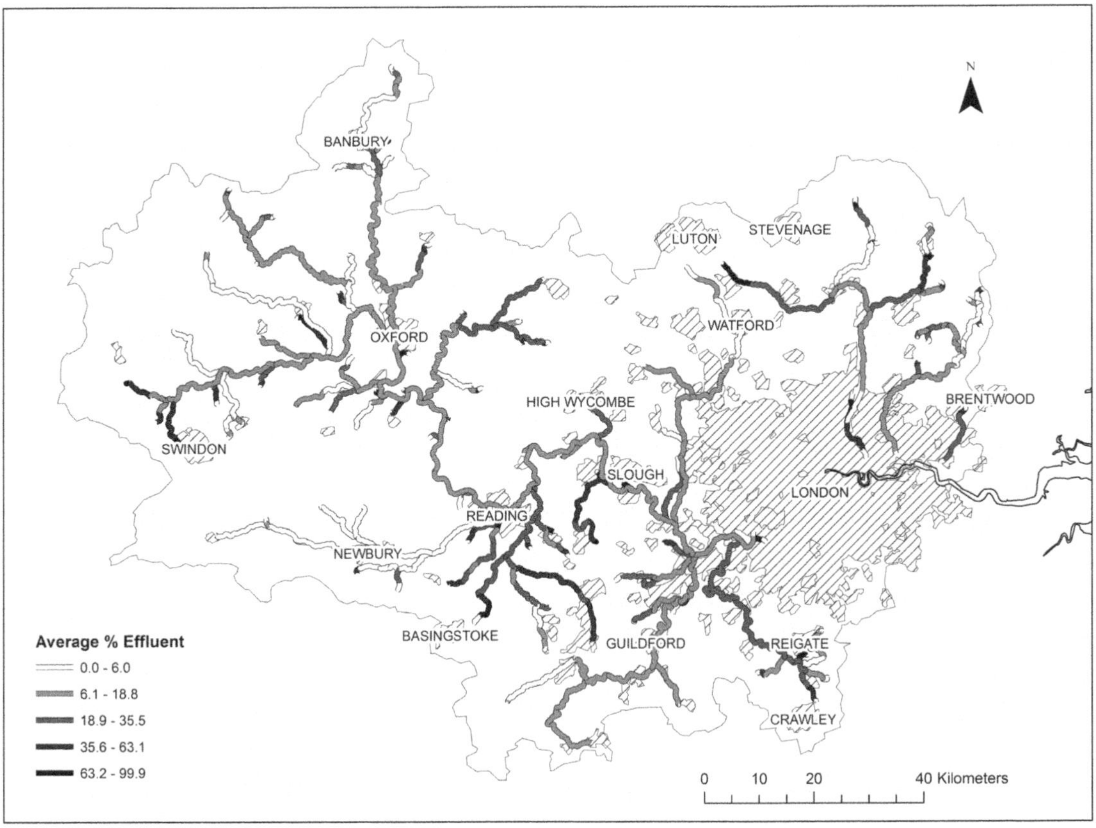

Fig 8.13: Average levels of effluent by volume in the Thames Basin downstream of major sewage treatment works based on flows from 1961–1990. Map compiled by CEH, based on data supplied by Environment Agency.

exposure levels of a range of emerging contaminants of a range of emerging contaminants including pharmaceutical and personal care products (e.g. Johnson et al, 2013; Johnson et al., 2015).

8.6.4 WATER QUALITY MODELLING AND MANAGEMENT

As demand for water resources increases and the variety of pollutants becomes more diverse, there is an increasing conflict between the use of rivers for water supply, and as 'sewers' for carrying away industrial and domestic waste. In fact downstream water users have been drinking upstream wastewater for millennia. Modern wastewater treatment provides water of a high quality (USEPA, 2012). In the UK, 30% of all drinking water is of a recycled nature (Lester, 1990). Figure 8.13 shows the pattern of increasing reuse of water moving down the River Thames basin. Considerable effort has been applied to developing techniques to improve decision making in water quality problems, and mathematical models can be used to assess alternative control measures to operate water and wastewater treatment facilities. Such models may be used to develop operating rules in river basin management to take into account factors such as flood control, water supply, sewage disposal, fisheries, recreation and amenity activities.

There are many different types of mathematical models used for water quality management (e.g. Chapra, 2008; Whitehead et al, 2009a). Early water quality models focussed on urban waste discharges, particularly dissolved oxygen levels in streams. The advent of digital computers in the 1960s led to major advances in models and the way they were applied. Rather than being constrained to local effects, they could view the whole of a drainage basin and address more general concerns about the aquatic environment, and particularly eutrophication. From the 1980s much effort was put into modelling the effects of 'acid rain' and potential remediation strategies. Subsequently attention turned to modelling problems of toxic substances and the importance of their adsorption and transport by sediments. Models are widely used to balance the need to control pollution and protect the environment, while maintaining economic growth. It is outside the scope of this chapter to discuss them in detail, but two widely used models are MAGIC and INCA. MAGIC - Model for Acidification of Groundwater In Catchments, is a process-oriented dynamic model by which long-term trends in soil and water acidification can be reconstructed and predicted at the catchment scale (e.g. Wright et al, 2006). INCA is a process-based dynamic model representing plant/soil system dynamics and instream biogeochemical and hydrological dynamics. It has been used to assess environmental change issues including land use change, climate change and changing pollution environments (Whitehead et al, 2009a).

Information on common river water quality determinands in Great Britain is analysed and collated as part of a Harmonised Monitoring Network. Table 8.3 gives data from a subset of basins showing median values of samples taken at approximately two–week intervals over a 20–year period. Differences between the sites can be related to differences in catchment characteristics. The Nene and Stour, for example, contain much intensively farmed and fertilized land, with high rates of NO_3–N loss. In contrast the Exe, Carron and Spey have only poor farmland, so the water is of high quality with low solute concentrations and a high level of dissolved oxygen. The Trent and Aire are examples of catchments with large inputs of industrial effluent and domestic sewage, and as a result have high concentrations of orthophosphate, chloride, NO_3–N and BOD. The Avon has an intensively farmed catchment and a high sewage effluent input which together give a median NO_3–N level close to the EU limit of 11.3 mg l^{-1}.

Recent advances in field autoanalyser and probe technology that enable very high resolution monitoring in time are providing new insights into nutrient sources and dynamics, which were not captured by conventional lower-frequency monitoring programmes. They reveal a wealth of flow related, flow independent, diurnal, seasonal and annual fluctuations (Halliday et al., 2012). The work indicates a complex structure of catchment functioning and provides new insights into hydrogeochemical functioning and links directly to the internal functioning of catchments. Studies of the upland Plynlimon catchments in mid-Wales, indicated a very variable soil and

RIVER, LOCATION	PH UNITS	CONDUCTIVITY (μS CM^{-1})	SUSPENDED SOLIDS	DISSOLVED OXYGEN
Nene, Wansford	8.0	955	14	10.6
Trent, Nottingham	7.8	904	14	10.2
Stour, Langham	8.2	911	10	10.9
Thames, Teddington	7.9	587	13	10.0
Gt. Stour, Horton	7.9	698	7	10.9
Itchen, Gatersmill	8.1	492	8	10.6
Aire, Fleet	7.5	680	17	8.0
Almond, Craigehall	7.7	595	10	9.7
Axe, Whitford	8.0	393	6	10.9
Ribble, Samlesbury	7.8	407	8	10.2
Exe, Thorverton	7.5	163	5	11.2
Avon, Evesham	8.0	937	16	11.0
Dee, Overton	7.2	164	4	11.1
Leven, Linnbrane	7.1	68	3	11.0
Spey, Fochabers	7.1	77	2	11.4
Carron, New Kelso	6.7	42	1	11.3

Table 8.3: Median values of selected determinands in river water for the 20-year period 1975–94. Units are mg l^{-1}, unless otherwise specified. (Data collected as part of the Harmonised Monitoring Network). Unfortunately the monitoring programme was later curtailed.

groundwater chemistry owing to localised inputs within a highly heterogeneous catchment system, a wide variation of water residence times, and advection and dispersion processes (Kirchner et al., 2001). A key point here is that by allowing variable water residence times within the catchments and incomplete water mixing, the predicted catchment volumetric water storage may be relatively small and much closer to hydrological estimates for these impermeable catchment than the unrealistically large storage volumes previously predicted by standard water quality models. As well as helping to reconcile hydrological and hydrochemical understanding the continuous monitoring studies, have provided insights that indicate that recovery rates from pollutant inputs may take a lot longer than previous modelling work might imply (Kirchner et al., 2000).

BOD O	AMMONIACAL NITROGEN N	NITRATE N	ORTHOPHOSPHATE P	CHLORIDE CL	ALKALINITY $CaCO_3$
2.7	0.13	9.4	0.98	76	209
3.0	0.25	8.6	1.50	99	163
2.1	0.07	7.2	0.59	68	250
2.3	0.23	7.1	1.08	42	190
2.3	0.12	6.2	0.93	53	223
1.8	0.09	5.3	0.35	22	239
7.0	1.49	4.9	1.08	77	126
2.9	0.95	3.7	0.45	61	120
1.6	0.06	3.6	0.23	23	140
2.4	0.15	3.4	0.31	30	120
1.6	0.05	2.3	0.08	17	38
2.7	0.16	10.4	1.60	74	199
1.1	0.03	1.0	0.05	18	25
1.8	0.02	0.3	0.01	10	15
0.9	0.02	0.3	0.01	10	25
0.9	0.01	0.1	0.01	10	5

8.7 IN CONCLUSION

The quality of water is of great importance in determining the uses to which the water can be put. Increasing demand for water and increasing sources of pollution (amounts and types) mean that there is greater potential for environmental damage than ever before, and emphasizes the need for public and political pressure to protect and manage our common resource. Climate change, may lead to longer summer periods of low flow in rivers which will reduce the dilution of pollutants. With increasing demand for water, and pressures for its more efficient use, more of the water we use will have been recycled. After it is treated it is put it back into the rivers and this reuse happens throughout the world.

In scientific terms the study of water quality is more than simply an 'add–on' to our existing knowledge of water quantity fluxes and stores. The use of tracers, both natural and artificial, can provide very revealing information on the flow paths of water, and the use of natural tracers such as ^{18}O has led to a questioning of some of the traditional hydrological theories of storm runoff generation. Water quality can also affect the behaviour and movement of subsurface water and high solute concentrations can lower the evaporation from water bodies as a result of the reduced saturation vapour pressure. Increasingly the hydrologist must consider both the quantity and the quality of water, and in the future the latter may pose the greater challenges.

Increasing **dissolved organic carbon (DOC)** is a big issue for the water companies (Whitehead et al, 2009b). Water colour is correlated with DOC, so as DOC increases water colour will be increasingly brown. Whilst colour per se is not a public health issue, the chlorination processes at water treatment plants generate by-products, such as trihalomethanes, which are carcinogens (Gough et al, 2014)

The Water Framework Directive (WFD) uses biological indicators to provide a fuller picture of the health of a river (Quevauviller, et al, 2012). Species present at a site are compared with those which would be expected in the absence of pollution, thereby allowing for different environmental characteristics in different areas to define stream ecological status. But this approach has drawbacks. Many large rivers have unknown reference conditions due to their history of use, impoundments, pollution etc. Also an alien invader may wipe out many species in a stream depleting its ecology even if it is not polluted.

REVIEW PROBLEMS AND EXERCISES

8.1 Discuss the reasons for the growing importance of water quality.

8.2 Why has water been described as a maverick compound? Compare its properties to those of some other commonly occurring substances.

8.3 Define the following terms: atom, molecule, ions, solvent, moles, valency, acidity, strong acid.

8.4 How are the techniques of chemical thermodynamics and kinetics used in the study of chemical processes in natural waters?

8.5 Describe the various processes by which acid deposition occurs, and discuss the problems of their measurement.

8.6 Discuss the concept of critical loads; in what situations may this approach not be appropriate?

8.7 Define and distinguish the following terms: advection, dispersion, convection, reaction and adsorption.

8.8 Discuss the likely differences in chemical and sediment loads of waters that have travelled via different flow pathways.

8.9 Discuss the challenges inherent in modelling the solute–water soil system.

8.10 Contrast the arguments that pesticides are 'elixirs of death' or 'lifesaving chemicals'. Discuss the importance of a proper hydrological understanding of the pathways and residence times of water movement, in any strategic planning to minimise environmental damage from the use of pesticides and nitrate fertilisers in a predominantly agricultural catchment.

WEBSITES

Dinking Water Inspectorate (England and Wales):
http://www.dwi.gov.uk/

Environment Agency Water Quality archive:
http://environment.data.gov.uk/water-quality/view/landing

European Environment Agency data:
http://www.eea.europa.eu/data-and-maps/

European Union Water Framework Directive:
http://ec.europa.eu/environment/water/water-framework/index_en.html

PHREEQC model for aqueous geochemical calculations:
http://wwwbrr.cr.usgs.gov/projects/GWC_coupled/phreeqc/

Piper trilinear digram:
http://training.usgs.gov/TEL/WQPrinciples/Lesson-13-Piper_Diagram_Form.zip

Effects of Air Pollution on Rivers and Lakes (ICP Waters):
www.icp-waters.no

UK Water quality and abstraction statistics:
https://www.gov.uk/government/collections/water-quality-and-abstraction-statistics

9

HYDROLOGY IN A CHANGING WORLD

"Water is the medium for life on Earth, and the thermostat of our planet".

KIRKBY, 2011

9.1 INTRODUCTION AND GLOBAL DRIVERS OF CHANGE

Having discussed the basic principles and processes in the earlier chapters, this final chapter considers a number of real-world water-related challenges including land use and climate change. There is then a discussion of the way that hydrology may evolve in the future – in terms of the science itself and in the light of wider social and economic issues.

The unity of the hydrological cycle was introduced in chapter 1 and emphasised frequently in the descriptive analyses of individual processes in the following chapters. Chapters 2 to 8 are intentionally focussed largely on individual components of the hydrological cycle and water quality, whereas in this concluding chapter the approach adopted is one of *synthesis*, to illustrate the interaction between hydrological processes, and in particular to show the very real relevance of hydrology to current and future challenges in a continually changing world. This chapter consequently takes a less didactic form than the others, introducing some topics that still being actively investigated and may be open to subjective opinion/interpretation, thus giving readers the opportunity to research the latest literature and to make up their own minds from the developing areas of science. While there is a consensus amongst climate scientists that anthropogenic emissions of greenhouse gases has altered global climate relative to the pre-industrial period, there remains uncertainty about the detailed regional pattern of projected future climate change. In some situations, such as how best to meet the needs of a water-scarce area, the choices and constraints of alternative options may be dictated by social or political as much as – or even more than – hydrological factors. The chapter considers alternative approaches to some of these issues, e.g. 'hard' vs 'soft' engineering solutions to flooding, the use of 'virtual' as opposed to 'real' large scale water transfers, the trading of water rights and payments for water and ecosystem 'services'. There is also discussion of the changing public perception of environmental issues, and opportunities as well as possible dangers posed by the trend towards increasing public scrutiny and participation.

Hydrology plays an increasing role in helping guide environmental policy at different levels, from wetland protection (and restoration), development of flood resilience (i.e. "working with nature") due to increasing recognition that there are limitations to hard engineering, and assessment of the implications of climate change predictions on the world's water resources.

As a subject, hydrology has advanced through technical developments, such as instrumentation and data handling, just as much – or even more so than by advances in theory (Sheffield et al, 2009), and as a subject it has necessarily become more comprehensive. It has broadened considerably within a single human lifespan from a subject that was still very closely linked to hydraulic engineering, to become recognised as one of the essential environmental sciences, with close links to such disciplines as ecology, geology, meteorology and chemistry.

Two decades ago, the majority of hydrological studies were at the catchment or sub-catchment scale (e.g. Baird, 1997). Several major interdisciplinary issues are now leading to a growing interest in the operation of hydrological processes at the global or major regional scale. This is turn emphasises the extraordinary range of spatial scale embraced by hydrology and the need for further work addressing scaling problems in the transfer of theory and of data from one spatial scale to another – an active area of research.

There are two principal forces of *global* change putting increasing stresses on water resources:

a. **Population growth**, which has trebled from 1950s to 2010s, and increasing per capita water usage has led to greater water demand for domestic, industrial and agricultural uses, causing changes in the way that the land is managed, and putting pressures on environmental use and water quality (Figure 9.1a),
b. **Climate change**, due to global warming, intensifying the hydrological cycle and the risk of extreme events, as well as placing additional pressures on the supply side of providing water resources for the growing population (Figure 9.1b).

Fifty years ago much of river and water management was based on the assumption that hydrological variability varied around a relatively stable long-term average, '**stationarity**'. Now that water management decisions can no longer be based on the assumption that the future will closely resemble the past, it is increasingly important to accept that although the fundamental *principles* of hydrology remain unchanged, we have a greater need than ever to further explore and to better understand the *hydrological processes* that bring about the changes to our water environment that are discussed in this book.

Continuing population growth and urbanisation, rapid industrialisation, and expanding and intensifying food production are all putting pressure on water resources (UN, 2010) at the very time when those resources are threatened by climate change. Entire books have been written about these topics so the remainder of this chapter is thus necessarily selective and concentrates briefly on scientific issues reflecting the role of hydrology including:

- Drivers of change – population growth and global warming;
- Hydrological impacts of the resulting changes in climate and land use to meet increased demand for food production; the increasing need for water and resulting stresses on water supply infrastructure and security - e.g. access to clean water).

In these ways we hope to reinforce our view that, although the principles and processes of hydrology may be conveniently taught and studied in a thematic, analytical way, their application to the solution of water-based problems will involve situations for which there may be no single 'right' answer, and so be best achieved through open-minded, integrated, studies at a wide range of scales, and in conjunction with other disciplines in the engineering, environmental and social sciences.

(a)

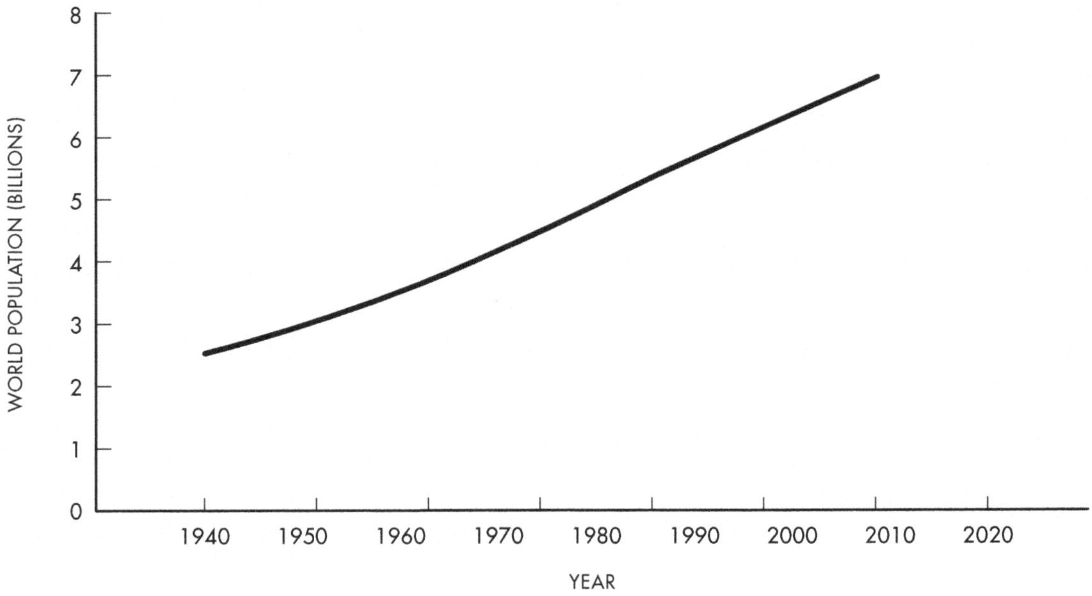

(b)

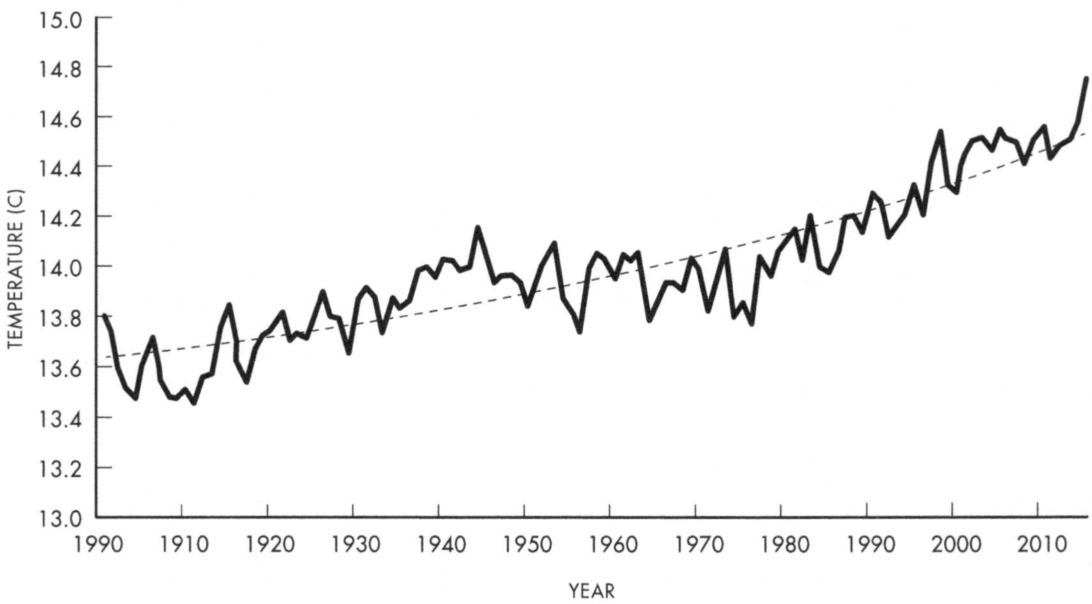

Fig 9.1: a) Global population growth (Source: FAOSTAT, downloaded 15 May 2016); b) Global temperature change (1900 - 2015) (Source: Climate Research Unit)

9.1.1 GLOBAL POPULATION

The estimated world population in 1700 was under 1 B people, and by 1900 it had increased to about 2 B. Since then the population has more than tripled, passing 7 B in 2011, and it is projected to reach 9.7 B in 2050 and exceed 11 B by 2100 (UN, 2015a). All these people will require food, goods and services, which all need water. Yet the amount of land and water available has not changed. The key point is that today's human numbers are a very recent phenomenon. "All environmental problems become harder – and ultimately impossible – to solve with ever more people" (Sir David Attenborough, www.populationmatters.org). To take just one example; when Napoleon invaded Egypt in 1797 a population survey was conducted and reported that the population was under 3 million - a figure that had probably changed little over several thousand years (Hillel, 1991). By 2015 the population had risen to almost 90 million, a 30-fold increase, supported by the same amount of water and land, and despite the exceptional productivity of the Nile Delta Egypt is now the world's biggest importer of grain.

Although the Earth's stock of fresh water is finite and remains relatively constant, with rapid population growth, water withdrawals have tripled over the last 50 years and have increased by more than 7-fold from 1900 to 2010, almost double the rate of population growth, due to rising living standards (FAO, 2016). Much of this increase has been associated with the rapid expansion in the world's irrigated area from an estimated 50 M ha in 1900 to about 250 M ha by the end of the 20th Century (www.fao.org). Demand for fresh water is doubling every 20 years, but the available supply has remained broadly constant over the last five millennia.

At the end of the 18th Century, when the global population was only about 1 B, and the Industrial Revolution had barely begun, Thomas Malthus (1798) was the first to draw attention to the relationship between population growth and resources, warning that population growth would sooner or later be checked by the finite resources available, resulting in mortality by famine and disease. The pace of technological developments in the latter half of the 19th Century due to the "Industrial Revolution" resulted in rising living standards and led many people to forget these early warnings. Furthermore, in the early and middle parts of the 20th Century the "Green Revolution" of intensified agricultural production by the development of high yielding crops and the use of chemical fertilisers, pesticides and large-scale irrigation made it seem possible that human ingenuity could outpace the demands of a rising population. The warnings of Norman Borlaug, the Nobel laureate, who has been called the 'father of the Green Revolution', were forgotten or ignored. Accepting a Nobel peace prize in 1970 he cautioned: "I have only bought you a 40-year breathing space to stabilise your populations."

By the mid-20th Century there were growing concerns in some quarters about the overuse of the world's natural resources, and the damage to the natural environment by agri-businesses. Such concerns were typified by the influential book *Limits to Growth* (Meadows, et al 1972). The authors used a computer simulation of exponential economic and population growth with finite resources to demonstrate broadly similar conclusions to Malthus and possible timescales. Subsequently, concerns were expressed in the 'Brundtland Report' (UN, 1987) published by World Commission on Environment and Development (WCED). This Report coined the term **Sustainable Development** – i.e. 'development that meets the needs of the present without compromising the ability of future generations to meet their own needs'. The UN Earth Summit in 1992 stressed this again, and sustainable development became a sort of catch-phrase or talisman for projects promoted by many intergovernmental and non-governmental bodies. But the concept of sustainable development has increasingly come into question as there is a fundamental tension between a desire to exploit now, and an obligation to protect for the future, as rising living standards clash with environmental conservation. There are conflicting goals, for example between the rich who may focus on 'quality of life' whilst the poor focus on survival itself (Bailey, 1997).

Population growth affects the hydrological cycle at global and local scales in two principal ways. Firstly, by directly increasing the demand for water and secondly as agriculture is the main human use of water it will alter land cover and land use by leading to a need to produce ever increasing amounts of food.

Family planning policy can only be successfully implemented in areas of the world where there are not strong religious objections, and there is good public health. The latter in turn depends upon safe water supplies through its effect on maternal and infant mortality. Both water quantity as well as quality are important since there must be not only sufficient for drinking water but also for washing and general sanitation. It is also widely accepted that educating young women, and providing them with equal opportunities is key to a global shift to smaller families.

Water quality and health

Water-borne diseases are the single greatest threat to global health, with diarrhoea, jaundice, typhoid, cholera, polio, and gastroenteritis spread by contaminated water. According to UN (2010) over half of the world's hospital beds are occupied by patients suffering from water-borne diseases, and more people die as a result of polluted water than are killed by all forms of violence including wars.

Hydrology can have a profound impact on mortality rates, especially for infants and small children, through its role in the provision of safe drinking water, and this rather than medical advances was responsible for the steep decline in death rates from the 19th to the 20th Centuries (Cutler and Miller, 2005) and has direct relevance to so-called emerging or **developing countries** today. Access to safe drinking water and basic sanitation was identified as one of the UN Millennium Development Goals for reducing poverty and improving lives. In 2000 World leaders agreed to halve the proportion of people without sustainable access to drinking water and sanitation by 2015, and the period 2005 to 2015 was declared the International Decade for Action: "Water for Life". The first goal (access to water) was achieved, principally due to rising living standards in China and

India in particular, rather than by successful policy, whilst the second goal (sanitation) was missed (UN, 2015b).

The decision to base the goals on the proportion of people rather than an absolute number was crucial; during the earlier International Drinking Water Supply and Sanitation Decade in the 1980s, clean water was brought to about 1 B people and 750 million people gained access to better sanitation, but due to population growth and rapid urbanization the number without proper sanitation actually increased (Bailey, 1997).

The Millennium Development Goals Review (World Bank, 2016; UN, 2015b), showed much progress towards better access to improved sources of water and sanitation, although due to population growth substantial numbers of people still lack such basic needs (Table 9.1). In fact the number of people without access to basic sanitation in 2015 was greater than the total world population in 1940.

As a successor to the Millennium Development Goals, the Sustainable Development Goals (SDGs) are an intergovernmental set of aspirational goals up to 2030 which aim for a combination of economic development, environmental sustainability, and social inclusion. Goal 6 calls for everyone to have access to clean water and sanitation. This includes sub-components to improve water quality by reducing pollution, increase water-use efficiency, implement integrated water resources management at all levels, including transboundary cooperation.

9.1.2 GLOBAL CLIMATE CHANGE

At the global scale it is known that the burning of fossil fuels has caused an increase in the carbon dioxide content of the atmosphere. This is one of a number of **greenhouse gases** that allow the passage of visible radiation from the sun but absorb part of the outgoing long–wave infrared radiation from the Earth, which would otherwise be lost into space. Without this natural greenhouse effect the mean temperature of the Earth would be about -18°C, some 30°C cooler than it actually is. Increases in greenhouse gases will absorb more of this outgoing radiation and therefore lead to a warming of the atmosphere. This in turn may result in changes to circulation patterns, with a poleward migration of climatic zones and spatially variable effects on precipitation.

With rising temperatures air can hold more water. Basic theory, observations and climate model results all show that a

	1990	2015	NUMBER STILL WITHOUT
Access to clean water	76%	90%	Over 0.6 B, or 1 in 10
Basic sanitation	54%	68%	Over 2.4 B, or 1 in 3

Table 9.1: From 1990–2015 than 2.5 billion people gained access to improved water sources, and almost 1.9 billon people to sanitation facilities (UN, 2015b).

warmer atmosphere can hold about 6% more water vapour for every 1°C of warming of the atmosphere leading to a likely increasing frequency of extreme climate events, with more intense rainstorms and more droughts. Increased greenhouse gases raise temperatures, which in turn causes greater evaporation and therefore greater rainfall. More water vapour – which is itself a greenhouse gas – amplifies the warming effect of increased atmospheric levels of carbon dioxide – a "positive feedback."

Water vapour is the main greenhouse gas' (Kiehl and Trenberth, 1997); with increasing temperature there is an exponential increase in the equilibrium vapour pressure of water. Thus, with rising temperatures more water is evaporated from the oceans which intensifies the greenhouse effect and further warms the surface.

Most scientists now accept that climate change is occurring. Observational evidence of global change includes (IPCC, 2014):

- Rising global atmospheric concentrations of CO_2 and methane concentrations to levels unprecedented in at least the last 800,000 years.
- Increasing global average near-surface air temperature of nearly 1°C between 1880 and 2012,
- Decrease in snow and ice cover in many parts of the world,
- Increase in length of the growing season in central England (defined as days >5°C) from 244 days (1861–90) to 251 (1961–90) to 280 (2006–15),
- Warming of the oceans and rising CO_2 concentrations in seawater increasing its acidity and resulting in mortality of coral reefs,
- Sea levels have risen by 20 cm since 1900, largely due to warming and expansion of the upper layers of the oceans.

There is increasing evidence that past periods of climate variation, such as aridity and wetness, have influenced human progress in particular parts of the world leading in some cases to the rise or fall of civilisations (see Chapter 1). Now, it seems that due to industrial development and rising populations the reverse has become possible and humans can alter climate. Furthermore, with climate change we are experiencing for the first time a human effect that is *global* – this has never happened before. Previously adverse effects, such as pollution, land use change, or overgrazing causing soil compaction and enhanced runoff, were just felt locally or regionally.

Long-term records show increasing concentrations of CO_2 and an increase in global temperatures in the 20th Century. CO_2 concentrations in the atmosphere at Mauna Loa on the island of Hawaii in the Pacific Ocean, far from urban centres, rose during the 19th Century from about 280 ppm to 380 ppm. The concentration of CO_2 is now about 40% higher than 200 years ago. Air bubbles trapped in ice cores show that concentrations were nearly constant during most of the last Millennium, then rose to levels that have not been experienced for at least 800,000 years (Luthi et al, 2008). The increase in CO_2 is due both to the burning of fossil fuels and the steady destruction of forests.

In response to growing international concern the Intergovernmental Panel on Climate Change (IPCC) was founded in

1988. It was set up by the World Meteorological Organization (WMO) and the United Nations Environment Program (UNEP). It does not conduct its own research, but rather periodically reviews and assesses the most recent scientific information on all aspects of climate change and its impacts. This is provided to world governments to assist them in formulating realistic response strategies for mitigation and adaptation to climate change. Successive IPCC reports have shown increasing certainty in the reality of anthropogenic climate change – increasing from 'little evidence' in its First Assessment Report published in 1991 to 'discernable' by 1995, to 'likely' and then 'very likely' in the Third and Fourth reports published in 2001 and 2007 respectively, to 'extremely likely', and 'unequivocal' by the Fifth Assessment (IPCC, 2014).

By the 1992 Earth Summit there was growing scientific consensus about the reality of climate change. In 1998 the British Prime Minister, Margaret Thatcher, acknowledged the problem and addressed a meeting at the Royal Society of London on global warming saying that humanity had "unwittingly become a massive experiment with the system of the planet itself". There was also growing political opposition since remedial action would interfere with continuing industrial development that was perceived to have brought so many benefits to humankind. There is intense lobbying against climate change science by multinational companies, particularly in the USA, where prominent public figures, including Donald Trump, have repeatedly called climate change science a 'hoax' (Time, 2016), and there has been a campaign of misinformation (Oreskes and Conway, 2010; Lewandowsky et al, 2015).

The IPCC publishes Assessment Reports at approximately 5-year intervals on the state of the atmosphere and the likely impact of increased levels of carbon dioxide and other greenhouse gases. The initial sessions of IPCC failed to recognise that the most serious impacts of climate change are those that affect the hydrological cycle and water resources. However, the more recent IPCC Technical Paper on Climate Change and Water (Bates et al., 2008) highlighted these impacts, making clear the vital nature of the relationship between climate change and water resources.

The IPCC (2014) Fifth Assessment Report (AR5) stated:

- Warming of the climate system is unequivocal, and it is extremely likely that human influence has been the dominant cause of the observed warming since the mid-20th Century'
- Recent anthropogenic emissions of greenhouse gases are the highest in history, and human influence has been detected in warming of the atmosphere and the ocean, in changes in the global water cycle, in reductions in snow and ice, in global mean sea level rise, and in changes in some climate extremes.
- Since 1901 precipitation amounts averaged over mid latitudes in the northern hemisphere have increased, and there were increases in the frequency and intensity of heavy rain events in Europe and North America.

Evidence of global temperature change comes from the analysis of the instrumental surface temperature. Scientists at the

Climate Research Unit (CRU) studied temperature data from 4,138 stations, and for each station calculated the mean temperature for 1961–1990 and the subsequent temperature anomalies relative to that period. They then arranged this data into a 5x5 degree grid to allow for the uneven distribution of stations (Morice et al, 2012).

Unlike the steady growth in global population the rise in temperature has not been smooth. Short-term oscillations lasting a few years are superimposed on the trend, just as troughs in temperature occurred in the late 19th and early 20th Centuries, and most recently in the 1960s (Figure 9.1b). The Central England Temperature series (Parker et al, 1992) begins in 1659 and shows periods of warming and cooling. The levelling out in the period from 1950s to 1980s is believed to have been caused by increasing air pollution such as sulphur and black carbon aerosols that reduced the amount of solar radiation reaching the Earth's surface (Wild, 2012). Following reductions in atmospheric pollution due to clean air acts in many regions (See Chapter 8) the temperature began increasing again. There was then a second "pause" was rising surface temperatures from about 2000–2009, before temperatures continued rising. A global network of profiling floats sampling ocean depths down to 2000 m indicates this 'pause' is due to the extra heat being stored in the deep oceans below normal measurement depths (Roemmich et al, 2015), but nevertheless it has still been used as evidence by some sceptics that there is no global warming.

The impacts of climate changes on the hydrological cycle are not expected to be uniform during the 21st Century. Contrasts between wet and dry regions and between wet and dry seasons are likely to increase. Such changes may have a profound impact on the water resources, the environment and world food production. Hydrologists will have a central role to play in understanding and managing resources in a changing environment. The World Bank has produced a Climate Change Knowledge Portal (CCKP) with information, data and reports about climate change around the world - See website lists at the end of this chapter.

9.2 TERRESTRIAL INTERACTIONS

In the past, the drainage basin was normally selected as the basic unit for research studies including modelling hydrological processes, water balance studies, chemical budgets and human impacts on hydrological systems. Each basin acts as an individual hydrological system, receiving quantifiable inputs of precipitation which are transformed into flows and storages and into outputs of evaporation and runoff. This is the appropriate level at which water-based issues affect local communities concerning land use changes, flood defence, pollution and ecology. Small experimental catchments in particular, in which variables could be more easily identified, controlled and measured, have made an important contribution to hydrology for more than 100 years (e.g. Rodda, 1976; Keller, 1988; Whitehead and Robinson, 1993; Herrmann et al, 2010). Significant specific examples include the pioneering basin experiments at Emmental, Switzerland (established in the 1890s), Wagon Wheel Gap, USA (1900s), Coweeta, USA (1930s), Cathedral Peak, South Africa (1940s) and Plynlimon, UK (1970s). There have also been studies of groups of basins for a common goal, most notably Flow Regimes from International Experimental and Network Data basins (FRIEND) (e.g. Servat et al, 2010).

Numerous examples in this chapter, and elsewhere in the book, illustrate that hydrology is studied, and hydrological processes operate, in the 'global village'. Evaporation from one country contributes to precipitation in another. Pollution in a headwaters of an international river may impact upon the economy of other countries in the lower part of the basin. In other words, it becomes ever clearer that decisions on water management (e.g. irrigation, river abstraction, or river diversion) or on the management of other resources which impact upon hydrological processes (e.g. deforestation), must in future be made with the fullest possible understanding of their potential wider impact.

9.2.1 BASIN SCALE – LAND USE CHANGE

Virtually every component of the drainage basin hydrological system may be modified by human activity. The most important of these alterations result from:

- Large-scale modifications of channel flow and storage, by means of surface changes such as deforestation, afforestation, and urbanisation, which affect surface runoff and the incidence or magnitude of flooding;
- The widespread development of irrigation and land drainage; and
- The large-scale abstraction of groundwater and surface water for domestic and industrial uses.

Other important modifications include artificial recharge of groundwater and inter-basin transfers of surface and groundwater.

Human activity (e.g. urbanisation, mining, forestry and agriculture) frequently acts as a flood-altering factor by modifying key hydrological variables such as water storage, infiltration, and transmissivity.

Irrigation

In absolute terms 80% of global cropland, producing 60–70% of the world's food is rainfed, without irrigation. But irrigation has a disproportionate impact on both water use and food production. Water withdrawals for agriculture, principally for irrigation, account for about 70% of global water use; this ranges from about 25% in Europe to over 80% in Africa, although in absolute terms the greatest withdrawals are in Asia (FAO, 2016). There has been a rapid increase in the world's irrigated area from an estimated 50 M ha in 1900, nearly doubling to 90 M ha by 1950 and then nearly tripling to about 250 M ha by the end of the 20th Century and reaching about 350 M ha by 2012 (www.fao.org). Although accounting for about 20% of the world's cropped area it produces about 40% of the food. The growth in irrigation has permitted the expansion of agriculture into arid regions, and in a favourable climate enables multiple cropping. Nearly 80% of the world's irrigated crop area is in Asia, which produces about 70% of the grain harvest in China and 50% in India.

The potential deleterious impacts of irrigation schemes include reduced flows downstream, due to evaporative loss and consumptive use in the irrigated area, changes in water quality including increased salinity as well as agricultural chemicals, with increased losses of nitrogen and phosphorus potentially leading to eutrophication (Tilman, 1999). Model simulations indicate that large-scale irrigation can alter climate significantly in some regions, cooling the near surface air by direct evaporative cooling effects and indirectly by increased cloud cover (Sacks et al, 2009). India, China and Pakistan and USA account for about half of total water withdrawals for irrigation, and worryingly, model studies suggest that over half of this water is taken from either non-renewable or nonlocal sources (Rost et al, 2008). It is known that groundwater 'mining' of water for irrigation is occurring widely across the globe (Scanlon et al, 2006).

Irrigation accounts for about 20% of global agricultural land, but produces 40% of the total food. To meet the increasing population's need for food by rain-fed agriculture alone would require an enormous expansion of agricultural land and resulting environmental damage.

Agricultural drainage

The substantial growth in the area of drained agricultural land, for example in Europe, which occurred after 1939, generated interest in the effects of **agricultural drainage** on flood hydrology. However, early work failed to recognise either the inadequacy of much of the available data or the importance of soil type. It is now clear (e.g. Robinson and Rycroft, 1999) that the drainage of heavy clay soils, which are prone to prolonged surface saturation in their undrained state, generally results in a lowering of large and medium flood peaks, as drainage ameliorates their naturally 'flashy' response by greatly reducing surface saturation. On more permeable soils, which are less prone to surface saturation, the effect of drainage is usually to accelerate the speed at which water follows subsurface flowpaths, thereby tending to increase flood flows.

Agricultural subsidies have been used in the developed world to support farmers and reduce the need to import food. More recently, there has been a move to environmental payments to farmers, instead of production payments, since people value the landscapes they produce and like cheaper imported food. However, much of this food comes from developing countries, many of which are semi-arid and cannot sustain large scale food production (e.g. mining groundwater reserves). So, in the long term due to rising global populations, future food production will have to be in those parts of the world where it is most sustainable – the temperate zones, due to their wet climates. With the prospect of climate change putting additional pressures on semi-arid areas, temperate countries such as Britain may turn into global 'breadbaskets' and export agricultural surpluses to feed the populations in less favoured areas. This would call into question the current process of de-intensification of agriculture with repercussions for the protection of water, soil and the wider environment. Should such areas protect and safeguard their environments at the expense of growing food for hungry people in other parts of the world?

Governments have encouraged the expansion of **biofuel production** to meet mandated targets for 'renewable energy' but these may have unintended environmental and social consequences. Fast growing fuel crops may have very high water use requirements and compete for land with food production (de Fraiture et al, 2008; Jorgensen, 2011) as well as the destruction of natural ecosystems – as in the case of the Amazon, where rainforest is being replaced by fast-growing Eucalyptus.

Forestry

One of the primary drivers of deforestation is the conversion of forests into agricultural land to feed the world's growing population. The global forest area has declined by over 100 M ha since 1990. The largest net loss of forests has occurred in South America and Africa, of around 3.6 and 3.4 M ha yr^{-1}, respectively, from 2005 to 2010 (UN, 2015b).

The role of forestry in modifying river basin flood hydrology remains a controversial topic. From theoretical considerations the larger interception loss of rainfall (Chapter 3) and greater rooting depth of trees will lead to a reduction in soil water stores and a consequent reduction of runoff (Chapter 7). These effects are most significant for small storms and least significant for the largest storms (Figure 9.2).

Forestry is generally thought to reduce peak flows, although temporary increases of flooding may initially result from forest road construction or from pre-planting drainage. Deforestation may intensify river flooding by: adversely affecting soil structure and volume; reducing infiltration rates, either through the effects of diminished root mass or by facilitating the development of surface crusts; and reducing water storage, either in the soil profile or within the canopy. However, these influences are normally significant only during frequent low-magnitude storms. Their effect during increasingly severe flood-producing storms diminishes as prolonged heavy rainfall and/or melting fills available storage and creates widespread conditions of surface saturation and zero infiltration. Accordingly, their effects are also insignificant where initial storage values are low, e.g. swamplands and

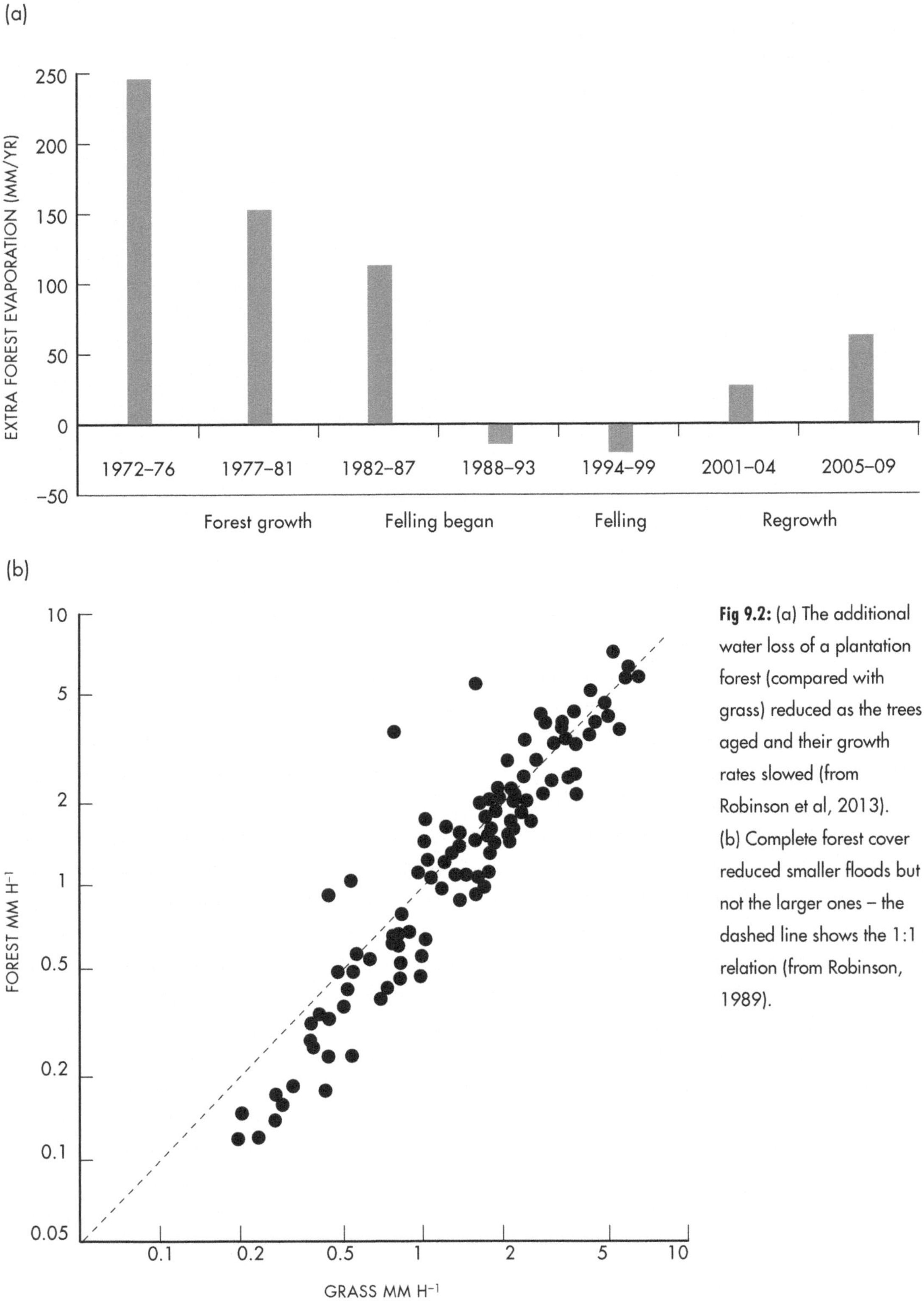

Fig 9.2: (a) The additional water loss of a plantation forest (compared with grass) reduced as the trees aged and their growth rates slowed (from Robinson et al, 2013).
(b) Complete forest cover reduced smaller floods but not the larger ones – the dashed line shows the 1:1 relation (from Robinson, 1989).

steep slopes with shallow soils, which will produce rapid quickflow irrespective of their vegetation cover. Furthermore, the effect on the flood hydrograph of an additional volume of quickflow diminishes with distance down the channel system, so that headwater deforestation becomes less significant as the flood peak moves downstream. These are important factors in high-rainfall areas such as the Himalayan foothills of the Indian sub-continent where forest clearance for agriculture has frequently been blamed for increased flooding many hundreds of kilometres downstream. Examination of several decades of hydrological data led Chawla and Mujumdar (2015) to conclude that there is no evidence that flooding on the Gangetic plain has increased. Hofer and Messerli (2006) found no evidence that forests in the Himalayas prevented large-scale floods in the Brahmaputra.

Urbanisation

The urban population is rapidly increasing throughout the world. Around 2005, the number of people living in urban areas exceeded those in rural areas, and it is expected that by 2030, about 60% of the world's population will live in cities (World Bank, 2016). China alone has over 130 cities with above 1 million inhabitants. Nearly all population growth will occur in cities, mostly in poor countries which by definition have the least resources and the lowest ability to dispose of wastes.

Cities may be economically efficient, but environmentally they suck in resources and emit wastes. Many originated as small settlements alongside a small stream, but as they grew into larger towns and cities the water sources that once served a few hundred people become no longer adequate.

The extent to which flood characteristics are modified by **urbanisation** depends very much on the nature of the modified urban surface, the urban hydrological system, and climate. Urban surfaces are less permeable than most of the surfaces which they replace. As a result they are effective source areas for quickflow and their flood hydrographs tend to have both higher and earlier peaks, reflecting the greater volume of quickflow and its rapid movement across the urban surface. Accordingly, urbanisation tends to increase downstream flood peaks and volumes. However, much depends on the permeability contrast between an urbanised area and the pre-urban surface, so that flood conditions are exacerbated more by the urbanisation of a high infiltration, sandy area than by the urbanisation of a low infiltration clay area. The additional quickflow produced by urban surfaces is routed through stormwater systems many of which, like those in the UK, are old and unable to cope with high-magnitude events, which may overflow and lead to widespread flooding.

Generally, the influence of urbanisation is smaller in winter than in summer and in wet climates than in dry ones. It specifically reduces with the severity of the flood-producing rain in the sense that, after prolonged and heavy rainfall, the infiltration characteristics of urban and saturated non-urban surfaces are very similar. Rising groundwater under UK and other European cities is primarily a recovery due to diminishing industrial abstraction from groundwater, while in arid areas such as the Middle East, rising groundwater may result from agricultural irrigation close to urban areas.

Flow abstractions

Substantial amounts of water are withdrawn for agricultural, domestic and industrial purposes. This is an important aspect that is usually not addressed, because of the amount of work involved in calculating **naturalised flows** and the necessary data are generally not available. Consequently, water balance calculations are usually performed using *measured* flow data, yet the distinction matters because it reveals the true magnitude of the underlying resource, and the sensitivity of is sustainable use to small changes in abstractions and returns.

One of the few basins with such data is the River Thames above Kingston, London, for which the Environment Agency has collated the licenced abstractions and returns as well as reservoir storages. This drains an area of 9,948 km^2 and the water balance is significantly affected by human activity.

The artificial factors which affect the flow of the Thames at Kingston include:

- Reservoir storage which diminishes river flows when the reservoirs are supplied from the drainage basin above the gauging station at Kingston,
- River regulation, especially by releases of water pumped from groundwater storage in the upper part of the basin,
- Abstraction of groundwater and river water for public water supplies,
- Effluent return, especially outflows from sewage treatment works, and
- Industrial and agricultural abstractions which result in a net reduction of natural flow.

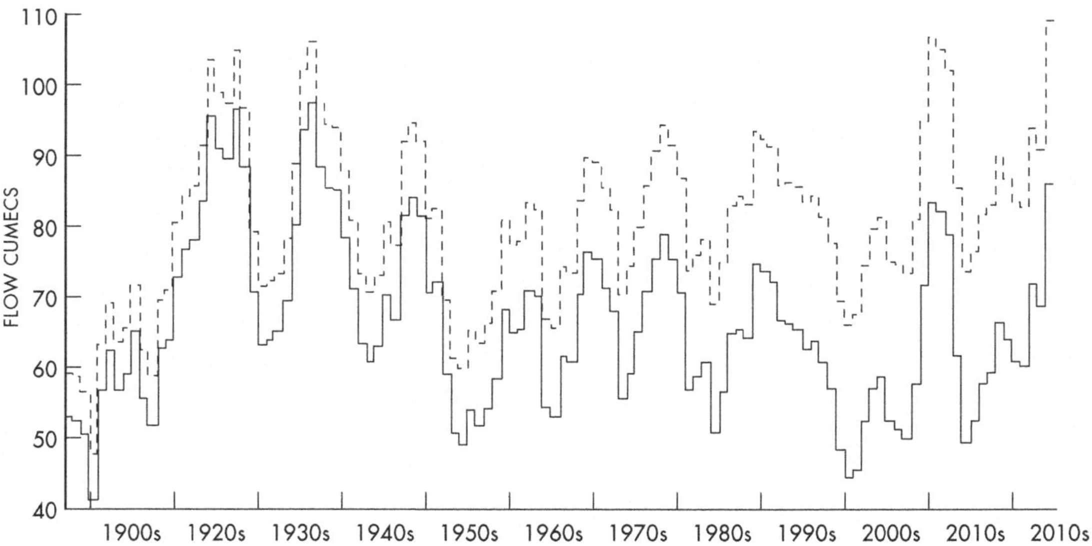

Fig 9.3: Time series of measured flow and net flow adjusted for abstractions and returns. River Thames 5-yr running mean annual flows at Teddington, London (cumecs) Solid line = gauged, Dashed = Naturalised). The difference shows increasing abstractions over time. (Data from NRFA).

The difference between the measured flow data, and the **naturalised flow values** corrected for human influences shows a steady increase reflecting the growing human influence in the catchment. In contrast, industrial withdrawals in the London region have decreased in recent years, resulting in a significant increase in water table levels (see also Section 5.4.3). Early, but less reliable, data at the beginning of the 20th Century show a reduction in mean annual flow of the Thames about 5 cumecs, or about 6% of flow, whereas from 2000 the difference is about four times that amount.

Thames Water, the UK's largest water and wastewater services company has over 13 million customers and supplies an average of 2,500 M litres of drinking water and treats nearly 3,000 million litres of sewage per day. The Water Resources Management Plan (WRMP14) attempts to forecast how the demand for water resources will change in the 25-year period from 2015 to 2040 (Thames Water, 2014). This uses forecasts of a total increase in population of between 2.0 million and 2.9 million people by 2040, and a central estimate of the impact of climate change from the UKCP09 forecasts. In addition, there must be reduction in existing abstractions from some rivers to ensure compliance with the EU Water Framework Directive (WFD).

Integrated catchment management

There are many often conflicting uses of water resources, and a need for a balance between humans and the impacts they have on ecosystems through continuing agricultural intensification, industrialization and urbanisation. This necessitates an **integrated catchment management** approach of coordinated land and water management to maximise economic and social benefits in an equitable manner without compromising the sustainability of vital ecosystems (e.g. Ferrier and Jenkins, 2010). But this ideal is often far from easy to achieve (Newson, 2009).

9.2.2 REGIONAL LAND COVER – CLIMATE INTERACTIONS

Although the drainage basin is an appropriate and useful scale for a wide range of hydrological investigations, major hydrological problems are increasingly being faced of a regional, continental, or global nature. What happens in one basin can influence the hydrology in other basins via rainfall and evaporative demand. Precipitation over an area derives partly from water vapour formed by local evaporation and partly from water vapour which moves into the atmosphere above that area by means of horizontal advection (see Figure 9.4, and Section 2.3).

For example, the vapour flux values shown in Figure 9.5 indicate that the Amazon basin is a sink, since more water vapour enters than leaves the atmosphere over the basin. The Amazon basin is not the only regional sink for water vapour, and it follows that other extensive areas of the Earth's surface must act as net sources of water vapour. The location, role and inter-relationships between global sources and sinks of water vapour currently constitute important issues in hydrology.

At the regional scale there has long been controversy as to whether vegetation cover could influence rainfall via its effect on evaporation losses. In the 19th Century,

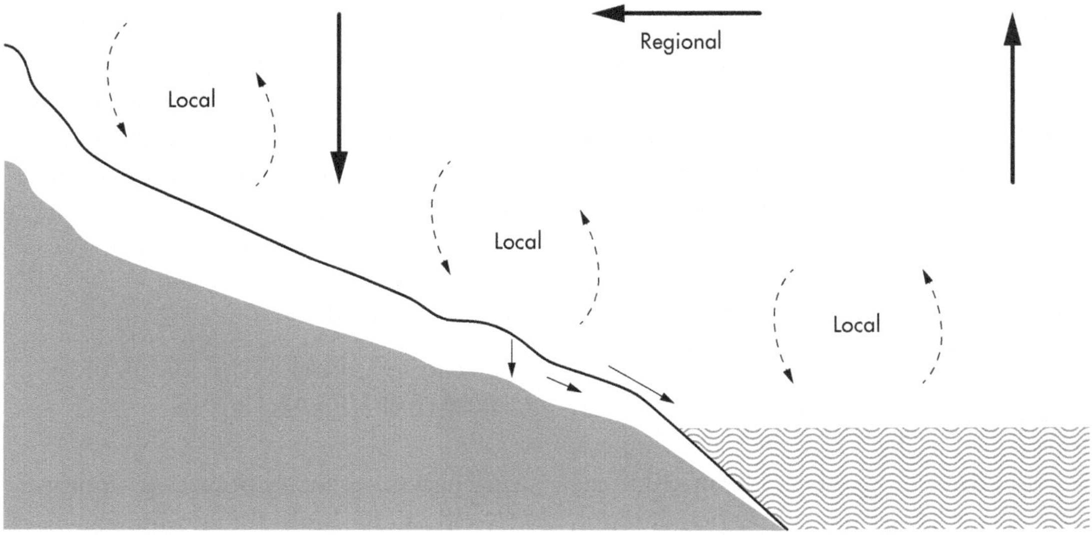

Fig 9.4: Local and regional water cycles.

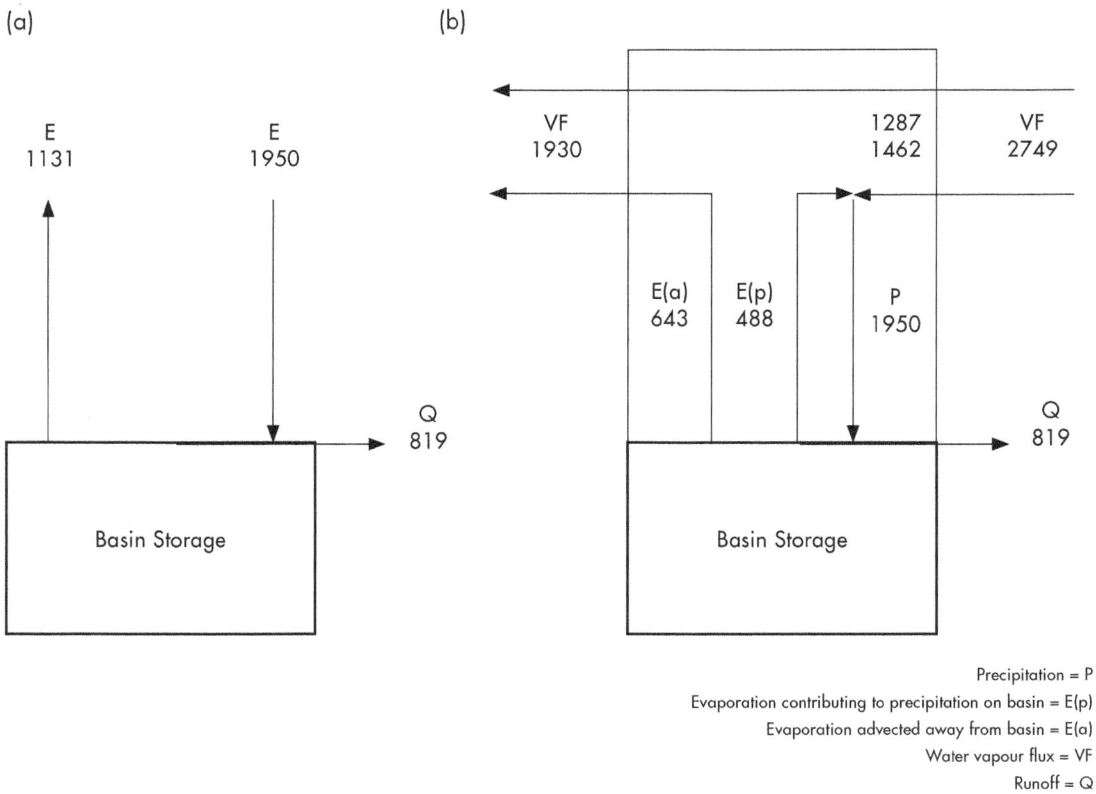

Fig 9.5: Components of the annual water balance of the Amazon basin (mm): (a) Terrestrial water balance, (b) Combined atmospheric and terrestrial water balance, illustrating the role of precipitation recycling (Based on data in Eltahir and Bras, 1994).

Humber (1876) attributed a period of reduced rainfall in parts of central USA to the felling of woodlands for the expansion of agriculture which was taking place at that time. Such notions were largely discounted by later research which emphasized the large–scale nature of water vapour transfer, with often great distances between the evaporation of water and its subsequent precipitation (Penman, 1963).

This view subsequently changed as a result of more recent studies. Deforestation can influence regional hydrology by decreasing evaporation and increasing runoff (see Chapter 3), and changes in the water and surface energy balance may result in a reduction in rainfall, and studies of atmospheric properties have found cooler and moister air downwind of forests (e.g. Butt et al, 2011; Spracklen et al, 2012).

The drainage basin of the River Amazon has been a major focus of hydrological interest for several reasons. Its gigantic area of 4.6 M km^2 (half the size of Europe), and its high average annual rainfall of nearly 2000 mm mean that the river contributes nearly 20% of the global runoff to the oceans. The basin contains the world's largest area of rainforest, which has been subjected in recent years to massive clearance operations, often involving burning, to facilitate the expansion of large-scale agriculture. Several studies have confirmed the significance of **precipitation recycling** within the Amazon basin, defined as the contribution of evaporation within an area to precipitation in that same area, expressed as a recycling ratio ρ , i.e. the proportion of precipitation which is contributed by local evaporation (Eltahir and Bras, 1994). This ratio is a function partly of the size of the area concerned and also on the vigour of the evaporation-rainfall process. Thus in tropical areas where the predominant movement of atmospheric moisture is vertical, it will be more important at a local scale than in mid-latitudes where large-scale horizontal transfer is predominant. Accordingly, a large equatorial drainage basin, like that of the Amazon, would be expected to have a large precipitation recycling ratio, and studies by Brubaker et al. (1993) and Eltahir and Bras (1994) calculated ρ = 0.25-0.35, i.e. at least one-quarter of all precipitation falling on the Amazon basin is derived from local evaporation. In such circumstances, the simple terrestrial water balance (Figure 9.5a) does not give a complete picture of the drainage basin hydrology. Although the input of precipitation (1950 mm) is balanced by the combined outputs of evaporation (1131 mm) and runoff (819 mm), the relationship between evaporation and precipitation is not apparent unless the combined terrestrial and atmospheric water balance is considered (Figure 9.5b). Then it is clear as noted in Section 2.3 that about half of the Amazon basin rainfall (43%, or 488 mm) originates from forest evaporation (Salati and Vose, 1984; Shuttleworth, 1988). This is significant since it means that continuing large–scale deforestation is likely to reduce evaporation, increase runoff, and ultimately to reduce precipitation in that region.

Water vapour recycling may have considerable hydrological implications for the impacts of land-surface changes. For example, in a drainage basin where virtually none of the precipitation results from 'local' evaporation, even large-scale changes may have little impact on drainage basin hydrology, whilst in a situation where all precipitation derives from local evaporation, changes of vegetation cover or other land-surface characteristics could have a significant effect on drainage basin hydrology and climate. Thus a high recycling ratio indicates a strong potential role for land surface hydrology (Eltahir and Bras, 1994) and a need for concern about the likely hydrological and climatological effects of any large-scale land cover change.

Another area where precipitation recycling is important is the **Sahel zone** of Africa where Brubaker et al. (1993) estimated $\rho = 0.35$. The Sahel region experienced a severe drought during the period 1970–2000, and although rainfall subsequently increased it was still below the long-term average and was associated with fewer and heavier daily rainfalls resulting in an increase in flood fatalities. It appears likely that the initial cause was sea surface temperature changes altering wind and moisture fluxes overland, and reducing precipitation. The resulting decline in soil moisture reduced the vegetation cover, enhancing the ground surface albedo by exposing more of the light-coloured sandy soil. This in turn would lower the ground surface temperatures, reducing the likelihood of convective precipitation, and so creating a positive feedback to drought conditions (Kucharski et al, 2013). It has also been suggested that overgrazing would contribute to the observed vegetation decline, reducing rainfall over the Sahel (Charney, 1975). Taylor and Lebel (1998) discovered a positive feedback between antecedent soil moisture and rainfall in a semi-arid West Africa. Previous storm patterns influenced the pattern of soil moisture and hence local evaporation rates; this in turn influenced subsequent convective rainfall patterns.

9.2.3 WATER CONFLICTS

Water conflicts have a very long history, particularly where there is a high dependency on water in a downstream country, or a history of antagonism between countries. Analysing recent data on river claims and conflicts over the 20th Century, Hensel et al (2006) confirmed that water scarcity increased the likelihood of militarized conflict over cross-border rivers, but noted that river institutions and regional or global organisations can be effective in reducing tension and aiding peaceful settlement of disputes by providing 'neutral' information (often hydrological).

More than 40 per cent of the world's population, including some of the poorest, depend upon water that originates in sources beyond their national borders. There are 260 'international river basins' shared by two or more countries.

One of the world's largest international basins is the River Nile which runs through 10 countries and drains 3.4 M km^2, 10% of the African continent. The Nile is fed by two main tributaries, the White Nile, originating at Lake Victoria in Uganda flows through the extensive Sudd swamps of southern Sudan and has a fairly constant flow, and the Blue Nile that starts in Lake Tana in Ethiopia. This has a much greater annual flow with a strong seasonal cycle due to monsoon rain on the highlands of Ethiopia and provides 70% of the annual flow at the Aswan Dam (See Section 1.2).

The Nile provides practically the sole source of water for Egypt and Sudan and little of its water has traditionally benefitted Ethiopia when it is needed most in the dry season. Consequently, Ethiopia decided to construct the Grand Ethiopian Renaissance Dam (GERD) to store water and produce up to 6,000 megawatts of electricity, which more than doubles Ethiopia's present capacity. This will raise the standard of living in Ethiopia, but it means Egypt giving up some of the water of the Nile, which has played such a pivotal role in its history, and Egypt has threatened possible military intervention. Conversely it may be argued that Egypt has long monopolized a large part of the Nile waters, so it is only fair that other countries be allowed to share the waters. The impact on downstream users will depend crucially on the way in which the dam is operated (Wheeler et al, 2016).

In Chapter 1, some of the negative aspects of the Aswan dam were mentioned. Biswas (Biswas and Tortajada, 2001; Biswas, 2002) provides a vigorous defence from a developing country point of view of the Aswan Dam, which has been so criticised by western experts. The reservoir has provided a store of water at times of drought when neighbouring countries such as Sudan and Ethiopia have suffered enormously. The dam also provides flood protection for communities living downstream, and the generation of hydroelectricity has brought power to many rural communities. While the dam construction reduced fish catches downstream, they have partially recovered, and the reservoir itself has become an important new source of fish.

With half of world's population dependent upon water shared by two or more nations it is essential that national governments are equipped to determine and enforce their legal entitlements and obligations ("right to water"). The UN 'Convention on the Law of Non-Navigational Uses of International Watercourses' was drafted to help conserve and manage water resources for present and future generations. And although still to be ratified by the majority of countries, it is regarded as an important step in establishing international water law. Wouter et al (2005) present a pragmatic, interdisciplinary methodology for interfacing water law and water science with the objective of assisting countries to develop effective national water policies.

There is an urgent need for a much clearer understanding of the physical, economic, social, and political consequences of large-scale water resources development and of major irrigation and flood-defence schemes which are designed to reduce the impact of drought and flood disasters.

Such concerns are understandable in the light of the channel and other geomorphological changes which followed the construction of the Aswan High Dam on the River Nile, and of the widespread and adverse ecological effects which have already resulted from some of the river diversions which have been carried out in the former USSR. Most notoriously, perhaps, was the diversion of almost all the waters of the rivers, which formerly drained to the Aral Sea in present day Kazakhstan and Uzbekistan, to irrigate cotton and other crops. This resulted in the drying out of the Aral Sea whose surface area shrank to less than half and whose salinity increased leading to a collapse of fisheries and of the regional economy (Micklin, 2007).

Global concern accompanied building of the enormous Three Gorges dam on the Yangtze River in central China. With an upstream catchment area of 1 million km^2 the dam created a 600 km long reservoir and displaced about 1.3 million people and flooded archaeological and cultural sites and is causing significant ecological changes. Conversely, it may be argued that the dam brought many benefits. It is the world's biggest hydroelectric power producer capable of generating 18.2 GW of 'clean' electric power, close to 10% of the whole country's electricity requirements, thereby cutting air pollution and greenhouse gas emissions. And by providing flood storage space it can significantly reduce the Yangzi's infamous floods which have resulted in over 1 million deaths in the past 100 years. Schemes such as this pose great social, environmental and ethical dilemmas, and hydrologists can provide essential factual and expert guidance to aid policy makers make difficult choices (Guo et al, 2012).

Successive UN conferences have placed providing water supply and sanitation for those without them high amongst their goals, but the environmental consequences have not always been recognised. Independent India's first prime minister, Jawaharlal Nehru called dams 'temples of modern India' in the sense that he was affirming a commitment to modernisation and socialism in post-Independence India. However, later he saw their negative side, with villages and farmland flooded and people displaced and their livelihoods destroyed. Amongst the exceptions was the Report of the World Commission on Dams (UNEP/IUCN, 2000) which has been widely praised for attempting to reach a compromise between promoters and opponents of dams through increased levels of transparency and participation.

9.2.4 WATER TRANSFERS

Globally there is enough water, but it is not evenly distributed or in the right place, and some of the earliest engineering works were for the storage and transfer of water. The Persian qanats dating from the early 1st Millennium BC pierced an aquifer and a subterranean channel then conveyed the water to villages and towns in desert areas, and Roman aqueducts were generally fed by springs.

Physical transfers – Engineering

Some of the highest growth in water demand is in areas where water is already in short supply. There have been many major water management proposals which, if implemented, could have had significant adverse effects on an international scale. For example, the North American Water and Power Alliance (NAWAPA) proposals envisaged diverting water from Alaskan rivers through the Rockies to the headwaters of the Colorado and Yellowstone rivers (Barr, 1975). The plan was abandoned since it would have caused widespread environmental damage and required enormous amounts of electricity to pump the water across the Rocky Mountains. to Mexico, using a major dam in the Rocky Mountain trench. Subsequently, there were Soviet proposals to divert the flow of some of the major Arctic rivers for use in regional irrigation projects in the Aral-Caspian basin

(Micklin, 1981). The reduction in freshwater inflows to the Arctic Ocean and resulting changes in its salinity would almost certainly have had significant ecological and climatological repercussions on a global scale. But in the event, economic and then political obstacles conspired to prevent further development of the proposals.

Water transfers deplete the water of the source area for agricultural production and ecosystems, and water transport is energy and carbon intensive (water is heavy) pumping water over long distances or up and over catchments divides between basins; water transmission is also vulnerable to disruption and needs ongoing maintenance (Grant et al, 2012). Water infrastructure projects also often involve storage - in Asia typically 80% of the runoff occurs from May to October, and water is often most needed in the dry season.

China has a very uneven distribution of water; two-thirds of its arable land and over 40% of its population are concentrated in the North, which has only 15% of the nation's water resource. The government began a major South-North inter-basin transfer project at more than 3-times the cost of the Three Gorges Project (INBO, 2014). This will transfer 50 M km^3 $year^{-1}$ of water from the Yangtze basin to the water-scarce north, although at great cost and to some minds a questionable departure from its successful 'virtual' water transfer policy (see below) using the Grand Canal (Allan, 2011).

Current issues in 'global hydropolitics' relating to the management of water in large international river basins and disputes between nations over shared rivers include: the Rio Grande and the Colorado River (USA and Mexico), Brahmaputra and Ganges (Bangladesh, Bhutan, Nepal and India), River Jordan (Israel, Jordan and the Palestinian territories), Tigris-Euphrates ((Turkey, Iraq and Syria) and the Nile (Ethiopia, Sudan and Egypt), In many cases tensions have arisen as a result of ill-judged actions and policies which have themselves been based on an incomplete understanding of drainage basin hydrology. Not surprisingly, therefore, attempts to improve the understanding of hydrological processes may be expected to have a key role in the solution of such problems (Garrick et al, 2014).

Virtual water transfers

Macroscale water transfer projects of this type may become less important as the potential of **'virtual water' trading** becomes more clearly understood. This approach regards water as one of a number of goods to be traded and recognises that it may be both easier and more environmentally friendly to move grain and other foodstuff into an area rather than develop capital-intensive programmes of crop irrigation. This is likely to be especially true of many arid and semi-arid areas, where the dis-benefits of irrigation, especially salinization, have long been recognised. Similarly, it makes more sense for economically wealthy, but water deficient, countries to purchase grain and other foodstuffs than to try to import vast quantities of water by major civil engineering projects.

As grain production requires large quantities of water (one ton of grain requires about 1,000 tons of water) it may be preferable for some water-scarce countries to limit their own grain production and import it instead. The water required for the same amount of agricultural product will vary between regions and generally crop yields and water efficiencies are greater in 'exporting' than in 'importing' countries. Then the amount of 'real water' used to produce a ton of grain is less than the amount of 'virtual water' saved by the grain importing country, so providing a net saving of global water use for other purposes. Virtual water trading is already a reality in the Middle East and North Africa region which has effectively run out of water. By 1980, about 20% of the region's water needs were being delivered as embedded in water-intensive commodities such as wheat, and despite the Aswan Dam and its programme of expansion of irrigated land Egypt depends heavily on 'imported' water in subsidised food from the industrialised countries of N America and Europe (Allan, 2011). International trade creates a direct link between the demand for water-intensive commodities in importing counties and the water used for the production of those commodities in the exporting countries. Thus, for example, Japanese consumers indirectly contribute to the mining of the Ogallala Aquifer and depletion of the Colorado River in the USA (Chapagain and Hoekstra, 2008).

Of course, trading and transporting water-hungry commodities is not new. The Grand Canal in China was built to transport grain from the agriculturally fertile Yangtze basin to feed the northern cities. Construction began in the 7th Century and it was extended several times, and is now over 1700 km long, the longest canal in the World. It is still in use today as an important transport link between northern and southern China, and it is currently being enlarged to take larger vessels.

9.2.5 WATER USE EFFICIENCY

There are a great number of ways to improve the efficiency of water usage; in thermal power generation and in some industries water may be recirculated and reused, and the more effective treatment of industrial wastes would reduce the pollution of downstream freshwaters. There are three general approaches to reducing water use (Grant et al, 2012). Firstly, make direct use of wastewater for some industrial and agricultural applications. Household use of potable water can be reduced by as much as 20% by 'rainwater harvesting' (from roofs) and using 'greywater' (from laundry, dishwashing and bathing) for non-drinking purposes such as toilet flushing and garden watering. Secondly, wastewater treatment may enable it to be used again. This may be full treatment and adding back to the water supply e.g. River Thames (see Section 8.6), or partial treatment such as coarse filtering and putting through a constructed reed bed to provide a biological filtration system encouraging micro-organisms to digest the organic pollutants. Thirdly, more efficient use can be made of water. As agriculture is the largest global user of freshwater water, even relatively small increases in water productivity can potentially result in substantial absolute water savings, although in practice savings may be hampered by a lack of technical knowledge and resources. There is, in principle, a large scope to reduce wastage in irrigation by leakage from transportation canals and the inefficient over-application of water which may be many times the actual crop requirement, and by better irrigation scheduling (such as in the evening or early morning rather than at the hottest time of the day, thus reducing non-productive evaporation of water from the soil). Water savings may be achieved by changing to lower water consumption crops, as well as reducing evaporation by fallowing and mulching. Pumping water for irrigation is also a major use of electricity in parts of the developing world.

Similarly, drinking water may be wasted by leakage from the distribution system to the consumer. The World Bank (Kingdom et al, 2006) estimates that about 30% of global water abstraction is lost through leakage, and that losses account for as much as 50% of the treated water in many developing countries. Even in richer countries pipeline losses can be substantial. Water companies in the UK are now required to reduce water mains leaks before new abstraction licences are issued, and they are required to publish annual figures on their leakage rates. It can be very difficult to accurately assess the true level of leakage and so companies often estimate it by measuring night flow when actual consumption levels will be very low and most water flow will be leakage. Leakage losses depend on the condition of the pipes, the pressures in the network and the efforts made by the companies to manage leakage. Ageing infrastructure leads to millions of litres of water being lost every day. In certain areas of Europe leakages are as high as 50%. In the US the Environmental Protection Agency estimates that US$ 300 B needs to be spent on waste water improvements and US$ 335 B on the country's water utilities infrastructure over the next 20 years. Domestic water use can be reduced by technical devices including dual flush toilets and more efficient washing machines.

In developed countries, mains leakage management and effluent recycling are the main means of reducing water use, while in many developing countries improvements in irrigation efficiency will make the greatest improvements.

Water pricing

One of the most effective means of controlling excessive demand and wastage of water is by the disincentive of water pricing. Rain is 'free' but its collection, treatment, storage and distribution are not. Delivering potable water costs money – especially for its treatment and purification, and the installation and maintenance of reservoirs and a network of pipes and pumps for its distribution. Nevertheless, in some Arab countries it is government policy to not charge for irrigation water, although they do charge for drinking water, and in rich Gulf countries such as the United Arab Emirates and Saudi Arabia, native citizens are not charged for domestic water (Ragab Ragab, pers. comm, 2016).

There is now an increasing recognition that water must be used more efficiently. One way is to regard water as both a natural resource and an economic commodity incurring costs of abstraction, treatment and transportation to where it is required. Thus, for example, water access rights may be traded, which can promote efficient water allocation as it provides incentives to reallocate water to higher value uses, although as with many aspects of economics it can be difficult to translate theory to reality as it may be hard to evaluate the value of the benefits, and to identify the beneficiaries and losers. One successful pioneering example is the water supply to New York City which is supplied unfiltered from the Catskill Mountains. It was found to be cheaper to protect the drinking water at its source than to treat the water at its destination (Pires, 2004; Grolleau and McCann, 2012). The cost of constructing a filtration facility would have been US \$6–10 B, with annual running costs of US \$300 M. Instead they established a voluntary watershed management program, providing financial incentives to landowners to modify their activities to reduce pollution, support conservation practices to reduce erosion and the establishment of buffer strips along watercourses. These, together with technical assistance, research and monitoring support have an annual cost equivalent to the running costs of the filtration plant. The water is collected in reservoirs and disinfected before going directly into the public dinking supplies.

Despite the persuasiveness of water pricing as a mean of allocating water, there are still many disagreements about the best means by which to determine the price level and, in a society with economically disparate groups, there may be additional social, institutional and political considerations (Johansson et al, 2002).

9.2.6 ECOSYSTEM HEALTH

As human pressure on the World's resources increases there is growing concern for the state of the environment. The subject of **hydroecology** integrates hydrology and ecology for the management and protection of water and ecosystems – i.e. water resources for people and the environment.

There is a growing desire to conserve or restore the ecological health of river systems that may have undergone centuries of management and manipulation. Hydrologists have a key role to play in balancing these often competing objectives, and working in conjunction with other river scientists can help define the **environmental flows** i.e. the flow regime that best meet the needs either to recreate the original 'natural' state, or in rivers where a return to natural conditions is no longer feasible, to sustain a particular 'desired ecosystem' (e.g. Acreman, 2016). This may involve physically habitat modelling to represent the functional link between river flows and ecology. The implementation of flow management may involve measures such as restrictions on abstractions and controlled releases from reservoirs.

The Instream Flow Incremental Methodology (IFIM) is a widely adopted methodology quantifying the incremental differences in instream habitat that result from alternative flow regimes. A channel is surveyed at representative reaches and an empirical relation established to predict how the amount of physical habitat available changes with flow. One of the key IFIM tools is Physical Habitat Simulation (PHABSIM) (Bovee, 1982; Bovee et al, 1998). Habitat is used instead of numbers or river biodiversity as it is more easily quantified and less subject to change (e.g. pollution incidents). Of course many other factors including water quality will impact on the ecological health of a river (see Section 8.6)

Natural flood protection

Towards the end of the 20th Century, as the infrastructures were completed to a large extent in the developed world, engineering started to lose importance, and hard engineering solutions to existing problems began to give way to more environmentally friendly solutions.

Flooding is a severe global problem and traditional methods of reducing flood risks have focused on a combination of flood defences and embankments along rivers and channel dredging. These techniques are expensive to build and need to be maintained, and by eliminating floodplain storage they may move the water faster downstream and actually make flooding worse further down the river. An alternative 'soft' engineering option is to use floodplains to store water for flood management purposes (e.g. Morris et al, 2005). This may mean that some previously protected floodplain land is returned to its previous unprotected condition, with compensation payments to farmers for loss of crops and perhaps additional payments to encourage environmental benefits. Examples in England of **washland storage**, where an area of land adjoining a river is allowed to flood, or is deliberately flooded, to provide flood protection for urban areas downstream include Lincoln, Shrewsbury and Leeds. Floodplain forests may also provide some additional flood storage due to woody debris dams.

There is growing interest in developing a twin-track approach using land management schemes to reduce flood peaks into the stream network alongside flood defences to control flows once in rivers. This work is informed by the results of catchment studies into the impacts of land cover change on runoff (see Section 9.2.1), often combined with the perceived environmental benefits of the alternative land use.

Claims are often made, particularly by foresters, that planting trees will reduce flooding downstream, although the actual evidence for this is disappointingly weak. Rather, a forest cover may help reduce flow peaks in smaller rainfall events but they are not effective for reducing very large floods (see Figure 9.3b). The slightly higher interception capacity of a forest canopy (Table 3.1) will make little difference during a major rainstorm. More significant may be the effect of soil differences; the biggest increases in flood peaks in deforestation studies being attributed to soil erosion rather than removal of the biomass itself. Similarly, Marshall et al (2014) measured lower runoff and much higher infiltration rates in tree shelter belts protected from grazing than on heavily grazed and compacted pasture. The impact of forests, although limited to smaller events, depends critically on the land to which it is being compared. Greater benefits will arise from planting trees on degraded land with thin soil than on parts of a catchment with a dense grass cover and permeable soil.

Flooding is an important process in the natural environment and cannot be entirely prevented. Even in tropical areas with the maximum possible biomass there are still massive flood peaks in extreme rainfall events. Changes to land use and management can make small hydrological changes, but only engineering structures can make a really big difference in extreme events - and even they have a design limit and can be overwhelmed.

With increasing human pressures on food and water resources there needs to be better coordination of the overall land and water management of a basin without compromising the sustainability of the ecosystem, sometimes called **Integrated Water Resource Management (IWRM)**. The example given earlier of the Catskill water catchment (Section 9.2.5) illustrated how protection of a natural or semi natural environment, such as a healthy forest cover may help to maintain a stable soil profile and moderate flood risk. Such **environmental** or **ecosystem services** are of value to people downstream and they may be willing to pay the upstream land owners to manage their land in a way to preserve and protect this service. Such an argument has been pioneered by ecologists wishing to protect, or perhaps restore, their preferred ecosystem by appealing to an audience beyond their traditional supporters, and generate a source of income to achieve that.

9.3 GLOBAL – CLIMATE CHANGE

There is almost universal agreement amongst scientists that climate change is occurring; and a warmer climate means the atmosphere can hold more water and has more energy, which will lead to more intense rainfall and more floods (the exponential increase in saturated vapour pressure with temperature implies more evaporation and hence more rain), as well as an increased likelihood of summer heat waves. What is less certain is how large the change will be for a given rise in CO_2 and other greenhouse gases, and what effects will be felt at a local level. Models embody the collective wisdom of many meteorologists and represent the best tool to predict what impacts climate change will have, so that we can take action to respond to mitigate the likely changes.

9.3.1 CLIMATE MODELLING

Climate change modelling comprises several distinct steps. Firstly, running large **Global Climate Models** (GCMs) which are typically large grids (~150+km across) and within them 'nest' finer scale regional models. They can be used to run different scenarios with/without rising CO_2, so the difference reflects the role of climate change. A **scenario** is not a forecast in the sense of predicting what will happen. We do not know all the parameter values and understand their interactions. Rather it is a 'plausible future' and a number of scenarios can be made using different assumptions. These indicate bigger increases in temperatures over land than over oceans due to heat storage lag, and globally increasing rain, but predictions for rain are less robust than for temperature increase, and regional rain patterns are far from clear. Increasing computer power is giving climate scientists more and more spatial detail, but there is still a need for downscaling to bridge the gap to regional and local scale processes. The predicted climate changes then have to be assessed for their likely impacts e.g. will there be an increased risk of floods?

9.3.2 CLIMATE CHANGE IMPACTS

Climate change will result in an intensification of the hydrological cycle and alter the spatial and temporal patterns of precipitation and hence water availability, resulting in changes to river flow regimes, soil water reserves and groundwater recharge (IPCC, 2014). Average annual runoff is expected to increase in high latitudes and in some wet tropical areas, but reduce in dry regions in the mid-latitudes and dry tropical regions. However, it is the extremes e.g. floods and droughts that will impact most severely economically and socially. The extent and location of the changes are uncertain, so that it may never be possible to say that a particular deluge or drought was caused by climate change; rather that it was made *more likely*. The problem for climate scientists is 'attribution' – determining the extent to which climate change drives any particular extreme from heavy rains to drought. The IPCC scientists anticipate that arid and semi-arid subtropical regions are likely to be subject to increased drought risk, while wet areas become wetter and flood risk will intensify, particularly in southern and eastern Asia, as well as tropical Africa and South America.

The impacts of climate change on water resource security may be two-fold. Firstly the direct change to water availability, and secondly the resulting changes in human demand for water - a drier summer means not just less rain but greater evaporative demand and increased water usage such as crop irrigation. We probably better understand and predict the physics of the atmosphere than predict future human emission scenarios that depend upon political and economic factors. Stakhiv (2011) offered some practical steps that can be taken by water managers and planners. These include institutional changes to strengthen interagency collaboration and emergency management planning, stress-testing existing infrastructures, applied research for adapting to climate change, and improved representation of extreme events in flood frequency analyses.

The Stern Report (2006) into the economics of climate change, written by a distinguished development economist and former chief economist at the World Bank, made the case for urgent action, saying that hundreds of millions of people could suffer from hunger and flooding as a result, and noted that "the impacts [of climate change] on people will be felt mainly through water". Agriculture is by far the most vulnerable sector to climatic change, and drives the economies of many developing countries. Historic records show a clear link between weather patterns and food production, with famines often caused by heavy rain and flood events affecting harvests (e.g. Stratton et al, 1978). However, there is no absolute link between weather and harvests, since the severity of the impact will depend upon the time of the year as well as on the particular crop. Human sensitivity to weather has reduced over time, particularly in more developed countries, as a result of plant breeding of higher yielding and disease resistant crops, land drainage to reduce waterlogging, and irrigation at times of water shortage, together with better transport infrastructure to distribute food surpluses to areas in need. So nowadays, although less 'in tune' with nature, people are generally better buffered against extreme events, given adequate financial resources.

9.4 FUTURE CHALLENGES IN HYDROLOGY

There are different levels of challenge in the future: increasing pressures on water supplies and dealing with extremes, the development of new hydrological tools and models for a non-stationary world, the spatial heterogeneity of hydrological systems and difficulties associated with transferring measurements and our understanding of hydrological processes from one scale to another, as well as inadequate measurement and estimation techniques, and the decline of land-based observation networks.

The European Commission brought together experts from 24 countries to take a strategic look at the research needs of water resource management, and their Report (Water JPI, 2016) identified the need for advances in:

- Understanding and managing ecological flows,
- Understanding the causes, and predicting the occurrence of droughts,
- Better water management to mitigate the harmful impacts of extreme events,
- Understanding and predicting the environmental behaviour of emerging pollutants, by-products and pathogens,
- Implementing efficient water use, and technologies for safe reuse of wastewater,
- Understanding of hydrological processes at different scales,
- Monitoring systems including new sensors and remote sensing.

The National Research Council (NRC, 2012) conducted a review of the current status of hydrology in the USA and identified areas to advance hydrological sciences to improve human welfare and the health of the environment. These included: better understanding of the water cycle and the processes linking its components and their sensitivity to environmental change; better understanding of the interconnection between water and the environment; further research into contaminant movement, their impacts on humans and ecosystems, and the impact on water quality of climate and land use change.

9.4.1 SCALE AND TRANSPOSITION

Some of the big remaining science challenges in hydrology are related to issues of **scaling**. Specifically, is it possible to transpose findings reliably between spatial and temporal scales? Many of the classical hydrological theories were developed at the scale of a few centimetres, and it has proved difficult to apply relationships developed at the laboratory scale to spatial units the size of even a small drainage basin (Beven, 1989; Kalma and Sivapalan, 1995). At the simplest level, even basic data may be difficult to extrapolate from one spatial scale to another. For example, measurements of hydraulic conductivity in soils and rocks often increase with the scale of measurement because of the incorporation of increasingly large and even more extensive macropore and fracture systems.

The problem was often addressed as that of **upscaling**, i.e. whether we have the understanding to aggregate from small scale to large scale, since much of the development of hydrological theory has been based upon observation and experimentation at the micro- and meso-scale. Increasingly, however, the problem is couched in terms of **downscaling** due to the growing use of global climate models, especially in relation to studies of climate change and of the hydrological impacts of climate change. Hydrologists are now interested in the extent to which GCM output may be downscaled for use in drainage basin investigations.

In a somewhat analogous manner to the problem of scale is the question of the application of a solution developed in one situation or region of the world to another new location where circumstances may be different – the 'one size fits all' principle. Techniques or solutions that have been developed in Europe or North America, may not be appropriate in Africa or Asia. Energy access is increasingly seen as a vital catalyst for wider social development, including better health and education, and is a critical input for the achievement of many of the Millennium Development Goals. Countries such as Germany and France have developed nearly all of their economically feasible hydropower potential, while over 70% of potential hydropower resources in the developing world are yet to be realized (World Bank, 2013). Hydropower is the largest source of affordable renewable energy for many developing countries, but the construction of new large dams is opposed by many environmentalists. This in turn has led to dilemmas on the appropriate scale of development for water resources in other areas of the world, i.e. to what extent should the developed nations try to influence or impose their solutions on the appropriate level of development in poorer less industrialised countries?

In an attempt to achieve the best balance between conflicting aims, the World Bank decided to renew its support for those dam construction schemes that meet environmental and social safeguard standards (World Bank, 2013).

9.4.2 INSTRUMENTATION AND DATA COLLECTION

The growth of interest in global climate change has stimulated great advances in the availability of satellite-based remote-sensed data of key hydrological information at a global or major regional scale (e.g. Neale and Cosh, 2012; Box 1.1). Such data are usually well suited to large-scale spatial integration, enabling progress in our understanding of the global scale operation of hydrological processes. This has been aided by advances in the technology permitting the rapid assimilation of these data into regional, continental, or even global GIS and database systems. McCulloch (2007) foresaw that future research progress would be built on vastly improved instrumentation for sensing, logging and retrieving data from all corners of the Earth, manipulation of these data, together with the modelling of systems using computers with undreamed of capabilities.

Remote sensing observations cannot be used in isolation, and must be verified or calibrated by using 'ground-truth' observations (Sheffield et al, 2009) based on traditional point-scale instruments such as raingauges and soil water probes, together with catchment-scale water balance calculations. It is a paradox that many of those areas where the demand for water is growing fastest, have the least data and worst capabilities for acquiring and managing water data. This applies to surface water and groundwater, and to quantity and quality. Indeed, with the exception of Latin America, the reliability and availability of terrestrial observations have declined sharply since the mid-1980s particularly in Africa, in eastern Europe and around the Arctic (Shiklomanov et al., 2002), largely because networks have been degraded by lack of investment. Hydrological monitoring is particularly weak in Africa due to the low number and uneven geographical coverage of gauges, and long gaps often due to civil conflict (INBO, 2016). In spite of progress in some nations' water data infrastructure, our overall ability to verify status and trends of global water resources from ground based measurements is actually declining (Sene and Farquharson, 1998; Mishra and Coulibaly, 2009).

Many flow records are significantly affected by human activities such as the large number of dams (e.g. Syvitski et al, 2005; ICOLD, 2011) as well as having many abstractions and returns, but in very few cases are the resources and sufficient information available for naturalised run-off values to be calculated; the Thames (see Section 9.2.1) is a rare example.

9.4.3 PUBLIC ENGAGEMENT AND SCRUTINY OF SCIENCE

Unless scientists are able to convey their findings in a coherent and understandable fashion to the general public there is a danger that the extensive promotion of certain land uses, environmental threats, or engineering interventions by vested-interest groups may create a disparity between public perceptions and the actual scientific evidence. Arguments which are outdated or incorrect may be uncritically repeated by those who benefit from their message, underscoring the failure of effective dissemination of scientific results. Stirring statements laden with sensation and potential conflicts are the usual lifeblood of many environmental journalists and many conservation organizations (Calder et al, 2004).

It is all too easy to claim hydrological or other benefits of a particular land cover with little proof, i.e. 'make claims that are easy to say, but difficult to disprove' (Chris Baines, 2003 pers. comm.). The application of scientific rigour may be defined as the *systematic observation* and *experimental investigation* of phenomena, and the organization of this knowledge in the form of *testable explanations and predictions*. This gives hydrologists the tools to test statements and identify any limitations or false assumptions. The essential point is that it should be *evidence-led*. To be successful at getting their findings accepted, scientists must be prepared to respond to, and engage with, people who have a different points of view.

Various 'arguments' may be used to oppose observational evidence, for example:

- To dismiss evidence of an increasing trend, e.g. "floods have always occurred, and some in the past have been even larger than the current ones". In fact a long (historic) period is likely to contain more floods than a shorter (current) period, but the *probability* of a large flood occurring may have changed.
- By taking a fact out of context, or simply ignoring new evidence. Yet, changing your mind in the face of evidence is absolutely central to a civilised democratic society (Cox and Cohen, 2016).
- Simply dismiss predictions, e.g. "the world is too complex for a model to include all the pertinent factors, and since it does not include everything its forecasts cannot be believed". But, a reputable model will be based on expert opinion and process understanding. The future is indeed uncertain, but one way of assessing the relative merits of alternative options is by attempting to model future outcomes. Doing nothing may not be a risk-free option; delaying action until the environment has deteriorated noticeably may greatly increase the eventual cost.

The role of the hydrologist is to conduct rigorous studies, and make that information available to policymakers and the public *in a form that they can use* including, where appropriate, providing guidelines on mitigation measures and best practice solutions. This can take a number of forms. For example, Blackmore (2015) estimated the public availability of a national database of peak flood data delivered over £5 M per year net benefits to UK environmental consultants and regulators. Some major Open Source data websites are listed in Chapters 1 and 9. The *Hydrological Outlook* website (http://www.hydoutuk.net/) provides an insight into likely hydrological conditions across the UK for river flows and groundwater levels over the next three months. Websites may also be established aimed at the public to provide a clear and objective analysis of a controversial topic or to correct misinformation (e.g. http://skepticalscience.com/). The proliferation of journals and science articles offering sometimes conflicting results may also cause confusion. A **Systematic Review** has become a recognised standard for accessing, appraising and synthesising scientific information (CEE, 2013). It provides a rigorous, transparent and repeatable objective assessment of a controversial topic by systematically reviewing multiple publications according to strict criteria including relevance and reliability.

Similarly, a complex issue may be made more easily understood by developing a numerical index or indicator so that progress towards targets to be monitored. The **Water Poverty Index (WPI)** combines the physical availability of water, with its ease of access for human use, their ability to manage water, their uses for the water and its allocation for ecological services (Sullivan, 2002). **Citizen Science** is an increasingly popular approach for scientific monitoring, providing people with the opportunity to engage with science and their environment, at the same time as helping scientists to collect data form a wider area and in more detail than would otherwise be feasible. Resources are needed to support the volunteers with training and equipment, and provide ongoing feedback to sustain motivation. An early example is the British raingauge observers' network, established in 1860 and still operating today with coordination by the Met. Office.

Although so vital for human life, the successful and reliable supply of an unlimited quantities of good quality drinking water in developed countries means that there is now a real danger that it is so taken for granted by many people that it has lost its true worth, and become just another commodity, of equivalent importance as a *fast* broadband connection (e.g. Countryside Alliance, 2016).

9.4.4 HYDROLOGY AS A PROFESSION

The United Nations (UN, 2016) estimates that three out of four of the jobs worldwide are in some way water-dependent. These jobs may be either directly related to its management (collection, supply, infrastructure, wastewater treatment etc.) or in its various uses (such as agriculture, fishing, power, industry and health), through to its ultimate return to the natural environment.

Water is mankind's most vital resource, but shortage (droughts), excess (floods), and the way in which man uses water, all pose risks to the environment. Hydrological expertise is of fundamental importance in managing this resource, and minimising these risks. Specifically, hydrologists are needed to:

- Secure water supplies for public use,
- Reduce loss of life and damage from flooding,
- Protect people and the environment from pollution and over abstraction,
- Anticipate floods and droughts in time for us to reduce their impact,
- Improve aquatic habitats for wildlife,
- Increase recreational opportunities in and around water.

Hydrologists must be knowledgeable and innovative in handling environmental projects. Since hydrology is an applied science they need a broadly based knowledge in the environmental sciences and it is increasingly important to understand water policy and social sciences. And as in most professions they should be able to communicate effectively both with fellow professionals and with regulators, clients and stakeholders who may have little technical knowledge.

According to the UK Water Partnership over 60 UK universities now conduct water-focused research including flood risk management and modelling, water resources and land-use change, as well as freshwater ecology, urban drainage and water treatment (UKWRIP, 2014). In total about £120 M per year is directed to water-related research, principality selected by "scientific excellence". Comparatively little is spent by the water utility companies on research and development – about £20 million per year in England and Wales, or just 0.18% of their revenues, which is less than half the amount spent by French companies. Their report concludes that while the UK has an excellent record of research and some world-class consultants, its water industry is fragmented, without a long-term strategy, tending to look primarily to the short-term and their own individual commercial interests, and the UK government has not seen water technology as a priority.

9.5 EPILOGUE

Although macro-scale issues are likely to dominate the future of hydrology in the next couple of decades, many hydrological problems will continue to be approached and solved at the meso- or even micro-scale. Some are regionally circumscribed but global in their impact, such as the interactions of snow and ice on macro-hydrological processes, and others are quasi-global in their distribution but essentially localised in their impact; such as the increasingly important topic of urban hydrology. These and many other issues will continue to focus the attention of hydrologists on the important inter-relationships between processes operating at very different spatial and temporal scales. In turn, it is hoped that such activity may resolve some of the current difficulties of upscaling and downscaling hydrological processes.

The environmental challenges have changed or intensified over time. Thus, concerns about the health risks of nitrates in drinking water have been addressed by stricter controls and Nitrate Vulnerable Zones, and in turn been replaced by worries about newly developed chemical and drug residues; concerns about industrialisation causing 'acid rain' subsided following considerable work on emission controls, and attention is now centred on understanding climate change and its impacts on a wide range of environmental aspects. There is a growing need to study water security and the supply of water for crops to feed the growing world population, while at the same time trying to protect the wider environment.

Earlier discussions of hydrological modelling emphasise the problems caused by the heterogeneity of hydrological systems and the potential contributions of remote sensing and distributed modelling techniques to the solution of these problems. For large drainage basins, and certainly for regional and continental areas, the data complexity associated with spatial heterogeneity are probably capable of resolution only through spatially distributed models using remotely sensed data generated by satellite sensors calibrated from a network of terrestrial point observations and interpolated to grid networks.

There are important challenges which demand an urgent and effective response if hydrology is to link productively with current and growing interests in global climate, climate change and GCMs. This point was emphasised by Watts (1997), who argued that: "The ultimate scope of hydrological models extends to the development of global circulation models representing the entire hydrosphere."

Wilby (1997) went further still in suggesting that hydrology is in danger of being marginalised by other geophysical disciplines unless it can demonstrate a *unique* contribution to the growing knowledge data base and suggested that the emphasis should be '... on solving the water balance equation ... at a global or regional scale'.

During the last 70 years great advances have been made in hydrology, and in its growth in stature as a science. In the early 1960s it was still largely the preserve of engineers, and a small number of physical geographers, for whom it will continue to provide an important framework of theories and methods for solving specific problems. However, we believe that the future of hydrology lies in its potential role as a scientific discipline and in its closer integration with other disciplines (Rodda and Robinson, 2015). Much of the impetus for such a development has been, and will probably continue to be, provided by the interest in climate variability and climate change, particularly in respect of anthropogenic influences on climate. But if hydrology is to fulfil its scientific potential, it will have to mature as a discipline. This point was made effectively and succinctly by Chahine (1992): 'Hydrological science must adjust itself to become a discipline not unlike atmospheric science or oceanography. Rather than fragmented studies in engineering, geography, meteorology and agricultural science, we need an integrated program of fundamental research and education in hydrological science.'

We believe this is a realistic assessment for the future of hydrology and that such a development would greatly strengthen its ability to grapple with macroscale problems other than those involving climate variability and change.

The impacts of climate change and population growth will be felt mainly through water. They are likely to hit hardest the economies and populations of some of the world's poorest countries which are least able to cope.

Water is our most precious resource – the original elixir of life – and for the majority of the global population a better understanding of the principles and processes of hydrology may, literally, be a matter of life and death.

REVIEW PROBLEMS AND DISCUSSION TOPICS CONCERNING CHOICES

9.1 What are the ways in which hydrologists can help to meet the rising demand for water, given the finite amount of water available?

9.2 Discuss and evaluate the benefits and disadvantages of projects such as the Aswan Dam, Three Gorges and Grand Renaissance Dams. Why are there different perceptions within, and outside of the host country?

9.3 There are many international conflicts between upstream and downstream water users. Undertake a case study summarising the main hydrological and environmental considerations.

9.4 Discuss the pros and cons of 'hard' and 'soft' engineering for flood mitigation.

9.5 Outline some of the factors which have contributed to the growing interest in global hydrology.

9.6 Imagine that you have to present the following case to the following groups: a) foresters, b) environmental activists: "Planting trees and protecting forests may have many environmental benefits, but preventing large-scale floods is not one of them". Now prepare a rebuttal to that statement.

HYDROLOGY JOBS

An introduction to the work of hydrologists is available on the USGS website http://water.usgs.gov/edu/hydrology.html.

For examples of job descriptions in the UK see the British Hydrological Society website: http://www.hydrology.org.uk/careers.php, or for New Zealand see http://www.hydrologynz.org.nz/index.php/nzhs-jobs. Look at the websites of particular employer organizations such as the Environment Agency https://www.gov.uk/government/organisations/environment-agency, or commercial recruitment sites. Alternatively use an internet search engine such as www.google.com or advert- and tracking-free https://duckduckgo.com.

WEBSITE RESOURCES FOR FURTHER STUDY

Central England temperatures (1772 – present):
http://www.metoffice.gov.uk/hadobs/hadcet/index.html
http://hadobs.metoffice.com/hadcet/

Citizen Science (Environmental Observation Framework)
www.ukeof.org.uk

Climatic Research Unit
http://www.cru.uea.ac.uk/

Collaboration for Environmental Evidence:
http://www.environmentalevidence.org/

Environmental Indicators (European Union):
http://europa.eu./european-union/index_en

European Water Statistics:
www.ec.europa.eu/eurostat

FAO Global water information system AQUASTAT
http://www.fao.org/nr/water/aquastat/main/index.stm

Foundation for Water Research (FWR):
http://www.euwfd.com/html/fwr-news.html

Global Population (UN)
http://www.un.org/en/development/desa/population/

Global Temperature (1880-present):
https://www.ncdc.noaa.gov/cag/time-series/global/globe/land_ocean/ytd/11/
http://data.giss.nasa.gov/gistemp/

Global temperature anomaly data:
http://www.cru.uea.ac.uk/cru/data/temperature/

Hydrological Outlook for the UK:
http://www.hydoutuk.net/

Intergovernmental Panel on Climate Change (IPCC):
http://www.ipcc.ch/
http://ipcc.ch/organization/organization_history.shtml

International Network of Basin Organizations:
www.inbo-news.org

International Water Management Institute:
http://www.iwmi.cgiar.org/

IPCC Fifth Assessment Report (AR5) Climate Change Synthesis:
http://ar5-syr.ipcc.ch/

Millennium Development Goals:
http://www.un.org/millenniumgoals/

Sustainable Development Goals:
http://www.un.org/sustainabledevelopment/sustainable-development-goals/

Walker Institute for Climate System Research
http://www.walker.ac.uk/

Water Poverty Index and other environmental information
http://www.grida.no/

World Bank Climate Change Knowledge Portal (CCKP)
http://sdwebx.worldbank.org/climateportal/index.cfm

REFERENCES

Many landmark papers in the development of understanding of hydrological processes have been collated in the *International Association of Hydrological Sciences* (IAHS) *Benchmark Series in Hydrology*. See Chapter 1, page 33.

Abdul, A. S. and Gillham, R.W. (1984) Laboratory studies of the effects of the capillary fringe on streamflow generation. *Water Resources Research*, **20**: 691-8.

Abrahams, A.D., Parsons, A.J. and Wainwright, J. (1994) Resistance to overland flow on semiarid grassland and shrubland hillslopes, Walnut Gulch, southern Arizona. J. *Hydrology*, **156**: 431-446

Acreman, M. (2016) Environmental flows—basics for novices. *WIREs Water 2016*. doi: 10.1002/wat2.1160

Agassi, M., Morin, J. and Shainberg, I. (1985) Effect of raindrop impact energy and water salinity on infiltration rates of sodic soils. Proc. *Soil Sci. Soc. America*, **49**: 186-90.

Akan, A.O. (2006) *Open Channel Hydraulics*. Butterworth-Heinemann, Oxford 364 pp

Albergel, C., de Rosnay, P., Gruhier, P. and coauthors (2012) Evaluation of remotely sensed and modelled soil moisture products using global ground-based in situ observations. *Remote Sensing of Environment*, **118**: 215–226

Albrecht, K.A., Fenster D.F. and Van Camp, S.G. (1990) Reducing groundwater flow model uncertainties at Yucca Mountain. *IAHS Publ.*, **195**: 301-309

Allan, J.A. (2011) *Virtual Water: Tackling the Threat to our Planet's Most Precious Resource*. I.B. Tauris and Co. London. 368 pp.

Allen, R.G., Pereira, L.S., Raes, D and Smith M (1998) Crop Evapotranspiration: Guidelines for computing crop water requirements. *FAO Irrigation and Drainage Paper*, 56. Food and Agriculture Organization, Rome. 300pp.

Alley, W.M. (1984) The Palmer drought severity index: limitations and assumptions. *J. Climate and Applied Meteorology*, **23:** 1100–1109.

Allott, T (2010) *The British Rainfall network in 2010*. Royal Meteorological Society meeting: 'The 150th Anniversary of the Founding of British Rainfall Organization'. Zoological Society of London, 17 April 2010 http://www.rmets.org/sites/default/files/pdf/presentation/20100417-allott.pdf

Amboise, B. (2004) Variable 'active' versus 'contributing' areas or periods: a necessary distinction. *Hydrological Processes*, **18**: 1149-1155.

Anderson, J.M. and Spencer, T. (1991) Carbon, nutrient and water balances of tropical rain forest ecosystems subject to disturbance, *MAB Digest*, **7**, UNESCO, Paris, 95 pp.

Anderson, M. (2008) (Ed.) *Groundwater*. Benchmark Papers in Hydrology, Vol **3**: IAHS Press, Wallingford, 625 pp

Anderson, M.G. and Burt, T.P. (1978) The role of topography in controlling throughflow generation. *Earth Surface Processes*, **3**: 331-4.

Andersson, L. and Harding, R.J. (1991) Soil moisture deficit simulations with models of varying complexity for forest and grassland sites in Sweden and the UK. *Water Resources Management*, **5**: 25–46.

Arcement, G.J., and Schneider, V.R. (1989) Guide for selecting Manning's roughness coefficients for natural channels and flood plains. *US Geological Survey Water Supply Paper* **2339**.

Arnold, G.E. and Willems W.J. (1996) European groundwater studies, *European Water Pollution Control*, **6**: 11-18

Asdak, C., Jarvis, P.G., van Gardingen, P. and Fraser, A. (1998) Rainfall interception loss in unlogged and logged forest areas of Central Kalimantan, Indonesia. *J. Hydrology*, **206**: 237–244.

Ashton, D., Hilton, M. Thomas, K.V. (2004) Investigating the environmental transport of human pharmaceuticals to streams in the United Kingdom. *Science of the Total Environment*, **333**: 167–184.

Austin, B.N. Cluckie I.D., Collier C.G. and Hardaker P.J. (1995) *Radar-based Estimation of Probable Maximum Precipitation and Flood*. Meteorological Office, Bracknell, 124 pp.

Bachmair, S., Kohn, I., and Stahl, K. (2015) Exploring the link between drought indicators and impacts. *Natural Hazards and Earth System Sciences*, **15**: 1381-1397.

Badon-Ghijben, W. (1889) Nota in verband met de voorgenomen putboring nabij Amsterdam (Notes on the probable results of the proposed well drilling near Amsterdam), *Tijdschrift van het Koninklijt Inst. van Ingenieurs*, The Hague, pp. 8-22.

Bailey, R (1997) *An introduction to Sustainable Development*. Chartered Institution of Water and Environmental management, London, 83 pp.

Baird, A.J. (1997) Continuity in hydrological systems. In: Wilby, R.L. (Ed.) *Contemporary Hydrology*. Wiley, Chichester, pp. 25-58.

Baldocchi, D.D., Falge, E, Gu, L. and coauthors (2001) FLUXNET: A new tool to study the temporal and spatial variability of ecosystem-scale carbon dioxide, water vapour and energy flux densities. *Bulletin American Meteorological Society*, **82**, 2415-2434.

Ball, J.T., Woodrow, I.E. and Berry J.A. (1986) A model predicting stomatal conductance and its contribution to the control of photosynthesis under different environmental conditions. In: Biggins, J. (Ed.) *Progress in Modern Photosynthesis Research* Vol **4**, pp 221-224. Martinius Nijhof, The Netherlands.

Barker, J.A. (1991) Transport in fractured rock. In: Downing, R.A. and Wilkinson W.B (Eds.). *Applied Groundwater Hydrology*. Clarendon Press, Oxford, pp. 199-216

Barnes, H.H. (1967) Roughness characteristics of natural channels: *US Geological Survey Water Supply Paper* 1849, 213 pp. http://pubs.usgs.gov/wsp/wsp_1849/

Barnett, C. (2015) *Rain: A Natural and Cultural History*. Crown Publishing, New York.

Barr, L. (1975) NAWAPA: A Continental Water Development Scheme for North America? *Geography*, **60**: 111-119.

Barry, R.G. and Chorley R.J. (2010) *Atmosphere, Weather and Climate,* Routledge, London.

Battarbee, R.W., Shilland, E.M., Kernan, M. and coauthors (2014) Recovery of acidified surface waters from acidification in the United Kingdom after twenty years of chemical and biological monitoring (1988–2008). *Ecological Indicators*, **37**: 267–273.

Baumgartner, A. and Reichel, E. (1975) *The World Water Balance.* Translated by R. Lee. Elsevier, Amsterdam. 179 pp.

Beran, M.A. and Gustard, A. (1977) A study into the low-flow characteristics of British rivers. *J. Hydrology*, 35: 147-52.

Beran, M.A., Wiltshire, S.E. and Gustard, A. (1984) Report of the European Flood Study. Institute of Hydrology, Wallingford.

Bergström, J. (1989) Incipient earth science in the Old Norse mythology, *Geologiska Föreningens i Stockholm Förhandlingar*, **111**: 187-191

Betson, R.P. (1964) What is watershed runoff? *J Geophysical Research*, **69**: 1541-1552

Beven, K.J. (1987) Towards a new paradigm in hydrology. *IASH Publ.* **164**: 393-387.

Beven, K. (1989) Changing ideas in hydrology – the case of physically-based models. *J. Hydrology*, **105**: 157-172.

Beven, K. (1989) Interflow. In: Morel-Seytoux (Ed.) *Unsaturated Flow in Hydrologic Modelling*. Kluwer Academic Publishers. London. pp 191-219.

Beven, K. (1997) TOPMODEL: A critique. *Hydrological Processes*. **11**: 1069-1085.

Beven, K. (2007) (Ed.) *Streamflow Generation Processes.* Benchmark Papers in Hydrology Vol **1**: IAHS Press, Wallingford, UK. 432.

Beven, K. (2012) *Rainfall-Runoff modelling: The Primer*. 2nd Edn. Wiley Blackwell, Oxford. 457 pp.

Beven, K.J. (2014) Here we have a system in which liquid water is moving; let's just get at the physics of it' (Penman 1965). *Hydrology Research* **45**: 727–736. doi:10.2166/nh.2014.130

Beven, K.J. and Germann, P. (1982) Macropores and water flow in soils. *Water Resources Research*, **18**: 1311-25.

Beven, K.J. and Germann, P. (2013) Macropores and water flow in soils revisited. *Water Resources Research*, **49**: 3071-3092.

Binnie, G.M. (1981) *Early Victorian Water Engineers*, Thomas Telford Ltd, London, 310 pp.

Biswas, A.K. and Tortajada, C. (2001) Development and Large Dams: A Global Perspective. *International Journal of Water Resources Development*, **17**: 9-21.

Biswas, A.K. (1967a) Hydrologic engineering prior to 600 B.C. *J Hydraulics Division*, ASCE, **93**: 115-136

Biswas, A.K. (1967b) Hydrology during the Hellenic civilization. *IASH Bulletin*, **12**: 5-14.

Biswas, A.K. (1970) *History of Hydrology*, North–Holland Publishing Co., Amsterdam, 336 pp.

Biswas, A.K. (1996) Water for the developing world in the 21st century: Issues and implications, *ICID Journal*, **45**: 1-12.

Biswas, A.K. (2002) Aswan Dam revisited: The benefits of a much maligned dam. *Development and Cooperation*, **6**: 25-27.

Black T.A. and Kelliher, F.M. (1989) Processes controlling understorey evapotranspiration, *Philosophical Transactions of the Royal Society, London. B*, **324**: 207–231.

Blackmore, R. (2015) The economic impact of the NRFA peak flow database. Research Impact Consulting, Reading. 19pp.

Blanken, P., Rouse, W., Culf, A., and coauthors (2000). Eddy covariance measurements of evaporation from Great Slave Lake, Northwest Territories, Canada. *Water Resources Research* **36**, 1069–1077.

Bonell, M. and Gilmour, D.A. (1978) The development of overland flow in a tropical rainforest catchment. *J. Hydrology*, **39**: 365-82. Also reproduced in Beven, KJ (2007) pp 306-323.

Bonell, M. and Williams, J. (1986) The generation and redistribution of overland flow on a massive oxic soil in a eucalypt woodland within the semi-arid tropics of north Australia. *Hydrological Processes*, **1**: 31-46.

Bonell, M., Cassells, D.S. and Gilmour, D.A. (1983) Vertical soil water movement in a tropical rainforest catchment in northeast Queensland. *Earth Surface Proc. and Landforms*, **8**: 253-272

Bonell, M., Hendriks, M.R., Imeson, A.C. and Hazelhoff, L. (1984) The generation of storm runoff in a forested clayey drainage basin in Luxembourg. J. *Hydrology*, **71**: 53-77.

Bormann, F.H. and Likens, G.E. (1994) *Pattern and Processes in a Forested Ecosystem.* Springer, New York, 226 pp.

Bouma, J. (1977) *Soil Survey and the Study of Water in the Unsaturated Soil.* Soil Survey Paper **13**, Netherlands Soil Survey Institute, Wageningen, 106 pp.

Bouma, J. (1981) Soil morphology and preferential flow along macropores, *Agric. Water Managagement*, **3**: 235–50.

Bouma, J. (1986) Using soil survey information to characterize the soil-water state. J. *Soil Sci.*, **37**: 1-7

Bovee, K.D. (1982) *A guide to stream habitat analysis using the Instream Flow Incremental Methodology*. Instream Flow Information Paper, **12**. US Fish and Wildlife Service Report FWS/OBS-82/26. Fort Collins, Colorado. 248 pp

Bovee, K. D., Lamb, B. L, Bartholow, J. M. and coauthors (1998) *Stream habitat analysis using the instream flow incremental methodology*. U.S. Geologicial Survey, Biological Resources Division Information and Technology Report USGS/ BRD-1998-0004. 131 pp.

Bowen, I.S. (1926) The ratio of heat losses by conduction and by evaporation from any water surface, *Physical Review*, **27**: 779–87.

Bowes, M.J., Neal, C., Jarvie, H.P. and coauthors (2010) Predicting phosphorus concentrations in British rivers resulting from the introduction of improved phosphorus removal from sewage effluent. *Science of the Total Environment*, **408**: 4239-4250.

Boxall, A. (2012) New and Emerging Water Pollutants arising from Agriculture. 48 pp. In: Water Quality and Agriculture: Meeting the Policy Challenge. Organisation for Economic Co-operation and Development (OECD). Paris, France www.oecd.org/agriculture/water.

Bracq, P. and Delay, F. (1997) Transmissivity and morphological features in a chalk aquifer: a geostatistical approach of their relationship. J. *Hydrology*, **191**: 139-160

Brady, N.C. and. Weil, R R. (2007) *The nature and properties of soils.* 14th Edn., Prentice Hall, New York, 980 pp.

Brakensiek, D.L. and. Onstad, C.A (1977) Parameter estimation of the Green and Ampt infiltration equation. *Water Resources Research*, **13**: 1009-1012.

Brubaker, K.L., Entekhabi, D. and Eagleson P.S. (1993) Estimation of continental precipitation recycling. *J. Climate*, **6**:1077-1089.

Brugge R and Burt S (2015) *One hundred years of Reading weather*. Department of Meteorology, University of Reading. ISBN 978-0-9569485-1-9 200 pp

Bruijnzeel, L.A. (1990) *Hydrology of Moist Tropical Forests and Effects of Conversion: A state of knowledge review.* Free University, Amsterdam, 224 pp.

Brutsaert, W.H. (1982) *Evaporation into the atmosphere – theory, history and applications,* D. Reidel Publishing Co., Dordrecht, 299 pp.

BSI (2010) *Hydrometry Specification for a Reference Raingauge Pit.* BS EN **13798**, British Standards Institution, London

BSI (2012) *Acquisition and Management of Meteorological Precipitation Data from a Raingauge Network*: **Part 1**. Guide for the design, development and review of a raingauge network. **Part 4**: Acquisition and management of meteorological precipitation data from a raingauge network. Guide for the estimation of areal rainfall. *BS 7843,* British Standards institution, London 4 vols.

Buchan, S. (1965) Hydrogeology and its part in the hydrological cycle. Informal discussion of the Hydrological Group, *Proc. Institution Civil Engineers,* **31**: 428-31.

Buchter, B., Hinz, C. and Leuenberger, J. (1997) Tracer transport in a stony hillslope soil under forest. *J. Hydrology,* **192**: 314-320

Burnett WC, Bokuniewicz H, Huettel M, and co-authors (2003) Groundwater and pore water inputs to the coastal zone. *Biogeochemistry,* **66**: 3-33

Burnett, W.C., Aggarwal, P.K., Aurelic, A. and co-authors (2006) Quantifying submarine groundwater discharge in the coastal zone via multiple methods. *Science of the Total Environment,* **367**: 498–543

Burpee, R.W. and Lahiff, L.N. (1984) Area average rainfall variations on sea breeze days in S Florida. *Monthly Weather Review,* **112**: 520-534.

Burt, S. (2005) Cloudburst upon Hendraburnick Down: the Boscastle storm of 16 August 2004. *Weather,* **60**: 219-227.

Burt, S. (2012) *The Weather Observer's Handbook.* Cambridge University Press. 456 pp

Busenberg, E., and Plummer, L.N. (2000) Dating young groundwater with sulphur hexafluoride: Natural and anthropogenic sources of sulphur hexafluoride. *Water Resources Research,* **36**: 3011–3030

Butt, N., de Oliveira, P. A. and Costa, M. H. (2011) Evidence that deforestation affects the onset of the rainy season in Rondonia, Brazil. J.Geophys. Res.**116**,D11120.

Buttle, J.M. (1994) Isotope hydrograph separation and rapid delivery of pre-event water from drainage basins, *Progress in Physical Geography,* **18**, 16-41.

Cabral, O.M.R., Rocha, H.R., Gash, J.H.C. and coauthors (2010) The energy and water balance of a Eucalyptus plantation in southeast Brazil. *J. Hydrology,* **388**: 208–216.

Cai, W., Borlace S., Lengaigne M., and coauthors. (2014) Increasing frequency of extreme El Nino events due to greenhouse warming. *Nature Climate Change.* **5**: 132–137.

Calder, I.R. (1976) The measurement of water loss from a forested area using a 'natural' lysimeter. J. *Hydrology,* **30**: 311–25.

Calder, I.R. (1977) A model of transpiration and interception loss from a spruce forest in Plynlimon, central Wales. *J. Hydrology,* **33**: 247–265.

Calder, I.R. (1979) Do trees use more water than grass? *Water Services,* **83**: 11–14.

Calder, I.R. (1986) The influence of land use on water yield in upland areas of the UK. J. *Hydrology,* **88**: 201–211.

Calder, I.R. (1990) *Evaporation in the Uplands.* John Wiley, Chichester. 148 pp.

Calder, I.R. (1996) Rainfall interception and drop size – development and calibration of the two layer stochastic interception model. *Tree Physiology,* **16**: 727–732.

Calder, I. Amezaga, J.M., Aylward, B. and coauthors (2004) Forest and Water Policies. The need to reconcile public and science perceptions. *Geologica Acta,* **2**: 157-166.

Calder, I.R., Hall, R.L. and Adlard, P.G. (1992) *Growth and water use of forest plantations.* John Wiley, Chichester, 381 pp.

Calder, I.R, Hall R.L., Harding R.J. and Wright I.R. (1984) The use of a wet-surface weighing lysimeter system in rainfall interception studies of heather (Calluna vulgaris). *J. Climate and Applied Meteorology,* **23**: 461–473.

Calder, I.R., Harding, R.J. and. Rosier, P.T.W (1983) An objective assessment of soil water deficit models. J. *Hydrology,* **60**: 329–355.

Calder I.R. and Kidd, C.H.R. (1978) A note on the dynamic calibration of tipping bucket gauges. J. *Hydrology*, **39**: 383–386.

Calder, I.R. and Neal, C. (1984) Evaporation from saline lakes: a combination approach, *Hydrological Sciences J.*, **29**: 89–97.

Calder, I.R. and Newson, M.D. (1979) Land use and upland water resources in Britain – a strategic look. *Water Resources Bulletin*, **15**: 1628–1639.

Calder, I.R. and Rosier, P.T.W. (1976) The design of large plastic sheet net rainfall gauges. *J. Hydrology*, **30**: 403–405.

Calder, I.R., Swaminath, H.H., Kariyappa, G.S.and coauthors (1992) Deuterium tracing for the estimation of transpiration from trees. 1: Field calibration. J. *Hydrology*, **130**: 17–25.

Calder, I.R. and Wright I.R. (1986) Gamma ray attenuation studies of interception from Sitka spruce: some evidence for an additional transport mechanism. *Water Resources Research*, **22**: 409–417.

Calder I.R., Wright, I.R. and Murdiyarso, D. (1986) A study of evaporation from tropical rain forest – West Java. *J. Hydrology*, **89**: 13–31.

Cameron, C.S. Murray D.L, Fahey B.D., and coauthors (1997) Fog deposition in tall tussock grassland, S. Island New Zealand. *J. Hydrology*, **193**: 363–376.

Cape, J.N., Fowler, D., Kinnaird, J.W., and coauthors (1987) Modification of rainfall chemistry by a forest canopy. In: Coughtrey, P.J., Martin, M.H,. and Unsworth, M.H. (Eds.) *Pollutant Transport and Fate in Ecosystems.* British Ecological Society Special Publication **6**, pp. 155–69.

Cappus, P. (1960) Etude des lois de l'écoulement, Application au calcul et à la prévision des débits, *La Houille Blanche A*, **July-August** : 493-520. Also reproduced in Beven, KJ (2007) pp 51-72.

Cassells, D.S., Gilmour, D.A. and Bonell, M. (1985) Catchment response and watershed management in the tropical rainforests in north-eastern Australia. *Forest Ecology and Management*, **10**: 155-75.

CEE (2013) *Guidelines for systematic review and evidence synthesis in environmental management.* Collaboration for Environmental Evidence, 78pp. www.environmentalevidence.org/Documents/Guidelines/Guidelines4.2.pdf

Chahine, M.T. (1992) The hydrological cycle and its influence on climate. *Nature*, **359**: 373-380.

Chapagain, A.K. and Hoekstra, A.Y. (2008) The global component of freshwater demand and supply: an assessment of virtual water flows between nations as a result of trade in agricultural and industrial products. *Water International*, **33**: 19-32.

Chapra, S.C. (2008) Surface Water Quality Modelling. Waveland Press, 844 pp

Charney, J.G. (1975) Dynamics of deserts and drought in the Sahel. Q. J. Royal Meteorological Society, **101**: 193–202.

Chawla, I. and Mujumdar, P.P. (2015) Isolating the impacts of land use and climate change on streamflow *Hydrology and Earth System Sciences* , **19**: 3633-3651

Chebotarev, I.I. (1955) Metamorphism of natural waters in the crust of weathering, *Geochima et Cosmochimica Acta*, **8**: 22–48, 137–70, 198–212.

Chen, M., Tomás, R., Li, Z. and coauthors (2016) Imaging land subsidence induced by groundwater extraction in Beijing (China) using satellite radar interferometry. *Remote Sensing*, **16,** 8(6), 468; doi:10.3390/rs8060468

Childs, E. C. (1969) *An Introduction to the Physical Basis of Soil Water Phenomena*. John Wiley and Sons Ltd., London, 493 pp.

Comstock, J.P. (2002) Hydraulic and chemical signalling in the control of stomatal conductance and transpiration. *J. Experimental Botany*, **53**: 195-200.

Cook, P. G. and Herczeg, A. L. (2000) (Eds.) *Environmental Tracers in Subsurface Hydrology.* Kluwer Academic Publishers. 529 pp.

Cooper, J.D. (2016) *Soil Water Measurements: A Practical Handbook*. Wiley Blackwell, Oxford. 358 pp.

Corradini, C., Melone, F. and Smith, R.E. (1997) A unified model for infiltration and redistribution during complex rainfall patterns. J. *Hydrology*, **192**: 104-124

Countryside Alliance (2016) Rural broadband is as vital as water or electricity. Sarah Lee (Head of policy at the Countryside Alliance) writing in *Daily Telegraph*, 7 May 2016.

Cox, B. and Cohen, A. (2016) *Forces of Nature*. BBC Publications. 288 pp.

Craddock, J.M. (1976) Annual rainfall in England since 1725, *Quarterly J. Royal Meteorological Society*, 102: 823–40.

Crampon, N., Custodio, E., and Downing, R.A. (1996) The hydrogeology of Western Europe: a basic framework. *Q. J. Engineering Geology*, **29**: 163-180.

Crane, S.B. and Hudson, J.A. (1997) The impact of site factors and climate variability on the calculation of potential evaporation at Moel Cynnedd, Plynlimon, *Hydrology and Earth System Sciences*, **1**:429–445.

Crockford, R.H. and D.P. Richardson (1990) Partitioning of rainfall in a eucalyptus forest and pine plantation in southeastern Australia: III Determination of the canopy storage capacity of a dry sclerophyll eucalypt forest. *Hydrological Processes*, **4**: 157–167.

Cutler, D. and Miller, G. (2005) The role of public health improvements in health advances: the twentieth Century United States. *Demography*, **42**: 1-22.

Czikowsky, M.J., Fitzjarrald, D.R. (2009). Detecting rainfall interception in an Amazonian rain forest with eddy flux measurements. *J. Hydrology*, **377**: 92–105.

Dacre, H. F, Clark, P. A., Martinez-Alvarado, O., and coauthors (2015). How do Atmospheric Rivers form? *Bull. Amer. Meteor. Soc.*, **96**: 1243–1255.

Dai, A., Fung, I.Y. and Del Genio, A. D. (1997) Surface observed global precipitation variations during 1900–88. J. *Climate*, **10**: 2943–2962.

Daito, K. and Galloway, D. (2015) (Eds.) Prevention and mitigation of natural and anthropogenic hazards due to land subsidence. *Proc. IAHS*, **372**, 538 pp

Dalton, J. (1802a) Experimental essays on the constitution of mixed gases, *Manchester Literary and Philosophical Society Memo.*, **5**: 535–602. Also reproduced in Gash, J.H.C. and Shuttleworth, W.J. (2007) pp 121-141.

Dalton, J. (1802b) Experiments and observations to determine whether the quantity of rain and dew is equal to the quantity of water carried off by the rivers and raised by evaporation; with an enquiry into the origin of springs. *Memoirs Manchester Literary and Philosophical Society.*, **V, II**, 346 – 372.

Darcy, H. (1856) *Les Fontaines Publiques de la Ville de Dijon*, V. Dalmont, Paris. Also partly reproduced in Anderson, M (2008) pp 12-16.

David, T.S., Gash, J.H.C., Valente, F., and coauthors (2006). Rainfall interception by an isolated evergreen oak tree in a Mediterranean savannah. *Hydrological Processes* **20**: 2713–2726.

David, J.S., Valente, F. and Gash J.H.C. (2005) Evaporation of Intercepted rainfall. Chap 43 in Anderson, M.G. (Ed.) *Encyclopedia of Hydrological Sciences*. Vol **4**, Hydrometeorology. J Wiley and Sons Ltd, pp 627-634.

Davies, C.L., Surridge, B.W.J., Gooddy, D.C. (2014) Phosphate oxygen isotopes within aquatic ecosystems: Global data synthesis and future research priorities. *Science of the Total Environment*, **496**: 563–575

Davies, H.N. and Neal, C. (2007) Estimating nutrient concentrations from catchment characteristics. *Hydrol. Earth Sys. Sci.*, **11**: 550-558.

Davis, S.N. and De Wiest, R.J.M. (1966) *Hydrogeology*. J. Wiley and Sons, New York, 463 pp.

De Bruin, H.A.R. and Evans, J.G. (2010) Long path scintillometry: a brief review. *IAHS Publ.* **352**: 180-183.

de Fraiture C, Giordano M, Liao, Y (2008) Biofuels and implications for agricultural water use: blue impacts of green energy. *Water Policy*, **10**: 67–81.

De Ploey, J. (1982) A stemflow equation for grasses and similar vegetation. *Catena*, **9**: 139-152.

De Zeeuw, J.W. (1966) Hydrograph Analysis of areas with prevailing Groundwater Discharge. PhD thesis, Agricultural University, Wageningen. In Dutch with English summary.

Deming D. (2005) Born to trouble: Bernard Palissy and the hydrologic cycle. *Groundwater* **43**: 969–972.

Derksen, C. and Brown, R. (2012) Spring snow cover extent reductions in the 2008–2012 period exceeding climate model projections. *Geophysical Research Letters*, 39, L19504, doi:10.1029/2012GL053387

Dhar, O.N. and Nandargi, S. (1996) Which is the rainiest station in India – Cherrapunji or Mawsynram? *Weather*, **51**: 314–315.

Dixon, H., Hannaford, J. and Fry, M.J. (2013) The effective management of national hydrometric data: experiences from the United Kingdom. *Hydrological Sciences J.*, **58**: 1383-1399.

Dolman A.J. (1987) Summer and winter rainfall interception in an oak forest. Predictions with an analytical and a numerical simulation model. *J. Hydrology*, **90**: 1–9.

Dolman, A.J. and Gash, J.H. (2011). Evaporation in the Global Hydrological Cycle. Treatise on Water Science 79–87. In: Wilderer, P (Ed.) *Treatise on Water Science*, vol. 2, Elsevier: Rotterdam.

Domenico, P.A. and Schwartz, F.W. (1998) *Physical and Chemical Hydrogeology*, 2nd Edn. John Wiley, New York, 528 pp.

Dominingo, F., Sanchez, G., Moro, M.J. and coauthors (1998) Measurement and modelling of rainfall interception by three semi-arid canopies. *Agricultural Forestry Meteorology*, **91**: 275-292.

Doorenbos, J. and Pruitt, W.O. (1977) *Guidelines for Predicting Crop Water Requirements*, Irrigation and Drainage Paper 24, Food and Agriculture Organisation (FAO), Rome.

Douglas, J.T., Goss, M.J. and Hill, D. (1980) Measurement of pore characteristics in a clay soil under ploughing and direct drilling, including the use of a radioactive tracer (114C) technique. *Soil and Tillage Research*, **1**: 11-18.

Downing, R.A., Smith, D.B., Pearson, F.J. and coauthors (1977) The age of groundwater in the Lincolnshire Limestone, England and its relevance to the flow mechanism. J. *Hydrology*., **33**: 201-16.

Drever, J.I. (1997) *The Geochemistry of Natural Waters*. 3rd Edn. Prentice–Hall, Englewood Cliffs, New Jersey, 436 pp.

Drinan, J. E. And Spellman, F. (2012) *Water and Wastewater Treatment: A guide for the nonengineering professional*. 2nd Edn. CRC Press, Taylor & Francis Group, London. 300 pp.

Dunkerley, D (2000) Measuring interception loss and canopy storage in dryland vegetation: a brief review and evaluation of available research strategies. *Hydrological Processes*, **14**: 669-678.

Dunne, T. (1978) Field studies of hillslope flow processes. In: Kirkby, M.J. (Ed.), *Hillslope Hydrology*, Wiley, Chichester, pp. 227-293

Dunne, T. and Black, R.D. (1970) Partial area contributions to storm runoff in a small New England watershed. *Water Resources Research*, **6**: 1296-200. Also reproduced in Beven, K.J. (2007) pp 185-200.

Dunne, T., Moore, T.R. and Taylor, C.H. (1975) Recognition and prediction of runoff-producing zones in humid regions, *Hydrological Sciences Bull.*. **20**: 305-26.

Dwyer, I.J. and Reed, D.W. (1995) *Allowance for discretization in hydrological and environmental risk estimation*. Report **123**, Institute of Hydrology, Wallingford, UK.

Edmunds, W. M., Guendouz, A. H., Mamouc, A. and coauthors (2003) Groundwater evolution in the Continental Intercalaire aquifer of southern Algeria and Tunisia: trace element and isotopic indicators. *Applied Geochemistry*, **18**: 805–822.

Elsworth, D. and. Mase, C.R (1993) Groundwater in rock engineering, in J.A. Hudson (Ed.), *Comprehensive Rock Engineering: Principles, Practice and Projects*. Pergamon, pp. 201-226

Eltahir, E.A.B. and. Bras, R.L (1994) Precipitation recycling in the Amazon basin, *Quart. J. Roy. Met. Soc.*, **120**: 861-880

Essenwanger, O.M. (1986) Elements of statistical analysis. *World Survey of Climatology*, vol. **1B**, Elsevier, Amsterdam, 424 pp.

Fan Y., Li, H. and Miguez-Macho, G. (2013) Global patterns of groundwater table depth. *Science* **339**: 940-943

FAO (2011) *The state of the world's land and water resources for food and agriculture – managing systems at risk*. Food and Agriculture Organisation, Rome 285pp.

FAO (2014) *Facts and Figures about water withdrawal and pressure on water resources*. AQUASTAT. Food and Agriculture Organization, Rome.

FAO (2015) Renewable Water Resources Assessment – 2015 AQUASTAT methodology review. 8pp. Food and Agriculture Organisation, Rome. http://www.fao.org/3/a-bc818e.pdf

FAO (2016) AQUASTAT: FAO's online global water information system. Food and Agriculture Organisation, Rome. http://www.fao.org/nr/water/aquastat/main/index.stm

Farquharson F, Beran, M., Bromley, J. and coauthors (2015) Water resource security. Chap 6 in: Rodda, J.C. and Robinson, M. (Eds.) *Progress in Modern Hydrology*. Wiley Blackwell, Oxford. 183 – 215.

Federer, C.A. (1975) Evapotranspiration. *Reviews of Geophysics and Space Physics*, **13**: 442–5.

Ferguson, P. and Lee, J.A. (1983) Past and present sulphur pollution in the southern Pennines, *Atmospheric Environment,* **17**: 1131–1137.

Ferrier, R.C. and Jenkins, A. (Eds.) (2010) *Handbook of Catchment Management*. Wiley-Blackwell, Oxford. 540 pp.

Finch, J., and Calver, A. (2008) Methods for the quantification of evaporation from lakes. Report to the World Meteorological Organization's Commission for Hydrology. CEH, Wallingford. 41 pp. http://nora.nerc.ac.uk/14359/1/wmoevap_271008.pdf

Finch, J., and Gash, J.H.C. (2002) Application of a simple finite difference model for estimating evaporation from open water. *J Hydrology,* **255**, 253–259.

Finney, H.J. (1984) The effect of crop covers on rainfall characteristics and splash detachment. *J. Agricultural Engineering Research,* **29**: 337–343.

Foster, S.S D. and Young, C.P. (1981) Effects of agricultural land use on groundwater quality with special reference to nitrate. In: *A survey of British hydrogeology*. Royal Society, London, pp. 47–59.

Fowler, D. (1984) Transfer to terrestrial surfaces. *Phil. Trans. Roy. Soc. London, B,* **305**: 281–97.

Fowler, D. and Cape, J.N. (1984) The contamination of rain samples by deposition on rain collectors. *Atmospheric Environment,* **18**: 183–9.

Fox, T. (2013) *Global food waste not, want not.* Institution of Mechanical Engineers. London http://www.imeche.org/knowledge/themes/environment/global-food

Freeze, R.A. and Cherry, J.A. (1979) *Groundwater.* Prentice-Hall, Englewood Cliffs, N. J., 604 pp.

Freeze, R.A. and. Witherspoon, P.A (1966) Theoretical analysis of regional groundwater flow, 1. Analytical and numerical solutions to the mathematical model. *Water Resources Research,* **2**: 641-56.

Freeze, R.A. and Witherspoon P.A. (1967) Theoretical analysis of regional groundwater flow, 2. Effect of water-table configuration and subsurface permeability variation. *Water Resources Research,* **3**: 623-34. Also reproduced in Anderson, M (2008) (Ed.) pp 318-329.

Freeze, R.A. and Witherspoon P.A. (1968) Theoretical analysis of regional groundwater flow, **3**. Quantitative interpretation. *Water Resources Research,* **4**: 581-90.

Frei, A., Tedesco M., Lee S., and coauthors (2012) A review of global satellite-derived snow products. *Advances in Space Research,* **50**: 1007-1029.

Galloway, D.L., Jones, D.R., and Ingebritsen, S.E., (2000) Measuring land subsidence from space: *U.S. Geological Survey Fact Sheet* **051-00**, 4 pp. http://pubs.usgs.gov/fs/fs-051-00/pdf/fs-051-00.pdf

Gardner, C.M.K. and M. Field (1983) An evaluation of the success of MORECS, a meteorological model, in estimating soil moisture deficits. *Agric. Met.,* **29**: 269–84.

Gardner, W. R., Hillel, D., and Benyamini, Y. (1970) Post irrigation movement of soil water, 2. Simultaneous redistribution and evaporation. *Water Resources Research,* **6**:1148-53.

Garmouma, M., Teil, M.J., Blanchard, M. and Chevreuil, M. (1998) Spatial and temporal variations of herbicide (triazines and phenylureas) concentrations in the catchment basin of the Marne river (France). *Science of the total environment,* **224**: 93–107.

Garrick, D.E., Booth, W., Anderson, G.R.M. and coeditors (2014) *Federal Rivers: Managing Water in multi-layered Political Systems.* Edward Elgar Publishing and IWA Publishing. 361 pp.

Garven, G. (1985) The role of regional fluid flow in the genesis of the Pine Point deposit, western Canada sedimentary basin. *Economic Geology,* **80**: 307-324

Gash J.H.C., and Morton, A.J. (1978) An application of the Rutter model to the estimation of the interception loss from Thetford Forest. *J. Hydrology,* **38**: 49–58.

Gash, J.H.C. (1979) An analytical model of rainfall interception by forests. *Quart. J. Royal Met. Soc.,* **105**: 43–55. Also reproduced in Gash and Shuttleworth (2007) pp 316-328.

Gash, J.H.C. (1986) A note on estimating the effect of a limited fetch on micrometeorological evaporation measurements. *Boundary Layer Meteorology* **35**, 409-413. Also reproduced in Gash, J.H.C. and Shuttleworth, W.J. (2007) pp 98-102.

Gash, J.H.C. and Shuttleworth, W.J. (2007) *Evaporation*. Benchmark Papers in Hydrology Vol **2**. International Association Hydrological Sciences Press, Wallingford, UK. 521 pp

Gash, J.H.C., Lloyd, C.R. and Lachaud, G. (1995) Estimating sparse forest rainfall interception with an analytical model. *J. Hydrology*, **170**: 79–86.

Gash, J.H.C., Valente, F., and David, J.S., (1999). Estimates and measurements of evaporation from wet, sparse pine forest in Portugal. *Agricultural Forestry Meteorology*, **94**, 149–158.

Gillham, R.W. (1984) The capillary fringe and its effect on water-table response. *J. Hydrology*, **67**: 307-24. Also reproduced in Beven, K.J. (2007) pp 414-431.

Giosan, L., Clift, P. D., Macklin, M. G. and coauthors (2012) Fluvial landscapes of the Harappan civilization. *Proceedings of the National Academy of Sciences*, **109**: El 688-E 1694.

Goodison, B.E., Sevruk, B. and Klemm, S. (1989) WMO solid precipitation measurement intercomparison: objectives, methodology, analysis. *IAHS Publ.*, **179**: 57–64.

Gorham, E. (1958) Atmospheric pollution by hydrochloric acid. *Quarterly J. Royal Met. Soc.*, **84**: 274–6.

Goss, M. J., Howse, K.R. and Harris, W. (1978) Effects of cultivation on soil water retention and water use by cereals in clay soils. *J. Soil Sci.*, 29: 475-88.

Gough, R., Holliman, P.J., Willis, N. and Freeman, C. (2014) Dissolved organic carbon and trihalomethane precursor removal at a UK upland water treatment works. *Science of The Total Environment*, **468-9**: 228–239.

Grace, J. (1983) *Plant-atmosphere Relationships*, Chapman and Hall, London. 96pp

Grant, S.B., Saphores, J.-D.,Hamilton, A.J. and coauthors (2012) Taking the 'waste' out of 'wastewater' for human water security and ecosystem sustainability. *Science*, **337**: 681-686.

Green, W.H. and Ampt, G.A. (1911) Studies in soil physics, part 1. The flow of air and water through soils. *J.. Agric. Sci.*, **4**: 1-24.

Gregory, P.J. and Nortcliff, S. (Eds.) (2013) *Soil Conditions and Plant Growth.* Wiley-Blackwell, 472 pp

Grelle A., Lundberg, A., Lindroth, A. and, Moren A.S. (1997) Evaporation components of a boreal forest: variations during the growing season. *J. Hydrology*, **197**, 70–87.

Grismer, M., Orang, M., Snyder, R., and Matyac, R. (2002). Pan Evaporation to Reference Evapotranspiration Conversion Methods. *J. Irrig. Drain Eng.*, **128**: 180-184.

Groisman P.Y. and Legates, D.R. (1994) The accuracy of United States precipitation data. *Bulletin of the American Meteorological Society*, **75**: 215–226.

Groisman, P.Y. and Easterling, D.R. (1994) Variability and trends of total precipitation and snowfall over the United States and Canada. *J. Climate*, **7**: 184–205.

Grolleau, G. and McCann, L.M.J. (2012) Designing watershed programs to pay farmers for water quality services: case studies of Munich and New York City. *Ecological Economics*, **76**: 87-94.

Grust, K. and Stewart, D. (2012) UK trial of the OTT Pluvio2. *BHS Eleventh National Symposium. Hydrology for a Changing World.* Dundee. 7 pp

Gunston, H.M. and Batchelor, C.H. (1983) A comparison of the Priestley–Taylor and Penman methods for estimating reference crop evapotranspiration in tropical countries. *Agricultural Water Management*, **6**: 65–77.

Guo, H., Hu, Q., Zhang, Q., and Feng, S. (2012) Effects of the Three Gorges Dam on Yangtze River flow and river interaction with Poyang Lake, China: 2003–2008. *J. Hydrology*, **416–417**: 19–27.

Gustard, A. and Irving, K.M. (1994) Classification of the low flow response of European soils, in P. Seuna, A. Gustard, N.W. Arnell and G.A. Cole (eds), *FRIEND: Flow regimes from international experimental and network data*, IAHS Publ. No. **221**, IAHS Press, Wallingford, 113-117.

Gustard, A., Bullock, A. and Dixon, J.M. (1992) Low flow estimation in the UK. Institute of Hydrology Report **108**, Wallingford, 292 pp. http://nora.nerc.ac.uk/6050/1/IH_108.pdf .

Habermehl, M.A. (1985) Groundwater in Australia. In: *Hydrogeology in the Service of Man.* Memoires of the XVIIth Congress, Int. Assoc. Hydrogeologists, 31-52.

Habib, E., Lee G., Kim D., and Ciach G.J. (2013) Ground-based direct measurement. In: Testik F.Y. and Gebremichael M. (Eds.) *Rainfall: State of the Science.* Geophysical Monograph Series, Vol **191**. American Geophysical Union. pp 61-77.

Haines, A.T., Finlayson, B.L. and McMahon, T.A. (1988) A global classification of river regimes. *Applied Geography*, **8**: 255-272.

Hall, D.G. (1964) What is hydrology? How is it applied? *Proc. Inst. Civil Engrs.*, **27**: 662-664.

Hall, R.L. (1985) Further interception studies of heather using a wet surface weighing lysimeter system. *J. Hydrology*, **81**: 193–210.

Hall, R.L. (1987) Processes of evaporation from vegetation of the uplands of Scotland. *Trans. Roy. Soc. Edinburgh: Earth Sciences*, **78**: 327–334.

Hall, R.L. and Harding, R.J (1993) The water balance of the Balquhidder catchments: a process approach. *J. Hydrology*, **145**: 285–314.

Hall, R.L. and Hopkins, R. (1997) A net rainfall gauge for use with multi–stemmed trees. *Hydrology and Earth System Sciences*, **1**: 213–215.

Hall, R.L. and Roberts, J.M. (1990) Hydrological aspects of new broadleaf plantations. In: Moffat, A.J., Jarvis, M.B. (Eds.) *SEESOIL*. J. Southeast England Soils Discussion Group, **6**: 2–38.

Hall, R.L., Calder, I.R., Rosier, P.T.W. and coauthors (1992) Measurement and modelling of interception loss from a Eucalyptus plantation in southern India. In: Calder I.R.,. Hall R.L and Adlard P.G. (Eds.), *Growth and Water Use in Plantations*. J. Wiley, Chichester. pp 270–289.

Halliday, S.J., Wade, A.J., Skeffington, R.A., and coauthors (2012) An analysis of long-term trends, seasonality and short-term dynamics in water quality data from Plynlimon, Wales. *Science of the Total Environment*, **434**: 186-200.

Hannaford, J., and Marsh, T.J. (2008) High-flow and flood trends in a network of undisturbed catchments in the UK. *International J. Climatology*, **28:** 1325–1338.

Harding, R.J., Neal, C. and Whitehead, P.G. (1992) Hydrological effects of plantation forestry in North–Western Europe. In:. Teller, A, Mathy, P,. and. Jeffers, J.N.R (Eds.), *Responses of Forest Ecosystems to Environmental Changes*. Elsevier Applied Science, London. pp 445–455.

Haria, A.H., Hodnett, M.G. and Johnson, A.C. (2003) Mechanisms of groundwater recharge and pesticide penetration to a chalk aquifer in southern England. *J. Hydrology*, **275:** 122–137.

Harris, G.L., Bailey, S.W. and Mason, D.J. (1991) The determination of pesticide losses to water courses in an agricultural clay catchment with variable drainage and land management. In: *Proceedings of Brighton Crop Protection Conference - Weeds, 1991*. British Crop Protection Council, Farnham, Surrey, Vol **3**: 1271–1278.

Harris, N.M., Gurnell, A.M., Hannah, D.M. and Petts, G.E. (2000) Classification of river regimes: a context for Hydroecology. *Hydrological Processes*, **14**: 2831-2848.

Harrison, D.L., Norman, K, Pierce, C., and Gaussiat, N. (2012) Radar products for hydrological application in the UK. *Water Management*, **165**: 89-103.

Havas, M. (1986) Effects of acidic deposition on aquatic ecosystems,. In: Stern, A.C. (Ed.) *Air pollutants, their transformation, transport and effects*. 3rd Edn., Academic Press, New York, pp. 351–89.

Haygarth, P.M. and Jarvis, S.C. (2002) (Eds.) *Agriculture, Hydrology and Water Quality*. CABI Publishing, Wallingford. 502 pp.

Heath, R.C. (2004) Basic groundwater hydrology. *USGS Wat. Supply Paper* 2220. 10th printing, revised. 86 pp. https://pubs.er.usgs.gov/publication/wsp2220

Heath, R.C. (1982) Classification of groundwater systems of the United States. *Groundwater*, **20**: 393-401

Hem, J.D. (1985) Study and interpretation of the chemical characteristics of natural water, *USGS Water Supply Paper*, **2254**, 3rd Edn., 263 pp. http://pubs.usgs.gov/wsp/wsp2254/

Hendriks, M.R. (1993) Effects of lithology and land use on storm runoff in east Luxembourg. *Hydrological Processes*, **7**: 213-226.

Hensel, P.R., McLaughlin Mitchell, S, and Sowers, T.E. (2006) Conflict management in riparian disputes. *Political Geography*, **25**: 383-411.

Herbst, M., Roberts, J.M., Rosier, P., and Gowing, D. (2006) Measuring and modelling the rainfall interception loss by hedgerows in southern England. *Agricultural Forestry Meteorology*, **141**, 244–256.

Herbst, M., Rosier, P., McNeil, D., and coauthors, (2008). Seasonal variability of interception evaporation from the canopy of a mixed deciduous forest. *Agricultural Forestry Meteorology*, **148**: 1655-1667

Herrmann, A, Schumann, S, Holko L and coauthors (2010) Status and perspectives of hydrology in small basins. *IAHS Publication*, **336**. 313 pp

Herschy, R.W. (2009) *Streamflow Measurement*. Taylor and Francis. London, 507pp

Herwitz, S.R. (1985) Interception storage capacities of tropical rainforest canopy trees. *J. Hydrology*, **77**: 237–52.

Herzberg, B. (1901) Die Wasserversorgung einiger Nordseebader (The water supply to some North Sea resorts). J. *für Gasbeleuchtung und Wasserversorgung*, **44**: 815-19.

Hewitt C.N. and Jackson A.V. (2009) *Atmospheric Science for Environmental Scientists*. Wiley-Blackwell. Chichester. 300pp.

Hewlett, J.D. (1961a) Watershed management. In: *Report for 1961 Southeastern Forest Experiment Station*. US Forest Service, Ashville, N. Carolina.

Hewlett, J.D. (1961b) Soil moisture as a source of baseflow from steep mountain watersheds. *Southeastern Forest Experiment Station, Paper* **132**, US Forest Service, Ashville, N. Carolina.

Hewlett, J.D. (1969) Tracing storm base flow to variable source areas on forested headwaters. *Technical Report* **2**, School of Forest Resources, University of Georgia, Athens.

Hewlett, J.D. and Hibbert, A.R. (1963) Moisture and energy conditions within a sloping soil mass during drainage. *J. Geophysical Research*, **68**: 1081-7. Also reproduced in Beven, K.J. (2007) pp 81-87.

Hewlett, J.D. and Hibbert, A.R. (1967) Factors affecting the response of small watersheds to precipitation in humid areas. In: Sopper, W.E,. and Lull, H.W. (Eds.) *Forest Hydrology*, Pergamon. Oxford, pp. 275-290. Also reproduced in Beven, K.J. (2007) pp 116-131.

Hibbert, A.R. (1967) Forest treatment effects on water yield. In: Sopper, W.E,. and Lull, H.W. (Eds.), *Forest Hydrology*, Pergamon, Oxford, pp. 527–43.

Hibbert, A.R. (1971) Increases in streamflow after converting chaparral to grass. *Water Resources Research*, **7**: 71–80.

Hillel, D. (2003) *Introduction to Environmental Soil Physics*. Academic Press, New York, 494 pp.

Hillel, D.J. (1991) *Out of the Earth: Civilization and the Life of the Soil*. The Free Press, New York. 321 pp.

Hodge. A.T (2002) *Roman Aqueducts and Water Supply*. Duckworth Archaeology, London 504 pp

Hodnett, M.G. and Bell, J.P. (1986) Soil moisture investigations of groundwater recharge through black cotton soils, in Madhya Pradesh, India, *Hydrological Sciences J.*, **31**: 361-81.

Hodnett, M.G., Vendrame, I., Marques Filho, A.De O., and coauthors (1997) Soil water storage and groundwater behaviour in a catenary sequence beneath forest in central Amazonia. II. Floodplain water table behaviour and implications for streamflow generation. *Hydrology and Earth System Sciences*, **1**: 279-290.

Hofer, T. and Messerli, B. (2006) *Floods in Bangladesh: History, Dynamics and Rethinking the Role of the Himalayas*. United Nations University Press, Tokyo, Japan. 468 pp

Holden, J. (2005) Controls of soil pipe frequency in upland blanket peat. *J. Geophysical Research*, **110**: 2156-2202

Horton, J.H. and Hawkins, R.H. (1965) Flow path of rain from the soil surface to the water table. *Soil Sci.*, **100**: 377-83.

Horton, R.E. (1919) Rainfall interception. *Monthly Weather Review*, **47**: 603–23. Also reproduced in Gash and Shuttleworth (2007) pp 278-291.

Horton, R.E. (1933) The role of infiltration in the hydrologic cycle. *Trans. American Geophysical Union*, **14**: 446-60. Also reproduced in Beven, KJ (2007) pp 13-27.

Horton, R.E. (1939) Analysis of runoff plot experiments with varying infiltration capacity, *Trans. American Geophysical Union*, **20**: 693-711.

Hough, M.N. and. Jones, R.J.A (1997) The UK Meteorological Office rainfall and evaporation calculation system: MORECS version 2.0 – an overview. *Hydrology and Earth System Sciences*, **1**:227–239.

Houze, R.A. (2004), Mesoscale convective systems, *Rev. Geophysics*, **42**, RG4003, doi:10.1029/2004RG000150

Howden, N.J.K., Burt, T.P., Worrall, F. and coauthors (2011) Nitrate concentrations and fluxes in the River Thames over 140 years (1868–2008): are increases irreversible? *Hydrological Processes*, **23**: 2657-2662.

Hubbert, M.K. (1940) The theory of groundwater motion. *J. Geology*, **48**: 785-944. Also partly reproduced in Anderson, M. (2008) (Ed.) pp 74-103.

Hursh, C.R. (1944) Report of the sub-committee on subsurface flow. *Trans. American Geophysical Union*, **25**: 743-6.

Hursh, C.R. and Brater, E.F. (1941) Separating storm hydrographs from small drainage areas into surface and subsurface flow, *Trans. American Geophysical Union*, **22**: 863-70.

Huuskonen, A ,Saltikoff, E, and Holleman, I, (2014) The Operational Weather Radar Network in Europe. *Bull. Amer. Meteor. Soc.*, **95**: 897–907.

ICOLD (2011) *World Register of Dams*. International Commission on Large Dams. (http://icold-cigb.org)

IH (1980) *Low Flow Studies Report*, Institute of Hydrology, Wallingford.

IH (1997) *Scientific Report 1996-97*, Institute of Hydrology, Wallingford, 62 pp.

IH (1998) *Broadleaf woodlands: The implications for water quantity and quality*. Institute of Hydrology Report to the Environment Agency. Environment Agency Research and Development Publication No. **5**. Stationery Office, London. 33 pp.

IH (1999) *Flood Estimation Handbook*, Institute of Hydrology, Wallingford. 5 Vols.

IH/BGS (1996) *Hydrological data UK, 1995*. Institute of Hydrology/British Geological Survey. 176 pp. http://nrfa.ceh.ac.uk/hydrological-yearbooks

Iizumi, T. and Ramankutty, N. (2015) How do weather and climate influence cropping area and intensity? *Global Food Security*, **4**: 46-50.

INBO (2014) China: South-North water diversion. *INBO Newsletter*, **22**. International Network of Basin Organizations.

INBO (2016) International conference on hydrology in large river basins of Africa. *INBO Newsletter*, **24**. International Network of Basin Organizations.

IPCC (2014) Fifth Assessment Report. Intergovernmental Panel on Climate Change. http://ar5-syr.ipcc.ch/topic_summary.php

ISO (1999) Hydrometric determinations: Flow measurements in open channels using structures - Guidelines for selection of structure. ISO **8368**. Reviewed in 2011. International Organization for Standardization, Geneva, Switzerland.

ISSS (1976) Soil physics terminology. Report of the Terminology Committee (Chairman G. H. Bolt) of Commission I (Soil Physics), *International Soil Science Society Bulletin*, **49**: 26-36.

Jackson, I.J. (1975) The relationships between rainfall parameters and interception by tropical forest. *J. Hydrology*, **24**: 215–38.

Jackson, J.H. (2010) *Paris under Water: How the City Of Light Survived the Great Flood of 1910*. Palgrave Macmillan, New York. 263 pp.

Jacobs, A.F.G. and Verhoef, A. (1997) Soil evaporation from sparse natural vegetation estimated from Sherwood Numbers. J. *Hydrology*, **188**: 443–452.

Jacobson, R.L. and Langmuir, D. (1974) Dissociation constants of calcite and CaHCO3– from 0°C to 50°C. *Geochima et Cosmochimica Acta*, **38**: 301–18.

Jaeger, L. (1985) Eleven years of precipitation measurements above a small pole wood pine stand, In: Sevruk, B. (Ed.), *Correction of precipitation measurements*, Swiss Federal Institute of Technology, Zurich, 101–103.

Jarvie, H.P., Neal, C. and Withers, P.A., 2006. Sewage-effluent phosphorus: A greater risk to river eutrophication than agricultural phosphorus? *Science of the Total Environment*, **306**: 243-253.

Jarvie, H.P., Neal, C., Leach, D.V. and coauthors (1997) Major ion concentrations and the inorganic carbon chemistry of the Humber rivers. *Science of the Total Environment*, **194/195**: 285–302.

Jarvie, H.P., Sharpley, A. N., Spears, B., and coauthors (2013a). Water Quality Remediation Faces Unprecedented Challenges from "Legacy Phosphorus". *Environmental Science & Technology*, **47**: 8997-8998.

Jarvie, H.P., Sharpley, A.N., Withers, P.J.A., and coauthors (2013b) Phosphorus mitigation to control river eutrophication: murky waters, inconvenient truths, and "postnormal" science. *J. Environmental Quality* **42**: 295-304.

Jayatilaka, C.J. and Gillham, R.W. (1996) A deterministic-empirical model of the effect of the capillary-fringe on near-stream area runoff. 1. Description of the model. J. *Hydrology*, **184**: 299-315

Jensen, M.E., Burman, R.D. and Allen, R.G (1990) (eds.) *Evaporation and irrigation water requirements*, Manuals and reports on engineering practice No. **70**, American Society Civil Engineers, New York, 360 pp.

Jobling, S., Nolan, M., Tyler , C.R. and coauthors (1998) Widespread sexual disruption in wild fish, *Environmental Science and Technology*, **32**: 2498–2506.

Johansson, R.C., Tsur, Y., Roe, T.L. and coauthors (2002) Pricing irrigation water: a review of theory and practice. *Water Policy*, **4**: 173-199.

Johnson, A.C., Besien, T.J., Bhardwaj, C.L. and coauthors (2001) Penetration of herbicides to groundwater in an unconfined chalk aquifer following normal soil applications. *J. Contaminant Hydrology*, **53:** 101–117.

Johnson, A.C. and Williams, R.J. (2004) A model to estimate influent and effluent concentrations of estradiol, estrone, and ethinylestradiol at sewage treatment works. *Environmental Science & Technology*, **38**: 3649-3658.

Johnson, A.C., Keller, V., Dumont, E. and Sumpter, J.P. (2015) Assessing the concentrations and risks of toxicity from the antibiotics ciprofloxacin, sulfamethoxazole, trimethoprim and erythromycin in European rivers. *Science of the Total Environment*, **511**: 747–755.

Johnson, A.C., Oldenkamp, R., Dumont, E. and Sumpter, J.P. (2013) Predicting concentrations of the cytostatic drugs cyclophosphamide, carboplatin, 5-fluorouracil, and capecitabine throughout the sewage effluents and surface waters of Europe. *Environmental Toxicology and Chemistry*, **32**: 1954-1961.

Johnson, R.C. (1990) The interception, throughfall and stemflow in a forest in Highland Scotland and comparison with other upland forests in the UK. *J. Hydrology*, **118**: 281–287.

Jones, H.G. (2007) Monitoring plant and soil water status: established and novel methods revisited and their relevance to studies of drought tolerance. *J. Experimental Botany*, **58**: 119-130.

Jones, J.A.A. (1997) Pipeflow contributing areas and runoff response. *Hydrologcial Processes*, **11**: 35-41.

Jones, J.A.A. (2010) Soil piping and catchment response. *Hydrological Processes*, **24**:1548-1566.

Jordan, W.C. (2010) The Great Famine revisited. In: Bruce, S.G. (Ed.) *Ecologies and Economies in Medieval and Early Modern Europe*. K Brill, Leiden, The Netherlands, pp 45-62.

Jorgensen, U. (2011) Benefits versus risks of growing biofuel crops: the case of Miscanthus. *Current Opinion in Environmental Sustainability*, **3**: 24–30.

Joslin, J.D., Mays, P.A., Wolfe M.H. and coauthors (1987) Chemistry of tension lysimeter water and lateral flow in spruce and hardwood stands. J. *Environmental Quality*, **16**: 152–60.

Kabat, P., Dolman, A.J. and Elbers, J.A. (1997) Evaporation, sensible heat and canopy conductance of fallow savannah and patterned woodland in the Sahel. J. *Hydrology*, **188**: 494–515.

Kalma, J.D. and Sivapalan, M. (Eds.) (1995) *Scale Issues in Hydrological Modelling*. Wiley, Chichester. 489. pp

Karahan, H (2012) Predicting Muskingum flood routing parameters using spreadsheets. *Computer Applications in Engineering Education*, **20**: 280–286.

Karl, T.R. and Knight, R.W. (1998) Secular trends of precipitation amount, frequency and intensity in the United States. Bull*etin of the American Meteorological Society*, **79**: 231–241.

Kay, P. and Grayson, R. (2014) Using water industry data to assess the metaldehyde pollution problem. *Water and Environment J.*, **28**: 410–417.

Keller, H.M. (1988) European experiences in long-term forest hydrology research. In: Swank, W.T. and Crossley, D.A. (Eds.) *Forest Hydrology and Ecology at Coweeta*. Ecological Studies, **66**, Springer-Verlag, New York. pp 407–59.

Keller, V.D.J.,. Tanguy M., Prosdocimi I., and coauthors (2015) CEH-GEAR: 1 km resolution daily and monthly areal rainfall estimates for the UK for hydrological use. *Earth System Science Data Discussions*, **8**: 83–112

Kendall, C. and McDonnell, J.J. (1998) *Isotope Tracers in Catchment Hydrology*. Elsevier, Amsterdam. 839 pp.

Kendon, E.J., Roberts, N.M., Fowler, H.J., and coauthors (2014) Heavier summer downpours with climate change revealed by weather forecast resolution model. *Nature Climate Change* **4**:570-576

Kerfoot, O. (1968) Mist precipitation on vegetation. *Forest Abstracts*, **29**: 8–20.

Kidd, C. and Huffman, G. (2011) Global precipitation measurement. *Meteorological Applications*, **18**: 334-353.

Kidd, C. and Levizzani, V. (2011) Status of satellite precipitation retrievals. *Hydrology and Earth System Sciences*, **15**: 1109-1116.

Kingdom, B., Liemberger, R. and Marin, P. (2006) *The Challenge of Reducing Non-Revenue Water (NRW) in developing countries*. Water Supply and Sanitation Board Discussion Paper **8**. World Bank, Washington.

Kirchner, J.W., Feng, X. and Neal, C. (2000) Fractal stream chemistry and its implications for contaminant transport in catchments. *Nature*, **403**: 524-527.

Kirchner, J.W., Feng, X. and Neal, C. (2001) Catchment-scale advection and dispersion as a mechanism for fractal scaling in stream tracer concentrations. *J. Hydrology*, **254**:, 81-100

Kirkby, M.J. (1985) Hillslope hydrology. In: M.G. Anderson and T.P. Burt (Eds.), *Hydrological Forecasting*, Wiley, Chichester. pp 37-75

Kirkby, M.J. (2011) *Hydro-geomorphology, Erosion and Sedimentation*. Benchmark papers in Hydrology. Vol **6**, IAHS. 640 pp.

Kirkby, M.J. and Chorley, R.J. (1967) Throughflow, overland flow and erosion, *Int. Assoc. Hydrological Sciences Bulletin*, **12**: 5-21.

Klaassen, W., Bosveld, F. and de Water, E.(1998) Water storage and evaporation as constituents of rainfall interception. *J. Hydrology,* **212–213**: 36–50

Klimont, Z., Smith, S.J. and Cofala, J. (2013) The last decade of global anthropogenic sulfur dioxide: 2000–2011 emissions. *Environ. Res. Lett.* **8 014003** doi:10.1088/1748-9326/8/1/014003

Klute, A. (1986b) Water retention, laboratory methods. Chap. 26 in *Methods of soil analysis, I. Physical and mineralogical methods.* In: Klute, A. (Ed.), 2nd Edn., ASA/SSSA, Madison. pp. 635-62.

Konikow, L.F. and Kendy, E. (2005) Groundwater depletion: A global problem. *Hydrogeology J.*, **13**:317–320.

Koutsoyiannis D. (2014) Reconciling hydrology with engineering. *Hydrology Research*, **45**: 2–22,

Kramer, P.J. and Boyer, J.S. (1995) *Water Relations of Plants and Soils*, Academic Press, New York. 495 pp.

Krasovskaia, I. (1997) Entropy-based grouping of river flow regimes. *J. Hydrology*, **202**: 173-191.

Kucharski, F., Zeng, N. and Kalnay, E. (2013) A further assessment of vegetation feedback on decadal Sahel rainfall variability. *Climate Dynamics*, **40**: 1453-1466.

Kutiel, H., Maheras, P. and Guika, S. (1998) Singularity of atmospheric pressure in the eastern Mediterranean and its relevance to inter–annual variations of dry and wet spells, *International J. Climatology*, **18**: 317–327.

Kuusisto, E. (1986) The energy balance of a melting snow cover in different environments, *IAHS Publ.*, **155**: 37–45.

Lamb, J.C. (1985) *Water Quality and its Control*. J. Wiley and Sons, New York, 384 pp.

Landes, D.S. (1998) *The Wealth and Poverty of Nations*. Little, Brown and Co. London, 650pp.

Lanza L.G. and Vuerich, E. (2009) The WMO Field intercomparison of rain intensity gauges. *Atmospheric Research* **94**: 534-543.

Lavers, D.A., Allan, R.P., Wood, E.F., and coauthors (2011) Winter floods in Britain are connected to atmospheric rivers. *Geophysical Research Letters* **38**, L23803.

Law, F. (1958) Measurement of rainfall, interception and evaporation losses in a plantation of Sitka spruce trees. *IASH* Proc., **2**: 397–411. Also reproduced in Gash, J.H.C. and Shuttleworth, W.J. (2007) pp 240-254.

Lee, R. (1980) *Forest Hydrology*, Columbia University Press, New York, 349 pp.

Leeks, G.J.L., Neal, C., Jarvie, H.P. and coauthors (1997) The LOIS river monitoring network: strategy and implementation. *Science of the total environment*, **194/195**: 101–109.

Legates, D.R. and Willmott, C.J. (1990) Mean seasonal and spatial variability in gauge–corrected, global precipitation, *International J. Climatology*, **10**: 111–127.

Lerner, D.N. and Barrett, M.H. (1996) Urban groundwater issues in the United Kingdom, *Hydrogeology J.*, **4**: 80-89

Lester, J.E. (1990) Sewage and sewage sludge treatment. In: *Pollution: causes effects and control.* 2nd Edn., Royal Society of Chemistry, Cambridge, pp. 33–62.

Lettau H., Lettau K., Molion L.C.B. (1979) Amazonia's hydrologic cycle and the role of atmospheric recycling in assessing deforestation effects. *Monthly Weather Review*, **107**: 227–238

Lewandowsky, S., Oreskes, N., Risbey, J.S. and coauthors (2015) Seepage: Climate change denial and its effect on the scientific community. *Global Environmental Change*, **33**: 1–13.

Lhomme, J.P. (1997) Towards a rational definition of potential evaporation, *Hydrology and Earth System Sciences*, **1**: 257–264.

Lloyd, C.R. (1990) The temporal distribution of Amazonian rainfall and its implications for forest interception, *Quarterly J. Royal Meteorological Society*, **116**: 1487–1494.

Lloyd, C.R. and Marques, A.de 0. (1988) Spatial variability of throughfall and stemflow measurements in Amazonian rain forest. *Agricultural Forestry Meteorology*, **42**: 63–73.

Lloyd, C.R., Gash, J.H.C., Shuttleworth, W.J. and Marques, A.de 0. (1988) The measurement and modelling of rainfall interception by Amazonian rain forest. *Agricultural Forestry Meteorology*, **43**, 277–94.

Lloyd, C.R. and Oliver, S. (2015) The physics of atmospheric interaction. In: Rodda, J.C. and Robinson, M. (Eds.) *Progress in Modern Hydrology*. Wiley Blackwell, Chichester. pp 135-182.

Lohman, S. W. (1972) (Chairman) Definitions of selected groundwater terms - revisions and conceptual refinements. Report of the Committee on redefinition of groundwater terms, *USGS Wat. Sup. Pap.*, **1988**, 21 pp http://pubs.usgs.gov/wsp/wsp_1988/

Loudyi, D, Falconer, R., and Lin, B. (2014) MODFLOW: An insight into thirty years development of a standard numerical code for groundwater simulations. 11th International Conference on Hydroinformatics. *City University of New York (CUNY) Academic Works* http://academicworks.cuny.edu/cc_conf_hic/168 8 pp.

Lull, H.W. (1964) Ecological and silvicultural aspects. Sec. **6** in Chow, V.T. (Ed.), *Handbook of Applied Hydrology*, McGraw– Hill, New York.

Lundberg A., Calder, I.R. and Harding, R. (1998) Evaporation of intercepted snow: measurement and modelling. *J. Hydrology*, **206**: 151–163.

Luthi, D., Le Floch, M., Bereiter, B. and coauthors (2008) High-resolution carbon dioxide concentration record 650,000–800,000 years before present. *Nature* **453**: 379-382.

Lvovitch, M.I. (1973) The global water balance. *EOS*, **54**: 28-53.

MacDonald, A M, Bonsor, H C, Dochartaigh, B É Ó and Taylor, R G (2012) Quantitative maps of groundwater resources in Africa. *Environmental Research Letters*, **7**: 24009-24015(7)

Maddox, R.A. (1983) Large-scale meteorological conditions associated with mid-latitude Mesoscale convective complexes. *Monthly Weather Review*, **111**: 1475–1493

Malthus, T.R. (1798) *An Essay on the Principle of Population*, Selected and introduced by Winch, D. Cambridge University Press, Cambridge (1992).

Mankin, J.S., Viviroli, D., Hoekstra A.Y. and Diffenbaugh, N.S. (2015) The potential for snow to supply human water demand in the present and future. *Environmental Research Letters*, **10**. 114016 doi:10.1088/1748-9326/10/11/114016

Marsh, TJ, Cole, G and Wilby, R (2007) Major droughts in England and Wales, 1800–2006. *Weather*, **62**: 87–93.

Marsh, T.J., Monkhouse R.A., Arnell, N.W. and coauthors (1994) *The 1988-92 Drought*. Institute of Hydrology and British Geological Survey, Wallingford, 79 pp. http://nora.nerc.ac.uk/6952/1/HDUKdrought.pdf

Marsh, T., Moore, R., Dixon, H. and coauthors (2015) *Hydrological data acquisition and exploitation*. In: Rodda J.C. and Robinson M. (Eds.) (2015) Progress in Modern Hydrology. Wiley-Blackwell, Oxford. pp 324-365.

Marshall, J.S. and Palmer, W.M. (1948) The distribution of raindrops with size. J. *Meteorology*, **5**: 165–6.

Marty, C. and Blanchet, J. (2012) Long-term changes in annual maximum snow depth and snowfall in Switzerland based on extreme value statistics. *Climatic Change*, **111**: 705–721

Martyn, D. (1992) *Climates of the world*. Developments in Atmospheric Science, **18**, Elsevier, Amsterdam, 435 pp.

Massman, W.J. (1980) Water storage on forest foliage: a general model. *Water Resources Research*, **16**: 210–216.

Massman, W.J. (1983) The derivation and validation of a new model for the interception of rainfall by forests. *Agricultural Meteorology*, **28**: 261–286.

Mayes, J. (1996) Spatial and temporal fluctuations of monthly rainfall in the British Isles and variations in the mid–latitude western circulation, *International J. Climatology*, **16**: 585–596.

McCulloch, J.S.G. (1975) Hydrology – The science of water. *NERC News Journal* **12**:14-15.

McCulloch, J.S.G. (1988) Hydrology - Science or just technology? *Research Report 1984-87*, Institute of Hydrology, Wallingford.

McCulloch, J.S.G. (2007) All our yesterdays: a hydrological retrospective. *Hydrology and Earth System Sciences*, **11**:.3-11

McGowan, M., Blanch, P., Gregory, P.J and Haycock, D. (1984) Water relations of winter wheat 5. The root system and osmotic adjustment in relation to crop evaporation. J. *Agric. Sci.*, **102**:415–425.

McKeague, J.A., Wang C. and Topp, G.C. (1982) Estimating saturated hydraulic conductivity from soil morphology. *Proc. Soil Science Society America*, **46**:1239-44.

McKee, T.B., Doesken, N.J, Kleist, J. (1993) The relationship of drought frequency and duration to time scales. Proc. 8th Conference on Applied Climatology, 179-184. http://ccc.atmos.colostate.edu/relationshipofdroughtfrequency.pdf

McMillan, W.D. and Burgy, R.H. (1960) Interception loss from grass. J. *Geophysical Research*, **65**: 2389–94.

McNaughton, K.G. and. Black, T.A (1973) A study of evapotranspiration from a Douglas fir forest using the energy balance approach. W*ater Resources Research*, **9**: 1579–90.

McNaughton, K.G., Clothier, B.E. and Kerr, J.P. (1979) Evaporation from land surfaces, in Murray, D.L. and Ackroyd, P. (Eds.), *Physical Hydrology: New Zealand Experience*. Hydrological Society, Wellington, N.Z., pp. 97–119.

McNaughton, K.G. and Jarvis, P.G. (1983) Predicting effects of vegetation changes on transpiration and evaporation. In: Kozlowski, T.T. (Ed.), *Water Deficits and Plant Growth*, Vol. **III**, Academic Press, New York pp. 1 - 47.

Meadows, D .H., Meadows, D.L., Randers, J. and Behrens, W.W. (1972) *Limits to Growth*, Club of Rome.

Metcalfe, P., Beven, K., and Freer, J. (2015) Dynamic TOPMODEL: A new implementation in R and its sensitivity to time and space steps. *Environmental Modelling & Software*, **72**: 155–172.

Meteorological Office (1982) *Observer's Handbook*, 5th Edn., HMSO, London, 220 pp.

Michiles, A.A. and Gielow, R. (2008) Above ground thermal energy storage rates, trunk heat fluxes and surface energy balance in a central Amazonian rainforest. *Agricultural and Forest Meteorology*, **148**: 917-930.

Micklin, P. (1981) A preliminary systems analysis of impacts of proposed Soviet River diversions on Arctic sea ice. *Eos Transactions of the American Geophysical Union*, **62**: 489–493

Micklin, P. (2007) The Aral Sea disaster. *Annual Review of Earth and Planetary Sciences*, **35**: 47-72.

Miralles, D.G., de Jeu, R.A.M., Gash, J.H., Holmes, T.R.H., Dolman, A.J. (2011a) Magnitude and variability of land evaporation and its components at the global scale. *Hydrol. Earth Syst. Sci.*, **15**, 967–981. doi:10.5194/hess-15-967-2011

Miralles, D.G., Gash, J.H., Holmes, T.R.H., and coauthors (2010) Global canopy interception from satellite observations. *J. Geophys. Res.*, **115**, D16122. doi:10.1029/2009JD013530

Miralles, D.G., Holmes, T.R.H., de Jeu, R.A.M., and coauthors (2011b) Global land-surface evaporation estimated from satellite-based observations. *Hydrol. Earth Syst. Sci.* **15**, 453–469. doi:10.5194/hess-15-453-2011

Miranda, A.C., Jarvis, P.G. and Grace, J. (1984) Transpiration and evaporation from heather moorland, *Boundary–Layer Met.*, **28**: 227–243.

Mishra, A.K. and Coulibaly, P. (2009) Developments in hydrometric network design: a review. Reviews of Geophysics, **47**, RG2001, doi:10.1029/2007RG000243.

Mizutani, K. (1997) Applicability of the eddy correlation method to measure sensible heat transfer to forest under rainfall conditions. *Agricultural and Forest Meteorology*, **86:**, 193–203.

Mizutani, K., Yamanoi, K., Ikeda, T. and Watanabe, T. (1997) Applicability of the eddy correlation method to measure sensible heat transfer to forest under rainfall conditions. *Agricultural and Forest Meteorology*, **86**: 193-203.

Monastersky, M. (2015) Anthropocene: The human age. *Nature*, **519**: 44-147.

Monteith, J.L. (1965) Evaporation and the environment. Proc. *Symposium on Experimental Biology*, **19**: 205–234. Also reproduced in Gash, J.H.C. and Shuttleworth, W.J. (2007) pp 337-366.

Monteith, J.L. (1985) Evaporation from land surfaces: Progress in analysis and prediction since 1948. In: *Advances in Evapotranspiration.* American Society of Agricultural Engineers, pp 4–12.

Monteith, J.L. (1995) Fifty years of potential evaporation, in Keane, T. and Daly, E. (Eds.), *The balance of water – present and future*. Proc. AGMET Conference, Dublin, Sept 7–9, 1994. pp 29–45.

Monteith, J.L. and Szeicz, G. (1961) The radiation balance of bare soil and vegetation. *Quart. J. Royal Met. Soc.*, **87**: 159-170.

Monteith, J.L. and Unsworth, M.H. (2013) *Principles of Environmental Physics, 4th Ed*, Academic Press, London. 422 pp.

Morice, C.P., Kennedy, J.J., Rayner, N.A. and Jones, P.D. (2012) Quantifying uncertainties in global and regional temperature change using an ensemble of observational estimates: the HadCRUT4 dataset. *J. Geophysical Research,* **117**, D08101, doi:10.1029/2011JD017187

Morris, J., Hess, T.M , Gowing, D.J.G and coauthors (2005) A framework for integrating flood defence and biodiversity in washlands in England. *International J. River Basin Management*, **3**: 1–11.

Morton, F.I. (1985) The complementary relationship evapotranspiration model: how it works. In: *Advances in Evapotranspiration*, American Society Agricultural Engineers, 377–384.

Morton, F.I. (1994) Evaporation research – A critical review and its lessons for the environmental sciences. *Critical Reviews in Environmental Science and Technology*, **24**: 237–280.

Moser, H., Rauert, W., Morgenschweis, G. and Zojer, H. (1986) Study of groundwater and soil moisture movement by applying nuclear, physical and chemical methods, in *Technical documents in hydrology*, UNESCO, Paris, 104 pp.

Mosley, M.P. (1979) Streamflow generation in a forested watershed, New Zealand. *Water Resources Research*, **15**: 795-806. Also reproduced in Beven, KJ (2007) pp 338-349.

Mualem, Y. (1976) A new model for predicting the hydraulic conductivity of unsaturated porous media. *Water Resources Research*, **12**: 512-22.

Munger, J.W. and Eisenreich, S.J. (1983) Continental–scale variations in precipitation chemistry. *Environmental Science and Technology*, **17**: 32A–42A.

Murphy, C.E. and Knoerr, K.R. (1975) The evaporation of intercepted rainfall from a forest stand: An analysis by simulation. *Water Resources Research*, **11**: 273–280.

Muste, M., Yu, K. and Spasojevic, M. (2004) Practical aspects of ADCP data use for quantification of mean river flow characteristics; Part I: moving-vessel measurements; and Part II: fixed-vessel measurements. *Flow Measurement and Instrumentation*, **15**: 1–16 and 17–28.

Muzylo, A., Llorens, P., Valente, F., and coauthors (2009). A review of rainfall interception modelling. *J. Hydrology* **370**: 191-206

Nagel, J.F. (1956) Fog precipitation on Table Mountain. *Quarterly J. Royal Meteorological Society*, **82**: 452–460.

Nakai, Y., Sakamoto, T. and Terajima, T. (1994) Snow interception by forest canopies: weighing a conifer tree, meteorological observations and analysis by the Penman–Monteith formula. *IAHS Publ. No.* **233**: 227–236.

Narasimhan, T.N. (1998) Hydraulic characterization of aquifers, reservoir rocks and soils: A history of ideas. *Water Resources Research,* **34**: 33-46

Neale, C.M.U. and Cosh, M.H. (2012) Remote sensing and hydrology. *IAHS Publ.* **352**. 476 pp

NERC (1975) *Flood Studies Report*, Natural Environment Research Council, 5 vols. Reprinted by Institute of Hydrology (1993).

NERC (1991) Hydrogeology of hot dry rocks. *NERC News*, July 1991. pp 28-29.

Newson, M.D. (2009) *Land, Water and Development*. 3rd Edn. Routledge, Abingdon. 441 pp.

NIH (1990) *Hydrology in ancient India*. National Institute of Hydrology, Roorkee, India, 103pp.

Niu, G-Y, Seo, K-W, , Yang, Z-L and co authors (2007) Retrieving snow mass from GRACE terrestrial water storage change with a land surface model. *Geophysical Research Letters,* **34**, L15704, doi:10.1029/2007GL030413

Nordin, C.F. (1985) The sediment loads of rivers. In: Rodda, J.C. (Ed.) *Facets of Hydrology, 2*. J. Wiley and Sons, Chichester, pp. 183–204.

Nortcliffe, S. and Thornes, J.B. (1984) Floodplain response of a small tropical stream. In: Burt T.P. and Walling D.E. (Eds.), *Catchment Experiments in Fluvial Geomorphology*, GeoBooks, Norwich, pp. 73-86.

NRC (2012) *Challenges and opportunities in the hydrologic sciences*. National Research Council. National Academies Press, Washington, DC. http://www.nap.edu.

NWSCA (1984) An index for low flows. *Streamline* **24**. Water Directorate. National Water and Soil Conservation Authority. Wellington New Zealand.

Ogallo, L.J. (1988) Relationships between seasonal rainfall in East Africa and the Southern Oscillation. J. *Climatology*, **8**: 31–43.

Oke, T.R. (1987) *Boundary layer climates*. 2nd Edn., Routledge, London, 435 pp.

Oki, T., Valeo, C. and Heal, K. (Eds.) (2006) Hydrology 2020: An Integrating Science to Meet World Water Challenges. *IAHS Publ.* **300**. 190 pp

Oliver, H. and Oliver, S. (2003) Meteorologist's profile – John Dalton. *Weather*, **58**: 206–211.

O'Loughlin, E.M. (1981) Saturation regions in catchments and their relations to soil and topographic properties. J. *Hydrology*, **53**: 229-46

O'Loughlin, E.M. (1990) Modelling soil water status in complex terrain. *Agricultural and Forest Meteorology*, **50**: 23-38

Oreskes, N. and Conway, E.M. (2010) *Merchants of Doubt*. Bloomsbury Press, London. 368 pp.

Oroud, I.M. (1998) The influence of heat conduction on evaporation from sunken pans in hot, dry environment. J. *Hydrology*, **210**: 1–10.

Ortloff, C.R. (2016) Water Management in Ancient Peru. In: Selin, H. (Ed.) *Encyclopaedia of the History of Science, Technology, and Medicine in Non-Western Cultures*, Springer Netherlands 14 pp.

Osborn, T.J., Hulme, M, Jones, P.D. and Basnett, T.A. (2000) Observed trends in the daily intensity of UK precipitation. *International J. of Climatology*, **20**: 347-364.

OST (2005) Access to water in developing countries. *Postnote* **178**. The Parliamentary Office for Science and Technology. London. www.parliament.uk/post/home.htm

Parker, D.E., Legg T.P., and Folland ,C.K. (1992) A new daily Central England temperature series, 1772-1991. *Int. J. Climatology*, **12**: 317-342

Parkhurst, D.L., and Appelo, C.A.J. (2013) Description of input and examples for PHREEQC version 3. A computer program for speciation, batch-reaction, one-dimensional transport, and inverse geochemical calculations. *U.S. Geological Survey Techniques and Methods.*, Book 6, Chapter A43, 497 pp. http://pubs.usgs.gov/tm/06/a43/.

Parris, K. (2011) Impact of Agriculture on Water Pollution in OECD Countries: Recent Trends and Future Prospects. *Water Resources Development*. **27**: 33–52.

Pazwash, H., Asce A.M. and Mavrigian, G. (1981) Millennial celebration of Karaji's hydrology. *Proc. Amer. Soc. Civil Engrs., J. Hydraulics Div.*, **107**, No. HY3: 303-309.

Pearce, A.J., and Rowe, L.K. (1981) Rainfall interception in a multi–storied, evergreen mixed forest: estimates using Gash's analytical model. *J. Hydrology*, **49**: 341–353.

Pearce, A.J., Rowe L.K. and O'Loughlin C.L. (1982) Hydrologic regime of undisturbed mixed evergreen forests, South Nelson, New Zealand. *J. Hydrology (N.Z.)*, **21**: 98–116.

Pearce, A.J., Rowe L.K. and Stewart J.B. (1980) Night-time, wet canopy evaporation rates and the water balance of an evergreen mixed forest. *Water Resources Research*, **16**: 955–959.

Pearce, A. J., Stewart, M.K. and. Sklash, M.G (1986) Storm runoff generation in humid headwater catchments. I: Where does the water come from? *Water Resources Research*, **22**: 1263-72.

Peixoto, J.P and Oort, A.H. (1992) *Physics of Clouds*. American Institute of Physics, New York. 520 pp.

Penman, H.L. (1948) Natural evaporation from open water, bare soil and grass. *Proc. Royal Society, Series A*, **193**: 120–145. Also reproduced in Gash and Shuttleworth (2007) pp 186-212.

Penman, H.L. (1952) Experiments on the irrigation of sugar beet. J. *Agricultural Science*, **42**: 286–292.

Penman, H.L. (1954) Evaporation over parts of Europe. *IAHS Publ.*, **3**: 168–176.

Penman, H.L. (1956) Evaporation: an introductory survey. *Netherlands J. Agricultural Science*, **1**: 9–29.

Penman, H.L. (1963) *Vegetation and Hydrology*. Technical Communication **53**, Commonwealth Agricultural Bureau, Harpenden.

Penman, H.L. (1967) In discussion of J. Delfs, Interception and stemflow in stands of Norway spruce and beech in West Germany. In: Sopper, W.E. and Lull, H.W. (Eds.), *Forest Hydrology*, Pergamon, Oxford, pp. 179–185.

Pereira, A.R and Paes de Camargo, A. (1989) An analysis of the criticisms of Thornthwaite's equation for estimating potential evapotranspiration. *Agricultural and Forest Meteorology*, **46**, 149-157.

Pereira, F.L., Gash, J.H.C., David, J.S., and Valente, F (2009a) Evaporation of intercepted rainfall from isolated evergreen oak trees. Do the crowns behave as wet bulbs? *Agricultural Forestry Meteorology*, **149**, 667-679.

Pereira, F.L., Gash, J.H.C., David, J.S., and coauthors (2009b) Modelling interception loss from evergreen oak Mediterranean savannas: Application of a tree-based modelling approach. *Agricultural Forestry Meteorology*, **149**, 680–688.

Perks, A, T. Winkler, T. and Stewart, B. (1996) *The adequacy of hydrological networks: a global assessment*, Technical Reports in Hydrology and Water Resources, **52**. World Meteorological Organisation, Geneva. 56 pp.

Perry, M., Hollis, D., and Elms, M. (2009) *The Generation of Daily Gridded Datasets of Temperature and Rainfall for the UK.* National Climate Information Centre Memorandum **24**. Met. Office, Exeter, 7 pp.

Peters, N.E. (1984) Evaluation of environmental factors affecting yields of major dissolved ions in streams in the United States, *USGS Wat, Sup. Pap.*, **2228**, 39 pp.

Pfister, L., Savenije H.H.G. and Fenicia, F. (2009) Da Vinci's water theory: on the origin and fate of water. *IAHS Special Publication* **9**. Wallingford. 92 pp.

Pham, H.Q., Fredlund, D.G. and Barbour, S.L. (2005) A study of hysteresis models for soil-water characteristic curves. *Canadian Geotechnical J.* **42**: 1548–1568

Philip, J. R. (1957) The theory of infiltration, 4. Sorptivity and algebraic infiltration equations. *Soil Sci.*, **84**: 257-64.

Philip, J. R. (1964) The gain, transfer and loss of soil water. In: *Water Resources Use and Management*, Melbourne University Press, pp. 257-75.

Pillsbury, A.F., Pelishek R.E.,. Osborn J.F and Szuszkiewicz T.E. (1962) Effects of vegetation manipulation on the disposition of precipitation on chaparral–covered watersheds. *J. Geophys. Research*, **67**: 695–702.

Piper, A.M. (1944) A graphic procedure in the geochemical interpretation of water analyses, *Trans. Amer. Geophysical Union*, **25**: 914–23.

Pires, M. (2004) Watershed protection for a world city: The case of New York. *Land Use Policy*, **21**: 161-175.

Plummer, L.N, and Busenberg, E. (2000) Chlorofluorocarbons. In: Cook, P.G. and Herczeg, A.L. (Eds.) *Environmental Tracers in Subsurface Hydrology*. Springer, New York. pp 441-478.

Poland, J.F. (1984) Guidebook to studies of land subsidence due to groundwater withdrawal. *Studies and Reports in Hydrology*, No. 40, UNESCO, Paris, 305 pp.

Pool, D. R., Winster, D. and Cole, K.C. (2000) Land-subsidence and groundwater storage monitoring in the Tucson Active Management Area, Arizona. *USGS Fact Sheet* **084-00**.

Price, M. (1987) Fluid flow in the chalk of England, in *Fluid flow in sedimentary basins and aquifers*. In: Goff, J.C. and Williams, B.P.J. (Eds.), Society of London Special Publication, **34**, pp. 141-56.

Price, M. (2004) *Introducing Groundwater*. 2nd Edn. Routledge, London, 304 pp

Price, M., Bird M.J., and Foster, S.S.D. (1976) Chalk pore-size measurements and their significance. *Water Services*, October, pp. 596-600.

Priestley, C.H.B. and. Taylor, R.J (1972) On the assessment of surface heat flux and evaporation using large scale parameters. *Monthly Weather Review*, **100**: 81–92. Also reproduced in Gash, J.H.C. and Shuttleworth, W.J. (2007) pp 213-224.

Prior, J. and Beswick, M. (2008) The exceptional rainfall of 20 July 2007. *Weather*, **63**, 261-267.

Prudhomme, C. and Williamson, J. (2013) Derivation of RCM-driven potential evapotranspiration for hydrological climate change impact analysis in Great Britain: a comparison of methods and associated uncertainty in future projections. *Hydrology and Earth System Sciences*, **17**: 1365-1377.

Pruppacher, H.R. and Klett, J.D. (1997) *Microphysics of clouds and precipitation*, Kluwer Academic Publishers, Dordrecht and London. 954 pp.

Querner, E.P. (1997) Description and application of the combined surface and groundwater model MOGROW. J. *Hydrology*, **192**: 158-188

Quevauviller, P., Barcelo D., Beniston, M. and coauthors (2012) Integration of research advances in modelling and monitoring in support of WFD river basin management planning in the context of climate change. *Science of the Total Environment*, **440**: 167–177.

Rab, M. A., Chandra, S., Fisher, P.D. and coauthors (2011) Modelling and prediction of soil water contents at field capacity and permanent wilting point of dryland cropping soils. *Soil Research*, **49**: 389–407

Ragab, R. and Cooper, J.D. (1993) Variability of unsaturated zone water transport parameters: Implications for hydrological modelling. 1. In situ measurements. J. *Hydrology*, **148**: 109-132

Ragab, R., Finch J. and Harding R. (1997) Estimation of groundwater recharge to chalk and sandstone aquifers using simple soil models. J. *Hydrology*, 190: 19-41

Ragan, R.M. (1968) An experimental investigation of partial area contributions. *IAHS Publ.*, **76:**, 241-9. Also reproduced in Beven, K.J. (2007) pp 142-152.

Rawls, W.J., Brakensiek, D.L. and Miller, N. (1983) Green-Ampt infiltration parameters from soils data. J. *Hydraulic Engineering*, **109**: 62-70.

Rawls, W.J., Brakensiek, D.L. and Saxton, K.E. (1982) Estimation of soil water properties, *Trans. ASAE*, **25**: 131 6-20.

Reed, D.W. (1995) Rainfall assessment of drought severity and centennial events, Chap. 16, *Proceedings of CIWEM Centenary Conference*, Chartered Institution of Water and Environmental Management, London. 17 pp.

Reed, D.W. (2002) Reinforcing flood risk estimation. *Phil Trans Royal Society London* **A 360**: 1373 – 1387.

Reis, S., Grennfelt, P., Klimonty, Z. and coauthors (2012) From acid rain to climate change. *Science*, **338**: 1153-1154.

Reynolds, E.R.C. and Henderson, C.S. (1967) Rainfall interception by beech, larch and Norway Spruce. *Forestry*, **40**: 165–185.

Rice, E.W. , Baird, R.B. , Eaton, A.D., and Clesceri, L. S. (Eds.) (2012) *Standard Methods for the Examination of Water and Wastewater*. 22nd Edn. Joint publication of: American Public Health Association , American Water Works Association, Water Environment Federation. 1496 pp.

Richards, K. (2004) *Rivers: Form and Process in Alluvial Channels*. Blackburn Press, 360 pp.

Richards, L.A. (1931) Capillary conduction of liquids through porous mediums. *Physics*, **1**: 318-33.

Rijtema, P.E. (1968) *On the relation between transpiration, soil physical properties and crop production as a basis for water supply plans.* Technical Bulletin 58, Institute of Land and Water Management Research.

Robert, A. (2003) *River Processes: An introduction to fluvial dynamics*. Routledge, 240 pp.

Roberts, J.M. (1999) Plants and water in forests and woodlands. In: Baird, A. and Wilby, R. (Eds.), *Ecohydrology: Plants and Water in Terrestrial and Aquatic Ecosystems*. Routledge, London, pp 181-236.

Roberts, J.M., Pymar, C.F., Wallace, J.S. and Pitman, R.M. (1980) Seasonal changes in leaf area, stomatal conductance and transpiration from bracken below a forest canopy. J. *Applied Ecology*, 17: 409–422.

Robinson, D.A., Dewey, K.F. and Heim, R.H. (1993) Global snow cover monitoring: an update. Bull*etin of the American Meteorological Society*, **74**: 1689–1696.

Robinson, M. (1989) Small catchment studies of man's impact on flood flows: agricultural drainage and plantation forestry. IAHS Publ. **187**: 299-308.

Robinson, M. (1999) The consistency of long-term climate datasets: Two UK examples of the need for caution. *Weather*, **54**: 2-9.

Robinson, M. and Armstrong, A.C. (1988) The extent of agricultural field drainage in England and Wales, 1971-80, *Trans. IBG*, **13**: 19-28.

Robinson, M. and Beven, K.J. (1983) The effect of mole drainage on the hydrological response of a swelling clay soil. J. *Hydrology*, **64**: 205-23.

Robinson, M. and Rycroft, D.W. (1999) The impact of drainage on streamflow. In: Skaggs, W. & van Schilfgaarde, J. (Eds.) *Agricultural Drainage*. Agronomy Monograph **38**. American Society of Agronomy, Madison, Wisconsin. pp 753-786.

Robinson, M., Grant, S.G. and Hudson, J.A. (2004) Measuring rainfall to a forest canopy: an assessment of the performance of canopy level gauges. *Hydrology and Earth System Sciences* **8**: 327-333.

Robinson, M., Rodda, J.C. and Sutcliffe, J.V. (2013) Long-term environmental monitoring in the UK: Origins and achievements of the Plynlimon catchment study. *Transactions of the Institute of British Geographers*, **38**: 451–463.

Robson, A.J. and Neal, C. (1997) A summary of regional water quality for eastern UK rivers. *Science of the Total Environment*, **194/195**: 15–37.

Roca R, Chambon P.-E., Jobard I., and coauthors (2010) Comparing satellite and surface rainfall products over West Africa at meteorological relevant scales during the AMMA campaign using error estimates. J. *of Applied Meteorology and Climatology*, **49**: 715 – 731.

Rockström, J. and Valentin, C. (1997) Hillslope dynamics of on-farm generation of surface water flows: The case of rain-fed cultivation of pearl millet on sandy soil in the Sahel. *Agric. Water Management*, **33**: 183-210.

Rodda ,J.C. (1976) Basin Studies. In: Rodda, J.C. (Ed.) *Facets in Hydrology*. Wiley, Chichester, pp. 257–297.

Rodda, J. C. and Dixon, H. (2012), Rainfall measurement revisited. *Weather*, **67**: 131–136.

Rodda, J.C., Downing, R.A. and Law, F.M. (1976) *Systematic Hydrology*, Newnes – Butterworths, London, 399 pp.

Rodda, J.C. and Marsh, T. (2011) *The 1975-76 drought: A contemporary and retrospective review.* Centre for Ecology and Hydrology, Wallingford. 58 pp.

Rodda J.C. and Robinson, M. (Eds.) (2015) *Progress in Modern Hydrology: Past, Present and Future.* Wiley-Blackwell. Oxford. 384 pp.

Rodda, J. and Marsh, T. (2015) 1975-76 *Drought: A contemporary and retrospective review.* Centre for Ecology & Hydrology. 58 pp. Available from www.ceh.ac.uk

Rodell, M., Beaudoing, H.K., L'Ecuyer, T.S, and coauthors (2015) The observed state of the water cycle in the early Twenty-First Century. *J. Climate*, **28**: 8289–8318. doi: http://dx.doi.org/10.1175/JCLI-D-14-00555.1

Roemmich, D, Church, J., Gilson, J. and coauthors (2015) Unabated planetary warming and its ocean structure since 2006. *Nature Climate Change*, **5**: 240–245.

Romkens, M.J., Prasad S.N. and. Whisler, F.D (1990) Surface sealing and infiltration. In: Anderson, M.G. and Burt T.P. (Eds.), *Process Studies in Hillslope Hydrology*. Wiley, Chichester, pp. 127-172

Rost, S., Gerten, D., Bondeau, A. and coauthors (2008) Agricultural green and blue water consumption and its influence on the global water system. *Water Resources Research*, **44**: W094505, 17 pp.

Rudloff, W. (1981) *World climates with tables of climatic data and practical suggestions*. Wissenschaftiiche Veriagsgeselischaft mbH, Stuttgart, 632 pp.

Rutter, A.J. (1963) Studies in the water relations of Pinus sylvestris in plantation conditions. *J. Ecology*, **51**: 191–203.

Rutter, A.J. (1967) An analysis of evaporation from a stand of Scots pine. In: Sopper, W. E. and Lull, H.W. (Eds.), *Forest Hydrology*, Pergamon, Oxford, pp. 403–17. Also reproduced in Gash and Shuttleworth (2007) pp 278-291

Rutter, A.J. (1975) The hydrological cycle in vegetation. In: Monteith, J.L. (Ed.), *Vegetation and the Atmosphere*, Vol. **1**, Academic Press, London, pp. 111 - 154.

Rutter, A.J. and Morton, A.J. (1977) A predictive model of rainfall interception in forests: III Sensitivity of the model to stand parameters and meteorological variables. *J. Applied Ecology*, **14**: 567–88.

Rutter, A.J., Kershaw, K.A., Robins, P.C. and Morton, A.J. (1971) A predictive model of rainfall interception in forests: I Derivation of the model from observations in a plantation of Corsican pine. *Agricultural Meteorology*, **9**: 367–84. Also reproduced in Gash and Shuttleworth (2007) pp 292-308.

Rutter, A.J., Morton, A.J. and Robins, P.C. (1975) A predictive model of rainfall interception in forests: II. Generalization of the model and comparison with observations in some coniferous and hardwood stands. *J. Applied Ecology*, **12**: 367–80.

Sacks, W J.; Cook, B I.; Buenning, N and coauthors (2009) Effects of global irrigation on the near-surface climate. *Climate Dynamics* **33**:159–175

Saeed, F., Hagemann, S. and Jacob, D. (2009) Impact of irrigation on the South Asian summer monsoon. *Geophysical Research Letters*, **36**, L20711. doi: 10.1029/2009GL040625.

Salati, E, Dall'Olio, A., Matsu, I.E. and Gat, A.R. (1979) Recycling water in the Amazon basin: an isotopic study. *Water Resources Research* **15**: 1250–1258.

Satterlund, D.R. and Haupt, H.F. (1970) The disposition of snow caught by conifer crowns. *Water Resources Research*, **6**: 649–52.

Scanlon, B.R., Keese, K.E., Flint, A.L. and coauthors (2006) Global synthesis of groundwater recharge in semiarid and arid regions. *Hydrological Processes*, **20**: 3335–3370.

Schmid, H. (2002) Footprint modelling for vegetation atmosphere exchange studies: a review and perspective. *Agricultural and Forest Meteorology* **113**, 159–183.

Schnoor, J.L. (1996) *Environmental modelling: Fate and transport of pollutants in water, air and soil*, J. Wiley and Sons, Chichester, 682 pp.

Scoging, H.M. and Thornes, J.B. (1979) Infiltration characteristics in a semiarid environment, *IAHS Publ* **128**: 159-168.

Sear, D, Wilcock, D., Robinson, M., and Fisher K. (2000) River channel modification in the UK. In: Acreman, M.C. (Ed.) *Hydrology of the UK*. pp 55-81.

Sebastianelli, S., Russo, F., Napolitano, F. and Baldine, L. (2013) On precipitation measurements collected by a weather radar and a raingauge network. *Natural Hazards and Earth System Sciences*, **13**: 605-623.

Sene, K.J. and Farquharson, F. (1998) Sampling errors for water resource design: The need for improving hydrometry in developing countries. *Water Resources Management*, **12**: 121-138.

Serafin R.J. and Wilson J.W. (2000) Operational Weather Radar in the United States: Progress and Opportunity. *Bulletin of the American Meteorological Society*, **81**: 501-518.

Servat E., Demuth S., Dezetter A., and Daniell T. (Eds.) (2010) Global Change : Facing risks and threats to water resources. Proc. Sixth World FRIEND conference. *IAHS Publ.* **340**. 698 pp.

Sevruk, B. (1982) *Methods of Correction for Systematic Error in Point Precipitation Measurement for Operational Use*. Operational Hydrology Report **21**. World Meteorological Organization, Geneva, 91 pp.

Sevruk, B. and Geiger, H. (1981) *Selection of Distribution Types for Extremes of Precipitation*. Operational Hydrology Report **15**. World Meteorological Organization, Geneva, 64 pp.

Sevruk, B. and Klemm, S. (1989) *Catalogue of National Standard Precipitation Gauges*. Instruments and observing methods Report **39**. World Meteorological Organization, Geneva, 50 pp.

Sevruk, B., Hertig, I.A. and Spiess, R. (1989) Wind field deformation above precipitation gauge orifices, *IAHS Publ.*, **179**: 65–70.

Sharpley, A., Jarvie, H.P., Buda, A. and coauthors (2013) Phosphorus legacy: Overcoming the effects of past management practices to mitigate future water quality impairment. *J. Environmental Quality*, **42**: 1308-1326.

Sheffield, J.S., Ferguson, C.R., Troy, T.J. and coauthors (2009) Closing the terrestrial water budget from satellite remote sensing. *Geophysical Research Letters*, **36**: L07403. 10.1029/2009GL037338

Shiklomanov, I.A. (1993) World fresh water resources. In: Gleick, P.H. (Ed.), *Water in Crisis*, OUP, New York, 13-24.

Shiklomanov, I.A. (1997) *Assessment of Water Resources and Water Availability in the World*. WMO, 88 pp.

Shiklomanov, I.A. and Rodda, J.C. (Eds.) (2003) *World Water Resources at the Beginning of the Twenty-First Century*. UNESCO International Hydrology Series. Cambridge University Press. 435 pp.

Shiklomanov, I.A., Lammers, R.B. and Vorosmarty, C.J. (2002) Widespread decline in hydrological monitoring threatens Pan-Arctic Research. *Eos Transactions of the American Geophysical Union*, **83**: 13–17.

Shuttleworth, W.J. (1977) The exchange of wind-driven fog and mist between vegetation and the atmosphere. *Boundary Layer Meteorology*, **12**: 463–489.

Shuttleworth, W.J. (1988) Evaporation from the Amazonian rainforest. Pr*oc. Phil. Trans. Royal Society*, London, B. **324**:299-334.

Shuttleworth, W.J. (1989) Micrometeorology of temperate and tropical forest. *Philosophical Transactions of the Royal Society, London. Series B*, **324**: 299–334.

Shuttleworth, W.J. (2012) *Terrestrial Hydrometeorology*. Wiley-Blackwell, Chichester, UK. 448 pp.

Shuttleworth, W.J. and Calder, I.R. (1979) Has the Priestley–Taylor equation any relevance to forest evaporation? *J. Applied Meteorology*, 18: 639–646.

Shuttleworth, W.J. and Wallace, J.S. (1985) Evaporation from sparse crops – an energy combination approach, *Quarterly J. the Royal Meteorological Society*, 111:839–855.

Shuttleworth, W.J., Gash, J.H.C., Lloyd, C.R and coauthors (1988) An integrated micrometeorological system for evaporation measurement, *Agricultural Forest Meteorology*, 143, 295–317.

Silar, J. (1990) Time and its meaning in groundwater studies. *IAHS Publ.*, **190**: 281-289

Singh, B. and Szeicz, G. (1979) The effect of intercepted rainfall on the water balance of a hardwood forest. *Water Resources Research*, **15**: 131–138.

Singh, P., Spitzbart, G., Hübl, H. and Weinmeister, H.W. (1997) Hydrological response of snowpack under rain-on-snow events: a field study. *J. Hydrology*, **202**: 1-20

Sklash, M.G. and Farvolden, R.N. (1979) The role of groundwater in storm runoff. *J. Hydrology*, **43**: 45-65. Also reproduced in Beven, KJ (2007) pp 352-372.

Sklash, M.G., Stewart, M.K. and Pearce, A.J. (1986) Storm runoff generation in humid headwater catchments, II. A case study of hillslope and low order stream response. *Water Resources Research*, **22**: 1273-82.

Smakhtin, V.Y. (2001) Low flow hydrology: a review. *J. Hydrology*, **240**: 147-186.

Smedema. L.K., Vlotman., W.F. and Rycroft, D.W. (2004) *Modern Land Drainage: Planning, Design and Management of Agricultural Drainage Systems*. 2nd Edn. CRC Press. 462 pp.

Smettem, K.R.J. (1986) Solute movements in soils. In: Trudgill, S.T. (Ed.), *Solute Processes*. J. Wiley and Sons, Chichester, pp. 141–65.

Smith, D.M. and Allen, S.J. (1996) Measurement of sap flow in plant stems. J. *Experimental Botany*, 305: 1833–1844.

Smith, K. (2013) *Environmental Hazards*. Routledge.

Smith, K. and Ward, R.C. (1998) *Floods: Physical processes and human impacts*. John Wiley and Sons, Chichester. 382 pp.

Smith, R.A. (1872) *Air and Rain: The Beginnings of Chemical Climatology*. Longmans, London, 600 pp.

Smith, R.C.G., Barrs, H.D. and Steiner, J.L. (1985) Relationship between wheat yield and foliage temperature: theory and its application to infrared measurements, *Agricultural Forest Meteorology*, **36**: 129–43.

Speidel, D.H. and Agnew, A.F. (1988) The world water budget. In: Speidel D. H., Ruedisili L. C. and Agnew, A. F. (Eds.), *Perspectives on Water*. Oxford University Press, New York, 27-36.

Sposito, G. (1994) *Chemical Equilibria and Kinetics in Soils*. Oxford University Press, Oxford. 280 pp.

Spracklen, D.V., Arnold, S.R. and Taylor, C.M. (2012) Observations of increased tropical rainfall preceded by air passage over forests. *Nature*, **489**: 282–285

Stakhiv, E.Z. (2011) Pragmatic approaches for water management under climate change uncertainty. *J. the American Water Resources Association*, **47**: 1183-1196,.

Stern, N (2006) *Stern Review on the Economics of Climate Change*. H M Treasury, London. 700 pp. www.sternreview.org.uk.

Stewart, E.J., Morris, D.G., Jones, D.A., and Svensson, C. (2012) A new FEH rainfall depth-duration-frequency model for hydrological applications. *IAHS Publ.*, **351**, 638-643.

Stewart, E.J. (1989) Areal reduction factors for design storm construction: joint use of raingauge and radar data, *IAHS Publ.*, **181**: 31–40.

Stewart, E.J., Beran, M., Farquharson, F. and coauthors (2015) Risks and Extremes. In: Rodda, J.C. and Robinson, M. (Eds.) *Progress in Modern Hydrology - Past, Present and Future*. Wiley-Blackwell, Oxford. pp 60-99.

Stewart, J.B. (1977) Evaporation from the wet canopy of a pine forest. *Water Resources Research*, **13**: 915–21. Also reproduced in Gash and Shuttleworth (2007) pp 309-315.

Stewart, J.B. (1988) Modelling surface conductance of pine forest, *Agricultural and Forest Meteorology*, **43**: 19–35.

Stewart, J.B. and Thom, A.S. (1973) Energy budgets in pine forest. *Quarterly J. Royal Meteorological Society*, **99**: 154–70.

Stone, A.T. and Morgan, J.J. (1990) Kinetics of chemical transformations in the environment. In: Stumm, W. (Ed.), *Aquatic Chemical Kinetics*. J. Wiley and Sons, Chichester, pp. 1–41.

Strangeways, I. (2003) *Measuring the Natural Environment*. 2nd Edn. Cambridge University Press. 534 pp.

Strangeways, I. (2004) Improving precipitation measurement. *International J. of Climatology*, **24**: 1443-1460.

Strangeways, I. (2007) *Precipitation: Theory, Measurement and Distribution*. Cambridge University Press. Cambridge and New York. 290 pp.

Stratton, J.M., Houghton Brown, J., and Whitlock, R. (1978) *Agricultural records AD 220 – 1977*. 2nd Edn. John Baker, London. 259 pp.

Stumm, W. and Morgan, J.J. (1996) *Aquatic Chemistry*. .3rd Edn. Wiley-Blackwell, New York. 1040 pp.

Sullivan, C. (2002) Calculating a Water Poverty Index. *World Development*, **30**: 1195–1210

Svensson, C. and Jones, D.A. (2010) review of methods for deriving areal reduction factors. *J Risk Management*, **3**: 232-245.

Sverdrup, H. and De Vries, W. (1994) Calculating critical loads for acidity with the simple mass balance method. *Water, air, soil pollution*, **72**: 143–162.

Swank, W.T. and Miner, N.H. (1968) Conversion of hardwood–covered watersheds to white pine reduces water yield. *Water Resources Research*, **4**: 947–54.

Swanson, R.H. (1994) Significant historical developments in thermal methods for measuring sap flow in trees. *Agricultural and Forest Meteorology*, **72**: 113–132.

Symons, G.J. (1867) Evaporators and evaporation. *British Rainfall*, **7**, 9-10.

Syvitski, J.P.M., Vorosmarty, C.J., Kettner, A.J. and Green, P. (2005) Impact of humans on the flux of terrestrial sediment to the global coastal ocean. *Science*, **308**, 376-380.

Szeicz, G., van Bavel, C.H.M. and Takami, S. (1973) Stomatal factor in the water use and dry matter production by sorghum, *Agric. Met.*, 12: 361–389.

Szeicz, G., Endrodi, G. and Tajchman, S. (1969) Aerodynamic and surface factors in evaporation. *Water Resources Research*, 5: 380–394.

Taha, A., Gresillon, J.M. and Clothier, B.E. (1998) Modelling the link between hillslope water movement and stream flow: application to a small Mediterranean forest watershed. *J. Hydrology*, **203**: 11-20

Tait, A.B. (1998) Estimation of snow water equivalent using passive microwave radiation data, *Remote Sensing of the Environment*, **64**: 286–291.

Takhar, H.S. and Rudge, A.J. (1970) Evaporation studies in standard catchments. J. *Hydrol.*, **11**: 329–362.

Tarnavsky, E., Grimes, D., Maidment, R., and coauthors (2014) Extension of the TAMSAT satellite-based rainfall monitoring over Africa and from 1983 to present. *J. Applied Meteorology and Climatology*, **53**: 2805–2822

Taylor, C.M. and Lebel, T. (1998) Observational evidence of persistent convective-scale rainfall patterns. *Monthly Weather Review*, **126**: 1597-1607.

Teatini, P., Strozzi, T., Tosi, L. and coauthors (2007) Assessing short- and long-time displacements in the Venice coastland by synthetic aperture radar interferometric point target analysis. J Geophysical Research, **112**: F01012, doi: 10.1029/2006JF000656.

Teklehaimanot, Z., and Jarvis, P.G. (1991) Direct measurement of evaporation of intercepted water from forest canopies. *J. Applied Ecology*, **28**: 603–618.

Teklehaimanot, Z., Jarvis, P.G. and Ledger, D.C. (1991) Rainfall interception and boundary layer conductance in relation to tree spacing. *J. Hydrology*, **123**: 261–278.

Ten Veldhuis, M.-C., Ten Veldhuis, J.A.E. Ochoa-Rodriguez, S. and coauthors (2014) High resolution radar rainfall for urban pluvial flood management: Lessons learnt from 10 pilots in North-West Europe within the RainGain project. *13th International Conference on Urban Drainage*, Sarawak, Malaysia, 7-12 September 2014. 9pp

Thames Water (2014) *Final Water Resources Management Plan 2015 – 2040*. Thames Water plc. Reading.

Thiessen, A.H. (1911) Precipitation averages for large areas, *Mon. Weather. Rev.*, **39**: 1082–4.

Thom, A.S. and Oliver, H.R. (1977) On Penman's equation for estimating regional evaporation, *Quarterly J. Royal Meteorological Society*, **103**: 345–358.

Thomas, D.M. and Benson, M.A. (1970) Generalization of streamflow characteristics from drainage-basin characteristics. Geological Survey Water-Supply Paper **1975**.

Thompson, N. (1982) A comparison of formulae for the calculation of water loss from vegetated surfaces, *Agric. Met.*, **26**: 265–272.

Thompson, N., Barrie, I.A. and Ayles, M. (1981) *The Meteorological Office Rainfall Evaporation Calculations System: MORECS*, Hydrological Memorandum **45**, Meteorological Office, Bracknell, 69 pp.

Thompson, W.R. and Hay, I.D. (2004). Complexity, diminishing marginal returns and serial Mesopotamian fragmentation. *J. World Systems*, **10**: 613-652.

Thornthwaite, C.W. (1944) A contribution to the Report of the Committee on transpiration and evaporation, 1943–44, *Trans. American Geophysical Union*, **25**: 686–693.

Thornthwaite, C.W. (1948) An approach towards a rational classification of climate, *Geographical Review*, **38**: 55–94.

Thornthwaite, C.W. (1954) A re-examination of the concept and measurement of potential evapotranspiration, *Publications in Climatology*, **7**.

Thurow, T.L., Blackburn, W.H., Warrren, S.D. and Taylor, C.A. (1987) Rainfall interception by midgrass, shortgrass and live oak mottes. *J. Range Management*, **40**: 455-460.

Tilman, D. (1999) Global environmental impacts of agricultural expansion: The need for sustainable and efficient practices. *Proc. National Academy of Sciences*, **96**: 5995–6000.

Time Magazine (2016) Donald Trump calls climate change a hoax, but worries it could hurt his golf course. *Time Magazine*, 23 May 2016.

Todd, D.K. (1980) *Groundwater Hydrology*. J Wiley and sons, New York. 535 pp.

Todd, D.K. and Mays L.W. (2004) *Groundwater Hydrology*, 3rd Edn. John Wiley and Sons, 656 pp

Tóth, J. (1962) A theory of groundwater motion in small drainage basins in central Alberta, Canada. J. *Geophysical Research*, **67**: 4375-87

Tóth, J. (1963) A theoretical analysis of groundwater flow in small drainage basins. J *Geophysical Research*, **68**: 4795-812. Also reproduced in Anderson, M. (2008) (Ed.) pp 298-315.

Tóth, J. (1995) Hydraulic continuity in large sedimentary basins. *Hydrogeology J.*, **3**: 4-16

Tóth, J. (1996) Reply to a comment by E. Mazor on Tóth (1995). *Hydrogeology J.*, **4**: 102-107

Tóth, J (2005) The Canadian School of hydrogeology: History and legacy. *Ground water*, **43**: 640 – 644.

Trenberth, K E, Smith, L, Qian, T. and coauthors (2007) Estimates of global water budget and its annual cycle using observational and model data. *J Hydrometeorology*, **8**: 758-769.

Trenberth, K.E. (1992) (Ed.) *Climate System Modelling*. Cambridge University Press, Cambridge 788 pp.

Troendle, C. A. (1985) Variable source area models, in *Hydrological forecasting*. In: Anderson, M.G. and. Burt, T.P (Eds.), John Wiley and Sons, pp. 347-403.

Turner, N.C. (1986) Crop water deficits: a decade of progress. *Advances in Agronomy*, **39**: 1–51.

Ubarana, V.N. (1996) Observations and modelling of rainfall interception at two experimental sites in Amazonia. In: Gash, J.H.C, Nobre, C.A., Roberts, J.M., and Victoria, R.L. (Eds.), *Amazonian Deforestation and Climate*. John Wiley and Sons, Chichester, 611 pp.

UKWRIP (2014) Tapping the potential: A fresh vision for UK water technology. UK Water Partnership 32pp. http://theukwaterpartnership.org/outputs/publications/

UN (1987) *Our Common Future*. Report of the World Commission on Environment and Development. United Nations, 247 pp.

UN (2010) *Sick Water? The Central Role of Wastewater Management in Sustainable Development*. United Nations Environment Programme, Nairobi. 85 pp.

UN (2014) *The United Nations World Water Development Report 2014: Water and Energy*. UNESCO, Paris. http://unesdoc.unesco.org/images/0022/002257/225741E.pdf

UN (2015a) *World Population Prospects: 2015 Revision*. United Nations. 20 pp. http://esa.un.org/unpd/wpp/

UN (2015b) *The Millennium Development Goals Report 2015*. United Nations.75 pp.

UN (2016) *Water and Jobs*. World Water Development Report 2016. Paris, UNESCO. 164 pp.

UNEP (1995) *Water Quality of World River Basins*, Environment library Series 14, United Nations Environment Programme. Nairobi, Kenya, 40 pp.

UNEP (2003) *Groundwater and its Susceptibility to Degradation: A Global Assessment of the Problem and Options for Management*. Nairobi, Kenya

UNEP (2008) *Global Glacier Changes: Facts and Figures*. World Glacier Monitoring Service. United Nations Environment Programme. Zurich, Switzerland 88pp

UNEP/IUCN (2000) *Dams and Development: A New Framework for Decision-Making*. Report of the World Commission on Dams. 356 pp.

UNESCO (1964) *Final Report, International Hydrological Decade*. Intergovernmental Meeting of Experts UNESCO/NAS/188. 51 pp. https://en.unesco.org/50-years-unesco-water-programmes/documents

UNESCO (2006) *Water, a Shared Responsibility*. UN World Water Development Report 2. UNESCO Publishing, Paris and Berghahn Books, New York. 584 pp.

UNESCO (2009) *Water in a Changing World*. UN World Water Development Report 3. UNESCO Publishing, Paris and Earthscan, London. 318 pp.

UNESCO (2015) *Water for a Sustainable World*. UN World Water Development Report 2015. UNESCO Publishing, Paris. 139 pp.

UNESCO/WMO/IASH (1974) *Three Centuries of scientific hydrology*. UNESCO, Paris. 123 pp.

USDA (2015) *Soil Survey Manual*. US Department of Agriculture. Published online: lulu.com. 318 pp

USEPA (2012) *Guidelines for Water Reuse*. EPA/600/R-12/618. US Environmental protection Agency.

Vachaud, G., Vauclin, M., Khanji, D. and Wakil, M. (1973) Effects of air pressure on water flow in an unsaturated stratified vertical column of sand. *Water Resources Research*, **9**:160-73.

Valente, F., David, J.S. and Gash, J.H.C. (1997) Modelling interception loss for two sparse eucalypt and pine forests in central Portugal using reformulated Rutter and Gash analytical models. *J. Hydrology*, **190**: 141–162.

Van der Kloet, P. and Lumadjeng, H.S. (1987) The development of an economic objective function for decision making in a water resource control problem. In: Carlsen, A.J. (Ed.) *Decision making in water resources planning*. Proceedings of UNESCO Symposium, May 5-7, 1986, Oslo, Norway, pp. 221-37.

Van der Leeden, F., Troise, F.L. and Todd, D.K. (1990) *The Water Encyclopedia*. Lewis Publishers, Michigan, 808 pp

van Dijk, A.I.J.M., and Bruijnzeel, L.A. (2001a). Modelling rainfall interception by vegetation of variable density using an adapted analytical model. Part 1. Model description. *J. Hydrology*, **247**: 230–238

van Dijk, A.I.J.M., Bruijnzeel, L.A. (2001b) Modelling rainfall interception by vegetation of variable density using an adapted analytical model. Part 2. Model validation for a tropical upland mixed cropping system. *J. Hydrology* **247**, 239–262.

van Dijk, A.I.J.M., Gash, J.H., van Gorsel, E. and coauthors (2015) Rainfall interception and coupled surface water and energy balance. *Agricultural and Forest Meteorology*, **214-215**: 402-415.

van der Tol, C., Gash, J.H.C, Grant, S.J and coauthors (2003) Average wet canopy evaporation for a Sitka spruce forest derived using the eddy correlation-energy balance technique. *J. Hydrology*, **276:** 12–19.

Van Genuchten, M.T. (1980) A closed-form equation for predicting the hydraulic conductivity of unsaturated soils. *Soil Science Society of America J.*, **44**: 892-898.

Van Stan, J.T., Levia D.F., Jenkins, R.B. (2015) Forest Canopy Interception Loss Across Temporal Scales: Implications for Urban Greening Initiatives. *The Professional Geographer*, **67**: 41-51

Viessman, W., Hammer, M.J., Perez, E.M. and Chadik, P.A. (2013) *Water Supply and Pollution Control*. 8th Edn. Pearson Education Ltd. 752 pp.

Vorosmarty, C.J. (2009) *The Earth's Natural Water Cycles*. In: UNESCO (2009), pp 166-180.

Vorosmarty, C.J., Leveque, C., Revenga C. and coauthors (2005) Fresh Water. In: Hassan, R. M., Scholes R. and Ash, N. (Eds.) *Ecosystems and Human Well-being: Current State and Trends*. Millennium Ecosystem Assessment, Island Press, Washington DC, pp 165-207.

Wallace J.G. (1997) *Meteorological Observations at the Radcliffe Observatory. Oxford: 1815–95*, School of Geography Research Paper **53**, University of Oxford.

Wallace, J.M. (2006) *Atmospheric Science: An introductory Survey*. Academic Press. London and New York. 504 pp.

Wallace, J.S., and Holwill, C.J. (1997) Soil evaporation from tiger-bush in south-west Niger. J Hydrology, **188-9**, 426-442.

Wallace, J.S., Gash, J.H.C., McNeil, D.D. and Sivakumar, M.V.K (1989) Evaporation from a sparse dryland millet crop in Niger, West Africa. In: Unger, P.W., Sneed, T.V., Jordan, W.R. and Jensen, R. (Eds.), *Proc. Internat. Conference on dryland Farming*, Texas Agricultural Experimental station, pp 325–327.

Wallace, J.S., Roberts J.M. and Roberts A.M. (1982) Evaporation from heather moorland in North Yorkshire, England. In: *Hydrological research basins and their use in water resources planning*. Proceedings of International Symposium, Bern, September, pp. 397–405.

Wallace, J.S., Roberts, J. and Shuttleworth, W.J. (1984) A comparison of methods for estimating aerodynamic resistance of heather (Calluna vulgaris (L.) Hull) in the field. *Agricultural and Forest Meteorology*, **32**: 289–805.

Wallen, C.C. (1970) *Climates of Northern and Western Europe*. World Survey of Climatology, Vol **5**, Elsevier, Amsterdam. 263 pp.

Walling, D.E. and Webb, B.W. (1983) The dissolved load of rivers: a global overview. *IAHS Publ.*, **141**: 3–20.

Walling, D.E. and Webb, B W. (1986) Solutes in river systems. In: Trudgill , S.T. (Ed.) *Solute processes*. J. Wiley and Sons, New York, pp. 251–327.

Wang, H.F. and Anderson, M.P. (1995) *Introduction to Groundwater Modelling*, Academic Press, London, 256 pp

Ward, H.C., Evans, J.G. and Grimmond, C.S.B. (2014) Multi-scale sensible heat fluxes in the suburban environment from large-aperture scintillometry and eddy covariance. *Boundary-layer meteorology*, **152**: 65-89.

Ward, R.C. and Robinson, M. (2000) *Principles of Hydrology*. McGraw-Hill Publishing Co., Maidenhead, 450 pp.

Warren, R.A., Kirshbaum, D.J., Plant, R.S. and Lean, H.W. (2014) A 'Boscastle-type' quasi-stationary convective system over the UK Southwest peninsular. *Quarterly J. of the Royal Meteorology Society*, **140**: 240-257.

Warrick, A.W. and Nielsen, D.R. (1980) Spatial variability of soil physical properties in the field. In: Hillel, D. (Ed.) *Applications of Soil Physics*, Academic Press, New York, pp. 319-44.

Warrick, A.W., Wierenga, P.J and Pan, L. (1997) Downward water flow through sloping layers in the vadose zone: analytical solutions for diversions. *J. Hydrology*, **192**: 321-337

Water JPI (2016) *An Introduction to the Strategic Research and Innovation Agenda 2.0*. Water Joint Programming Initiative. 46 pp. www.waterjpi.eu

Watts, G. (1997) Hydrological modelling in practice. In: Wilby, R.L. (Ed.) *Contemporary Hydrology*. Wiley, Chichester. pp 151-193

Wellings, S.R. (1984) Recharge of the upper chalk aquifer at a site in Hampshire, England: 1. Water balance and unsaturated flow. J. *Hydrology*, **69**: 259-73.

Wellings, S.R. and Bell, J.P. (1982) Physical controls of water movement in the unsaturated zone. *Quarterly J. Engineering Geology*, **1**: 235-41.

Werrell, C.E. and Femia, F. (Eds.) (2013). *The Arab Spring and Climate Change*. Climate and Security Correlations Series. Centre for American Progress. Washington. 68 pp

Weyman, D.R. (1973) Measurements of downslope flow of water in a soil. J. *Hydrology*, **20**: 267-88.

Wheeler, K.G., Basheer, M., Mekonnen, Z.T. and coauthors (2016) Cooperative filling approaches for the Grand Ethiopian Renaissance Dam. *Water International*, **12**: 1-24.

Whitaker, F.F. and. Smart, P.L (1997) Groundwater circulation and geochemistry of a karstified bank-marginal fracture system. South Andros Island, Bahamas. J. *Hydrology*, **197**: 293-315

White, R.E. (2005) *Introduction to the Principles and Practices of Soil Science*. 4th Edn., Wiley-Blackwell 384 pp.

White, W.B. (2002) Karst hydrology: recent developments and open questions. *Engineering Geology* **65**: 85–105

White, W.R. and Watts, J. (Eds.) (1994) *River Flood Hydraulics*. Wiley, Chichester, 604 pp.

Whitehead, P.G. and Robinson, M. (1993) Experimental basin studies – an international and historical perspective of forest impacts. *J Hydrology*, **145**: 217-230.

Whitehead, P.G., Neal C. and Neale R. (1986) Modelling the effects of hydrological changes on stream water acidity. J. *Hydrology*, **84**: 353–64.

Whitehead, P.G., Wade, A.J. and Butterfield, D. (2009a) Potential impacts of climate change on water quality and ecology in six UK rivers. *Hydrology Research*, **40**: 113-122

Whitehead, P.G., Wilby, R.L.. Battarbee, R.W. and coauthors (2009b) A review of the potential impacts of climate change on surface water quality. *Hydrological Sciences J.*, **54**: 101-123.

Whiting, P.J. and Pomeranets, M. (1997) A numerical study of bank storage and its contribution to streamflow. *J. Hydrology*, **202**: 121-136

WHO (2011) *Guidelines for Drinking Water Quality*. 4th Edn.. World Health Organization, Geneva.

Wiesner, C.J. (1970) *Hydrometeorology*, Chapman and Hall Ltd, London, 232 pp.

Wilby, R.L. (1997) Beyond the river catchment. In: Wilby, R.L. (Ed.) *Contemporary Hydrology: Towards Holistic Environmental Science*. Wiley, Chichester, 317-346.

Wild, M. (2012) Enlightening global dimming and brightening. *Bulletin of the American Meteorological Society*, **93**: 27-37

Wilhite, D.A. (1993) The enigma of drought, In: Wilhite, D.A. (Ed.), *Drought assessment, management and planning: theory and case studies*, Kluwer Academic Publishers, London, pp 5–15.

Williams R.J., Keller V.D.J., Johnson A.C., and coauthors (2009). A national risk assessment for intersex in fish arising from steroid estrogens. *Environmental Toxicology and Chemistry*, **28**: 220-230.

Williams, R.J., Brooke, D.N., Matthiessen, P., and coauthors (1995) Pesticide transport to surface waters within an agricultural catchment. J. *Institution of Water and Environmental Management*, **9**: 72–81.

Williams, R.J., Neal, C., Jarvie, H. and coauthors (2015). Water quality. In: Rodda, J.C. and Robinson, M. (Eds.) Progress in Modern Hydrology: Past, present and future. Wiley-Blackwell, Oxford. pp 240-266.

Winter, T.C. , Rosenberry, D.O. and LaBaugh, J.W. (2003) Where does the groundwater in small catchments come from? *Groundwater*, **41**: 989-1000

WMO (1975) *Manual on the observation of clouds and other meteors: International Cloud Atlas,* Volume **I** (Text), Report 407. World Meteorological Organization, Geneva.

WMO (1995a) *Manual on the observation of clouds and other meteors: International Cloud Atlas,* Volume **II** (Photos), Report 407. World Meteorological Organization, Geneva.

WMO (1995b) *INFOHYDRO Manual*, Operational Hydrology Report **28**. World Meteorological Organization, Geneva.

WMO (2008) *Guide to hydrological practices*, Report **168**. World Meteorological Organization, Geneva, http://www.whycos.org/hwrp/guide/index.php

WMO (2010) *Manual on Stream Gauging*. Vol 1: Fieldwork, and Vol 2: Computation of Discharge. WMO-No. **1044**. World Meteorological Organization, Geneva, Switzerland.

WMO 2009 Manual on Estimation of Probable Maximum Precipitation (PMP) WMO-No. **1045** World Meteorological Organization, Geneva. 291 pp. http://www.wmo.int/pages/prog/hwrp/publications/PMP/WMO%201045%20en.pdf

World Bank (2013) *Toward a Sustainable Energy Future for All: Directions for the World Bank Group's Energy Sector*. World Bank, Washington, 39 pp.

World Bank (2016) *World Development Indicators 2016*. World Bank Group, Washington. 180 pp.

Wouter, P., Vinogradov, S. , Allan, A. and coauthors (2005) *Sharing Transboundary Waters - An Integrated Assessment of Equitable Entitlement : The Legal Assessment Model.* Technical Documents in Hydrology, 74. UNESCO, Paris 142 pp

WRI (1996) *World Resources 1996-1997*. World Resources Institute, Washington DC,

WRI (2000) *World Resources 2000-2001*. World Resources Institute, Washington DC,

Wright, E. P., Benfield, A.C., Edmunds, W.M. and Kitching, R. (1982) Hydrogeology of the Kufra and Sirte basins, eastern Libya. *Quarterly J. of Engineering Geology,* **15**: 83-103.

Wright, E.P. (1992) The hydrogeology of crystalline basement aquifers in Africa. In: Wright, E.P. and Burgess, W.G. (Eds.) *The Hydrogeology of crystalline basement aquifers in Africa*. The Geological Society, London, pp. 1-27

Wright, R.F., Aherne, J., Bishop, K. and coauthors (2006) Modelling the effect of climate change on recovery of acidified freshwaters: Relative sensitivity of individual processes in the MAGIC model. *Science of The Total Environment,* **36**: 154–166.

Xue, Y-Q, Zhang, Y, Ye, S-J and coauthors (2005) Land subsidence in China. *Environmental Geology*, **48**: 713-720

Yamanaka, T., A. Takeda and J. Shimada (1998) Evaporation beneath the soil surface: some observational evidence and numerical experiments, *Hydrological Processes*, **12**: 2193–2203.

Youngs, E.G. (1991) Infiltration measurements - a review, *Hydrological Processes*, **5**: 309-320

Zaslavsky, D. and Sinai, G. (1981) Surface hydrology, I. Explanation of phenomena. *Proc. Amer. Soc. Civil. Engr. J. Hydraul. Div.*, **107**(HY1): 1-16.

Zhu, T.X., Cai, Q.G. and Zeng, B.Q. (1997) Runoff generation on a semi-arid agricultural catchment: Field and experimental studies. J. *Hydrology*, **196**: 99-118

INDEX

S

X

Z

ACKNOWLEDGEMENTS

The authors have worked together for many years in the field of hydrological education and this book is the culmination of work and discussions with many people who have (consciously or unconsciously) contributed to its strengths, as well as to the students at the Universities of Hull (RCW) Oxford and Reading (MR) who have been the recipients of our teaching and on whom we have tested our ideas.

In addition to the many published references quoted, both of us wish to thank friends and colleagues who have read draft versions of the manuscript or provided data and photos. Especial thanks to (in alphabetical order) John Bromley (Water Research Associates), David J Cooper (Ex-IH), Simon Dadson (Oxford University), John Davis (Environment Agency), Rosemary Fry (British Geological Survey), Mike Kendon (Met. Office), Harry Dixon (NRFA), John Gash (Ex-CEH), Jamie Hannaford (NRFA), Richard Harding (CEH), Olivia Hitt (NRFA), David MacDonald (British Geological Survey), Terry Marsh (CEH), Katie Muchan (NRFA), John Rodda (Ex-WMO), Dave Rylands (Environment Agency), Richard Skeffington (Reading University), Lisa Stewart (CEH), Ian Strangeways (TerraData Ltd), Rob Stroud (Environment Agency), Fernanda Valente (Technical University of Lisbon), Paul Whitehead (Oxford University) and Richard Williams (CEH). Their detailed comments and discussions have done much to improve the quality of this book. For any failings that remain, errors and misinterpretations, however, we must take responsibility.

When our previous publisher, McGraw-Hill (UK), changed priorities in the Environmental and Engineering field, this book project was welcomed and taken over by the International Water Association Publishing, and we are grateful to both organisations for the smooth transfer from the book formerly known as 'Principles of Hydrology' to this new book, with a wider remit for the modern age: 'Hydrology – Principles and Processes'.

The following figures are adapted from diagrams in Ward and Robinson, Principles of Hydrology, 4e, © 2000 Reproduced with the kind permission of McGraw-Hill Education. All rights reserved. 1.3, 2.1, 2.2, 2.3, 2.8, 3.1, 4.6, 4.8, 4.13, 4.14, 5.2, 5.10, 6.6, 7.2, 7.3, 7.4, 7.5, 7.6, 7.7, 7.13, 8.12 and 9.5. Figures 3.3b, 3.5, 3.6, 3.8, 4.10, 4.11, 5.5, 6.11, 6.12, 6.15, 8.7, 8.9 and 8.10 are republished with permission of J. Wiley and Sons Ltd; permission conveyed through Copyright Clearance Center, Inc. Figures 3.2, 3.9, 4.7, 6.14, 8.8 are reprinted with permission from Elsevier. Taylor & Francis Ltd, www.tandfonline.com gave permission for the use of figure 6.13, Springer Science & Bus Media, BV for 5.15, and Melbourne University Press for 6.10. We especially thank the Centre for Ecology and Hydrology for the provision of data and permission to use diagrams.

IWA Publishing's authorised EU representative for General Product Safety Regulations is Diane D'Arras, 15 rue Duret, 75116 Paris, France, e-mail: safety@iwap.co.uk.

Printed and bound by CPI Group (UK) Ltd, Croydon, CR0 4YY
12/05/2026
02108866-0014